THIRD EDITION

Electric Motors and Control Systems

Frank D. Petruzella

ELECTRIC MOTORS AND CONTROL SYSTEMS, THIRD EDITION

Published by McGraw-Hill Education, 2 Penn Plaza, New York, NY 10121. Copyright © 2020 by McGraw-Hill Education. All rights reserved. Printed in the United States of America. Previous editions © 2016 and 2010. No part of this publication may be reproduced or distributed in any form or by any means, or stored in a database or retrieval system, without the prior written consent of McGraw-Hill Education, including, but not limited to, in any network or other electronic storage or transmission, or broadcast for distance learning.

Some ancillaries, including electronic and print components, may not be available to customers outside the United States.

This book is printed on acid-free paper.

2 3 4 5 6 7 8 9 LKV 22

ISBN 978-1-260-25805-9 (bound edition)
MHID 1-260-25805-X (bound edition)
ISBN 978-1-260-43939-7 (loose-leaf edition)
MHID 1-260-43939-9 (loose-leaf edition)

Product Developer: *Tina Bower*
Marketing Manager: *Shannon O'Donnell*
Content Project Managers: *Jane Mohr*
Buyer: *Susan K. Culbertson*
Design: *Beth Blech*
Content Licensing Specialist: *Lori Hancock*
Cover Image: *Mr. B-King/Shutterstock*
Compositor: *Aptara, Inc.*

All credits appearing on page or at the end of the book are considered to be an extension of the copyright page.

Library of Congress Cataloging-in-Publication Data

Names: Petruzella, Frank D., author.
Title: Electric motors and control systems / Frank D. Petruzella.
Description: Third edition. | Dubuque : McGraw-Hill Education, 2020. |
  Includes index.
Identifiers: LCCN 2019025847 | ISBN 9781260258059 (hardcover) | ISBN 9781260439397 (spiral bound)
Subjects: LCSH: Electric motors. | Electric controllers. | Electric
  driving.
Classification: LCC TK2514 .P48 2019 | DDC 621.46—dc23
LC record available at https://lccn.loc.gov/2019025847

The Internet addresses listed in the text were accurate at the time of publication. The inclusion of a website does not indicate an endorsement by the authors or McGraw-Hill Education, and McGraw-Hill Education does not guarantee the accuracy of the information presented at these sites.

mheducation.com/highered

ABOUT THE AUTHOR

Frank D. Petruzella has extensive practical experience in the electrical motor control field, as well as many years of experience teaching and authoring textbooks. Before becoming a full-time educator, he was employed as an apprentice and electrician in areas of electrical installation and maintenance. He holds a Master of Science degree from Niagara University, a Bachelor of Science degree from the State University of New York College–Buffalo, as well as diplomas in Electrical Power and Electronics from the Erie County Technical Institute.

One unique feature with all of his texts is that they are all supported with the latest in related computer simulation software. Working in conjunction with National Instruments for Multisim, CMH Software for Constructor, and The Learning Pit for LogixPro, he has developed program files directly related to circuits explained in the text.

BRIEF CONTENTS

About the Author iii
Preface ix
Acknowledgments xi
Walk-through xii

Chapter 1 Safety in the Workplace 1
    PART 1 Protecting against Electrical Shock 1
    PART 2 Grounding—Lockout—Codes 9

Chapter 2 Understanding Electrical Drawings 16
    PART 1 Symbols—Abbreviations—Ladder Diagrams 16
    PART 2 Wiring—Single Line—Block Diagrams 24
    PART 3 Motor Terminal Connections 28
    PART 4 Motor Nameplate and Terminology 37
    PART 5 Manual and Magnetic Motor Starters 42

Chapter 3 Motor Transformers and Distribution Systems 47
    PART 1 Power Distribution Systems 47
    PART 2 Transformer Principles 57
    PART 3 Transformer Connections and Systems 62

Chapter 4 Motor Control Devices 72
    PART 1 Manually Operated Switches 72
    PART 2 Mechanically Operated Switches 80
    PART 3 Sensors 86
    PART 4 Actuators 98

Chapter 5 Electric Motors 105
    PART 1 Motor Principle 105
    PART 2 Direct Current Motors 110
    PART 3 Three-Phase Alternating Current Motors 122
    PART 4 Single-Phase Alternating Current Motors 131
    PART 5 Alternating Current Motor Drives 136
    PART 6 Motor Selection 139
    PART 7 Motor Installation 146
    PART 8 Motor Maintenance and Troubleshooting 151

Chapter 6 Contactors and Motor Starters 158
    PART 1 Magnetic Contactor 158
    PART 2 Contactor Ratings, Enclosures, and Solid-State Types 169
    PART 3 Motor Starters 175

Chapter 7 Relays 186
    PART 1 Electromechanical Control Relays 186
    PART 2 Solid-State Relays 191
    PART 3 Timing Relays 195
    PART 4 Latching Relays 203
    PART 5 Relay Control Logic 207

Chapter 8 Motor Control Circuits 211
    PART 1 NEC Motor Installation Requirements 211
    PART 2 Motor Starting 218
    PART 3 Motor Reversing and Jogging 231
    PART 4 Motor Stopping 238
    PART 5 Motor Speed 242

Chapter 9 Motor Control Electronics 245
    PART 1 Semiconductor Diodes 245
    PART 2 Transistors 251
    PART 3 Thyristors 259
    PART 4 Integrated Circuits (ICs) 265

Chapter 10 Adjustable-Speed Drives and PLC Installations 275
    PART 1 AC Motor Drive Fundamentals 275
    PART 2 VFD Installation and Programming Parameters 283
    PART 3 DC Motor Drive Fundamentals 297
    PART 4 Programmable Logic Controllers (PLCs) 304

Appendix 318
Index I-1

CONTENTS

About the Author iii
Preface ix
Acknowledgments xi
Walk-through xii

Chapter 1
Safety in the Workplace 1

PART 1 Protecting against Electrical Shock 1
   Electrical Shock 1
   Arc Flash Hazards 4
   Personal Protective Equipment 5
Machine Safety 7
   Safety Light Curtains 7
   Safety Interlock switches 7
   Emergency Stop Controls 8
   Safety Laser Scanners 8
PART 2 Grounding—Lockout—Codes 9
   Grounding and Bonding 9
   Lockout and Tagout 11
   Electrical Codes and Standards 12

Chapter 2
Understanding Electrical Drawings 16

PART 1 Symbols—Abbreviations—Ladder Diagrams 16
   Motor Symbols 16
   Abbreviations for Motor Terms 17
   Motor Ladder Diagrams 17
PART 2 Wiring—Single Line—Block Diagrams 24
   Wiring Diagrams 24
   Single-Line Diagrams 26
   Block Diagrams 26
   Riser Diagrams 27
PART 3 Motor Terminal Connections 28
   Motor Classification 28
   DC Motor Connections 28
   AC Motor Connections 30
PART 4 Motor Nameplate and Terminology 37
   NEC Required Nameplate Information 37
   Optional Nameplate Information 39
   Guide to Motor Terminology 41

PART 5 Manual and Magnetic Motor Starters 42
   Manual Starter 42
   Magnetic Starter 43

Chapter 3
Motor Transformers and Distribution Systems 47

PART 1 Power Distribution Systems 47
   Transmission Systems 47
   Unit Substations 48
   Distribution Systems 50
   Power Losses 51
   Switchboards and Panelboards 52
   Motor Control Centers (MCCs) 54
   Electrical Grounding 56
PART 2 Transformer Principles 57
   Transformer Operation 57
   Transformer Voltage, Current, and Turns Ratio 58
   Transformer Power Rating 60
   Transformer Performance 61
PART 3 Transformer Connections and Systems 62
   Transformer Polarity 62
   Single-Phase Transformers 63
   Three-Phase Transformers 65
   Instrument Transformers 67
   Transformer Testing 69

Chapter 4
Motor Control Devices 72

PART 1 Manually Operated Switches 72
   Primary and Pilot Control Devices 72
   Toggle Switches 73
   Pushbutton Switches 73
   Pilot Lights 77
   Tower Light Indicators 78
   Selector Switch 78
   Drum Switch 79
PART 2 Mechanically Operated Switches 80
   Limit Switches 80
   Temperature Control Devices 82
   Pressure Switches 83
   Float and Flow Switches 84

PART 3 Sensors 86
    Proximity Sensors 86
    Photoelectric Sensors 89
    Hall Effect Sensors 91
    Ultrasonic Sensors 92
    Temperature Sensors 93
    Velocity and Position Sensors 95
    Flow Measurement 96
    Magnetic Flowmeters 97

PART 4 Actuators 98
    Relays 98
    Solenoids 99
    Solenoid Valves 100
    Stepper Motors 101
    Servo Motors 102

Chapter 5
Electric Motors 105

PART 1 Motor Principle 105
    Magnetism 105
    Electromagnetism 106
    Generators 106
    Motor Rotation 107

PART 2 Direct Current Motors 110
    Permanent-Magnet DC Motor 110
    Series DC Motor 112
    Shunt DC Motor 113
    Compound DC Motor 114
    Direction of Rotation 115
    Motor Counter Electromotive Force (CEMF) 116
    Armature Reaction 117
    Speed Regulation 117
    Varying DC Motor Speed 118
    DC Motor Drives 119
    Brushless DC Motors 120

PART 3 Three-Phase Alternating Current Motors 122
    Rotating Magnetic Field 122
    Induction Motor 124
    Squirrel-Cage Induction Motor 124
    Wound-Rotor Induction Motor 128
    Three-Phase Synchronous Motor 129

PART 4 Single-Phase Alternating Current Motors 131
    Split-Phase Motor 131
    Split-Phase Capacitor Motor 133
    Shaded-Pole Motor 135
    Universal Motor 135

PART 5 Alternating Current Motor Drives 136
    Variable-Frequency Drive 136
    Inverter Duty Motor 139

PART 6 Motor Selection 139
    Mechanical Power Rating 140
    Current 140
    Code Letter 140
    Design Letter 140
    Efficiency 140
    Energy-Efficient Motors 141
    Frame Size 141
    Frequency 141
    Full-Load Speed 141
    Load Requirements 141
    Motor Temperature Ratings 142
    Duty Cycle 143
    Torque 143
    Motor Enclosures 143
    Metric Motors 144

PART 7 Motor Installation 146
    Foundation 146
    Mounting 146
    Motor and Load Alignment 146
    Motor Bearings 147
    Electrical Connections 148
    Grounding 149
    Conductor Size 149
    Voltage Levels and Balance 149
    Built-in Thermal Protection 150

PART 8 Motor Maintenance and Troubleshooting 151
    Motor Maintenance 151
    Troubleshooting Motors 152

Chapter 6
Contactors and Motor Starters 158

PART 1 Magnetic Contactor 158
    Switching Loads 159
    Capacitor Switching Contactors 162
    Contactor Assemblies 163
    Arc Suppression 166

PART 2 Contactor Ratings, Enclosures, and Solid-State Types 169
    NEMA Ratings 169
    IEC Ratings 170
    Contactor Enclosures 171
    Solid-State Contactor 172

PART 3 Motor Starters 175
    Magnetic Motor Starters 175
    Motor Overcurrent Protection 176
    Motor Overload Relays 178
    NEMA and IEC Symbols 182

Chapter 7
Relays 186

PART 1 Electromechanical Control Relays 186
    Relay Operation 186
    Relay Applications 188
    Relay Styles and Specifications 188
    Interposing Relay 190
PART 2 Solid-State Relays 191
    Operation 191
    Specifications 193
    Switching Methods 194
PART 3 Timing Relays 195
    Motor-Driven Timers 195
    Dashpot Timers 196
    Solid-State Timing Relays 196
    Timing Functions 197
    Multifunction and PLC Timers 201
PART 4 Latching Relays 203
    Mechanical Latching Relays 203
    Magnetic Latching Relays 204
    Latching Relay Applications 204
    Alternating Relays 204
PART 5 Relay Control Logic 207
    Control Circuit Inputs and Outputs 207
    AND Logic Function 207
    OR Logic Function 207
    Combination Logic Functions 208
    NOT Logic Function 208
    NAND Logic Function 208
    NOR Logic Function 209

Chapter 8
Motor Control Circuits 211

PART 1 NEC Motor Installation Requirements 211
    Sizing Motor Branch Circuit Conductors 212
    Branch Circuit Motor Protection 212
    Selecting a Motor Controller 215
    Disconnecting Means for Motor and Controller 215
    Providing a Control Circuit 216
PART 2 Motor Starting 218
    Full-Voltage Starting of AC Induction Motors 218
    Reduced-Voltage Starting of Induction Motors 223
    DC Motor Starting 229
PART 3 Motor Reversing and Jogging 231
    Reversing of AC Induction Motors 231
    Reversing of Single-Phase Motors 234
    Reversing of DC Motors 236
    Jogging 236

PART 4 Motor Stopping 238
    Plugging and Antiplugging 238
    Dynamic Braking 240
    DC Injection Braking 240
    Electromechanical Friction Brakes 241
PART 5 Motor Speed 242
    Multispeed Motors 242
    Wound-Rotor Motors 243

Chapter 9
Motor Control Electronics 245

PART 1 Semiconductor Diodes 245
    Diode Operation 245
    Rectifier Diode 246
    Zener Diode 249
    Light-Emitting Diode 249
    Photodiodes 250
    Inverters 251
PART 2 Transistors 251
    Bipolar Junction Transistor (BJT) 252
    Field-Effect Transistor 254
    Metal Oxide Semiconductor Field-Effect Transistor (MOSFET) 255
    Insulated-Gate Bipolar Transistor (IGBT) 257
PART 3 Thyristors 259
    Silicon-Controlled Rectifiers (SCRs) 259
    Triac 262
    Electronic Motor Control Systems 264
PART 4 Integrated Circuits (ICs) 265
    Fabrication 265
    Operational Amplifier ICs 266
    555 Timer IC 267
    Microcontroller 268
    Electrostatic Discharge (ESD) 270
    Digital Logic 270

Chapter 10
Adjustable-Speed Drives and PLC Installations 275

PART 1 AC Motor Drive Fundamentals 275
    Variable-Frequency Drives (VFDs) 276
    Volts per Hertz Drive 280
    Flux Vector Drive 281
PART 2 VFD Installation and Programming Parameters 283
    Selecting the Drive 283
    Line and Load Reactors 284
    Location 284

Enclosures 284
Mounting Techniques 285
Operator Interface 285
Electromagnetic Interference 285
Grounding 286
Bypass Contactor 286
Disconnecting Means 287
Motor Protection 287
Braking 288
Ramping 289
Control Inputs and Outputs 289
Motor Nameplate Data 292
Derating 292
Types of Variable-Frequency Drives 293
PID Control 294
Parameter Programming 294
Diagnostics and Troubleshooting 295

PART 3 DC Motor Drive Fundamentals 297
Applications 297
DC Drives—Principles of Operation 297
Single-Phase Input—DC Drive 299
Three-Phase Input—DC Drive 300
Field Voltage Control 300
Nonregenerative and Regenerative DC Drives 301
Parameter Programming 302

PART 4 Programmable Logic Controllers (PLCs) 304
PLC Sections and Configurations 304
Ladder Logic Programming 306
Programming Timers 309
Programming Counters 310
Troubleshooting 313

Appendix 318
Index I-1

PREFACE

This book has been written for a course of study that will introduce the reader to a broad range of motor types and control systems. It provides an overview of electric motor operation, selection, installation, control, and maintenance. Every effort has been made to present the most up-to-date information, reflecting the current needs of the industry.

The broad-based approach taken makes this text viable for a variety of motor and control system courses. Content is suitable for colleges, technical institutions, and vocational/technical schools as well as apprenticeship and journeymen training. Electrical apprentices and journeymen will find this book to be invaluable because of National Electrical Code references as well as information on maintenance and troubleshooting techniques. Personnel involved in motor maintenance and repair will find the book to be a useful reference text.

The text is comprehensive! It includes coverage of how motors operate in conjunction with their associated control circuitry. Both older and newer motor technologies are examined. Topics covered range from motor types and controls to installing and maintaining conventional controllers, electronic motor drives, and programmable logic controllers.

Features you will find unique to this motors and controls text include:

Self-Contained Chapters. Each chapter constitutes a complete and independent unit of study. All chapters are divided into parts designed to serve as individual lessons. Instructors can easily pick and choose chapters or parts of chapters that meet their particular curriculum needs.

How Circuits Operate. When understanding the operation of a circuit is called for, a bulleted list is used to summarize its operation. The lists are used in place of paragraphs and are especially helpful for explaining the sequenced steps of a motor control operation.

Integration of Diagrams and Photos. When the operation of a piece of equipment is illustrated by means of a diagram, a photo of the device is included. This feature is designed to increase the level of recognition of devices associated with motor and control systems.

Troubleshooting Scenarios. Troubleshooting is an important element of any motors and controls course. The chapter troubleshooting scenarios are designed to help students with the aid of the instructor to develop a systematic approach to troubleshooting.

Discussion and Critical Thinking Questions. These open-ended questions are designed to give students an opportunity to reflect on the material covered in the chapter. In most cases, they allow for a wide range of responses and provide an opportunity for the student to share more than just facts.

The following content has been added to the chapters listed below:

- Chapter 1 - Safety light curtains
  - Safety interlock switches
  - Emergency stop controls
  - Safety laser scanners
- Chapter 2 - Comparison of common motor NEMA and IEC symbols
  - Riser diagrams
  - Dual voltage three-phase motor connections
  - IEC three-phase motor connections
  - IEC 2-wire and 3-wire control circuits
- Chapter 3 - Motor control center three-phase full-voltage starter bucket
  - Electrical grounding
  - Transformer testing
- Chapter 4 - IEC break-make pushbutton control circuit
  - Two motor emergency stop control circuit
  - Signal light towers
  - Alternating pumping operation and control circuit
  - Comparison of the features and application of sensors
- Chapter 5 - DC brushless motor operation and applications

Chapter 6 - Capacitor switching contactor operation and applications
- DC inverter power contactors

Chapter 7 - Interposing relay operation and applications
- Analog-switching relay operation and applications
- Conveyor motor warning signal control circuit
- Timed and instantaneous relay timer contacts
- One-shot timer solenoid control circuit
- Symmetrical recycle timer flasher circuit

Chapter 8 - Three motor sequential motor starting interlocking circuit.
- Two motor sequential motor stopping interlocking circuit.
- Three-phase motor selector jogging circuit.
- Zero-speed switch operation.
- Antiplugging executed using time-delay relays.

Chapter 9 - Inverter applications and output waveforms.
- Building blocks of an electronic motor control system.
- Three-wire sourcing and sinking sensor connections.

Chapter 10 - Analog versus digital signals.
- 4–20 mA control loop.
- PLC processor module troubleshooting.
- PLC input module troubleshooting.
- PLC output module troubleshooting.

Ancillaries

- **Activities Manual for *Electric Motors and Control Systems*.** This manual contains quizzes, practical assignments, and computer-generated simulated circuit analysis assignments.

  Quizzes made up of multiple choice, true/false, and completion-type questions are provided for each part of each chapter. These serve as an excellent review of the material presented.

  Practical assignments are designed to give the student an opportunity to apply the information covered in the text in a hands-on motor installation.

  The Constructor motor control simulation software is included as part of the manual. This special edition of the program contains preconstructed simulated motor control circuits constructed using both NEMA and IEC symbols. The constructor analysis assignments provide students with the opportunity to test the motor control circuits discussed in the text. The constructor simulation engine visually displays power flow to each component and using animation and sound effects; each component will react accordingly once power is supplied.

  The constructor troubleshooting mode includes a Test Probe that provides an indication of power or continuity. The test probe leads are inserted into the circuit to determine common preprogrammed motor faults.

- **Instructor's Resources** are available to instructors who adopt *Electric Motors and Control Systems*. They can be found on the Instructor Library on Connect and include:

  Answers to the textbook review questions and the Activities Manual quizzes and assignments.

  PowerPoint presentations that feature enhanced graphics along with explanatory text.

  Instructional videos for text motor control circuits.

ACKNOWLEDGMENTS

The efforts of many people are needed to develop and improve a text. Among these people are the reviewers and consultants who point out areas of concern, cite areas of strength, and make recommendations for change. In particular, I would like to acknowledge Don Pelster of Nashville Community College. Don has done an impeccable job of performing a technical edit of the text as well as all the additional Instructor resources.

Electric Motors and Control Systems, 3e contains the most up-to-date information on electric motor operation, selection, installation, control, and maintenance. The text provides a balance between concepts and applications to offer students an accessible framework to introduce a broad range of motor types and control systems.

CHAPTER OBJECTIVES provide an outline of the concepts that will be presented in the chapter. These objectives provide a roadmap to students and instructors on what new material will be presented.

CHAPTER OBJECTIVES

This chapter will help you:
- Recognize symbols frequently used on motor and control diagrams.
- Differentiate between NEMA and IEC motor control symbols.
- Interpret and construct ladder diagrams.
- Interpret wiring, single-line, and block diagrams.
- Explain the terminal connections for different types of motors.
- Interpret connection schemes used for dual-voltage three-phase motors.
- Interpret information found on motor nameplates.
- Explain the terminology used in motor circuits.

Electric Motors and Control Systems provides...

CIRCUIT LISTS When a new operation of a circuit is presented, a bulleted list is used to summarize the operation. The lists are used in place of paragraphs to provide a more accessible summary of the necessary steps of a motor control operation.

The operation of the circuit can be summarized as follows:

- Three-wires are run from the start/stop pushbutton station to the starter.
- When the momentary-contact start button is closed, line voltage is applied to the starter coil to energize it.
- The three main M contacts close to apply voltage to the motor.

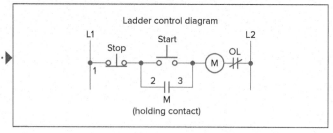

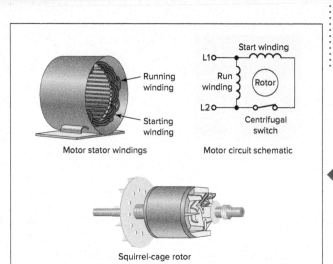

DIAGRAMS AND PHOTOS When the operation of a piece of equipment is illustrated, a photo of the device is included. The integration of diagrams and photos increases the students' recognition of devices associated with motor and control systems.

▶ an engaging framework in every chapter to help students master concepts and realize success beyond the classroom.

REVIEW QUESTIONS Each chapter is divided into parts designed to represent individual lessons. These parts provide professors and students the flexibility to pick and choose topics that best represent their needs. Review questions follow each part to reinforce the new concepts that have been introduced.

Part 1 Review Questions

1. Does the severity of an electric shock increase or decrease with each of the following changes?
   a. A decrease in the source voltage
   b. An increase in body current flow
   c. An increase in body resistance
   d. A decrease in the length of time of exposure
2. a. Calculate the theoretical body current flow (in amperes and milliamperes) of an electric shock victim who comes in contact with a 120 V energy source. Assume a total resistance of 15,000 Ω (skin, body, and ground contacts).
   b. What effect, if any, would this amount of current likely have on the body?
3. Normally a 6 volt lantern battery capable of delivering 2 A of current is considered safe to handle. Why?
4. Why is AC of a 60 Hz frequency considered to be potentially more dangerous than DC of the same voltage and current value?
5. What circuit fault can result in an arc flash?
8. State the piece of electrical safety equipment that should be used to perform each of the following tasks:
   a. A switching operation where there is a risk of injury to the eyes or face from an electric arc.
   b. Using a multimeter to verify the line voltage on a three-phase 480 volt system.
   c. Opening a manually operated high-voltage disconnect switch.
9. Outline the safety procedure to follow when you are connecting shorting probes across de-energized circuits.
10. List three pieces of personal protection equipment required to be worn on most job sites.
11. Explain the way in which safety light curtains operate.
12. a. Describe a typical example of point of operation light curtain control.
    b. Describe a typical example of perimeter access light curtain control.
13. What type of safety switch is used to monitor the

TROUBLESHOOTING SCENARIOS These scenarios are designed to help students develop a systematic approach to troubleshooting that is vital in this course.

Troubleshooting Scenarios

1. Heat is the greatest enemy of a motor. Discuss in what way nonadherence to each of the following motor nameplate parameters could cause a motor to overheat: (a) voltage rating; (b) current rating; (c) ambient temperature; (d) duty cycle.
2. Two identical control relay coils are incorrectly connected in series instead of parallel across a 230 V source. Discuss how this might affect the operation of the circuit.
3. A two-wire magnetic motor control circuit controlling a furnace fan uses a thermostat to automatically operate the motor on and off. A single-pole switch is to be installed next to the remote thermostat and wired so that, when closed, it will override the automatic control and allow the fan to operate at all times regardless of the thermostat setting. Draw a ladder control diagram of a circuit that will accomplish this.
4. A three-wire magnetic motor control circuit uses a remote start/stop pushbutton station to operate the motor on and off. Assume the start button is pressed but the starter coil does not energize. List the possible causes of the problem.
5. How is the control voltage obtained in most motor control circuits?
6. Assume you have to purchase a motor to replace the one with the specifications shown below. Visit the website of a motor manufacturer and report on the specifications and price of a replacement motor.

Horsepower	10
Voltage	200
Hertz	60
Phase	3
Full-load amperes	33
RPM	1725
Frame size	215T
Service factor	1.15
Rating	40C AMB-CONT
Locked rotor code	J
NEMA design code	B
Insulation class	B
Full-load efficiency	85.5
Power factor	76
Enclosure	OPEN

DISCUSSION TOPICS AND CRITICAL THINKING QUESTIONS These open-ended questions are designed to give students an opportunity to review the material covered in the chapter. These questions cover all the parts presented in each chapter and provide an opportunity for the student to show comprehension of the concepts covered.

Discussion Topics and Critical Thinking Questions

1. Why are contacts from control devices not placed in parallel with loads?
2. Record all the nameplate data for a given motor and write a short description of what each item specifies.
3. Search the Internet for electric motor connection diagrams. Record all information given for the connection of the following types of motors:
   a. DC compound motor
   b. AC single-phase dual-voltage induction motor
   c. AC three-phase two-speed induction motor
4. The AC squirrel-cage induction motor is the dominant motor technology in use today. Why?
5. In general, how do NEMA motor standards compare to IEC standards?

xiii

CHAPTER ONE

Safety in the Workplace

Banner Engineering

CHAPTER OBJECTIVES

This chapter will help you:

- Identify the electrical factors that determine the severity of an electric shock.
- Describe arc flash hazard recognition and prevention.
- List of general principles of electrical safety including wearing approved protective clothing and using protective equipment.
- Understand the application of different types of electrical machine safety devices.
- Explain the safety aspects of grounding an electrical motor installation.
- Outline the basic steps in a lockout procedure.
- Identify the functions of the different organizations responsible for electrical codes and standards.

Safety is the number one priority in any job. Every year, electrical accidents cause serious injury or death. Many of these casualties are young people just entering the workplace. They are involved in accidents that result from carelessness, from the pressures and distractions of a new job, or from a lack of understanding about electricity. This chapter is designed to develop an awareness of the dangers associated with electrical power and the potential dangers that can exist on the job or at a training facility.

PART 1 PROTECTING AGAINST ELECTRICAL SHOCK

Electrical Shock

The human body conducts electricity. Even low currents may cause severe health effects. Spasms, burns, muscle paralysis, or death can result, depending on the amount of the current flowing through the body, the route it takes, and the duration of exposure.

The main factor for determining the severity of an electric shock is the amount of electric current that passes through the body. This current is dependent upon the voltage and the resistance of the path it follows through the body.

Electrical **resistance** (R) is the opposition to the flow of current in a circuit and is measured in ohms (Ω). The lower the body resistance, the greater the current flow and potential electric shock hazard. Body resistance can be divided into external (skin resistance) and internal (body tissues and blood stream resistance). Dry skin is a good insulator; moisture lowers the resistance of skin, which explains why shock intensity is greater when the hands are wet. Internal resistance is low owing to the salt and moisture content of the blood. There is a wide degree of variation in body resistance. A shock that may be fatal to one person may cause only brief discomfort to another. Typical body resistance values are:

- Dry skin—100,000 to 600,000 Ω
- Wet skin—1,000 Ω
- Internal body (hand to foot)—400 to 600 Ω
- Ear to ear—100 Ω

Thin or wet skin is much less resistant than thick or dry skin. When skin resistance is low, the current may cause little or no skin damage but severely burn internal organs and tissues. Conversely, high skin resistance can produce severe skin burns but prevent the current from entering the body.

Voltage (E) is the pressure that causes the flow of electric current in a circuit and is measured in units called volts (V). The amount of voltage that is dangerous to life varies with each individual because of differences in body resistance and heart conditions. Generally, any voltage *above 30 V* is considered dangerous.

Electric **current** (I) is the rate of flow of electrons in a circuit and is measured in amperes (A) or milliamperes (mA). One milliampere is one-thousandth of an ampere. The amount of current flowing through a person's body depends on the voltage and resistance. Body current can be calculated using the following Ohm's law formula:

$$\text{Current} = \frac{\text{Voltage}}{\text{Resistance}}$$

If you came into direct contact with 120 volts and your body resistance was 100,000 ohms, then the current that would flow would be:

$$I = \frac{120 \text{ V}}{100,000 \text{ }\Omega}$$
$$= 0.0012 \text{ A}$$
$$= 1.2 \text{ mA } (0.0012 \times 1,000)$$

This is just about at the threshold of perception, so it would produce only a tingle.

If you were sweaty and barefoot, then your resistance to ground might be as low as 1,000 ohms. Then the current would be:

$$I = \frac{120 \text{ V}}{1,000 \text{ }\Omega} = 0.12 \text{ A} = 120 \text{ mA}$$

This is a lethal shock, capable of producing ventricular fibrillation (rapid irregular contractions of the heart) and death!

Voltage is not as reliable an indication of shock intensity because the body's resistance varies so widely that it is impossible to predict how much current will result from a given voltage. The amount of current that passes through the body and the length of time of exposure are perhaps the two most reliable criteria of shock intensity. Once current enters the body, it follows through the circulatory system in preference to the external skin. Figure 1-1 illustrates the relative magnitude and effect of electric current. It doesn't take much current to cause a painful or even fatal shock. A current of 1 mA (1/1000 of an ampere) can be felt. A current of 10 mA will produce a shock of sufficient intensity to prevent voluntary control of muscles, which explains why, in some cases, the victim of electric shock is unable to release grip on the conductor while the current is flowing. A current of 100 mA passing through the body for a second or longer can be fatal. Generally, any current flow *above 0.005 A, or 5 mA*, is considered dangerous.

A 1.5 V flashlight cell can deliver more than enough current to kill a human being, yet it is safe to handle. This is because the resistance of human skin is high enough to limit greatly the flow of electric current. In lower voltage circuits, resistance restricts current flow to very low values. Therefore, there is little danger of an electric shock. Higher voltages, on the other hand, can force enough current though the skin to produce a shock. The danger of harmful shock increases as the voltage increases.

The pathway through the body is another factor influencing the effect of an electric shock. For example, a current from hand to foot, which passes through the heart and part of the central nervous system, is far more dangerous than a shock between two points on the same arm (Figure 1-2).

AC (alternating current) of the common 60 Hz frequency is three to five times more dangerous than DC (direct current) of the same voltage and current value. DC tends to cause a convulsive contraction of the muscles, often forcing the victim away from further current exposure. The effects of AC on the body depend to a great extent on the frequency: low-frequency currents (50–60 Hz) are usually more dangerous than high-frequency currents. AC causes muscle spasm, often "freezing" the hand (the most common part of the body to make contact) to the

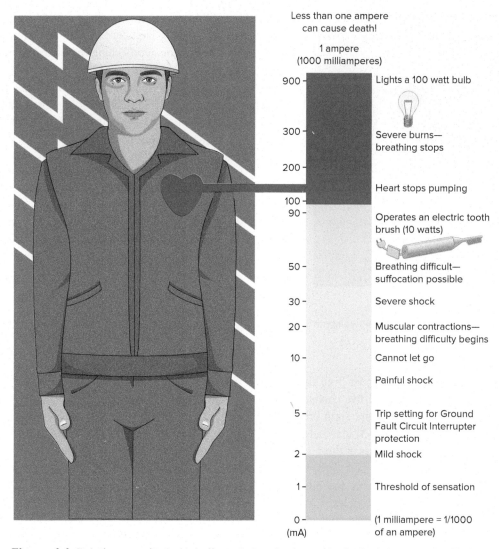

Figure 1-1 Relative magnitude and effect of electric current on the body.

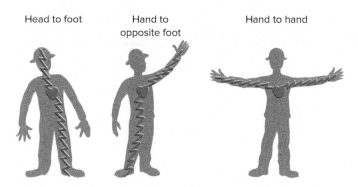

Figure 1-2 Typical electric current pathways that stop normal pumping of the heart.

circuit. The fist clenches around the current source, resulting in prolonged exposure with severe burns.

The most common electric-related injury is a burn. The major types of burns:

- **Electrical burns,** which are a result of electric current flowing through the tissues or bones. The burn itself may be only on the skin surface or deeper layers of the skin may be affected.
- **Arc burns,** which are a result of an extremely high temperature caused by an electric arc (as high as 35,000°F) in close proximity to the body. Electric arcs can occur as a result of poor electrical contact or failed insulation.
- **Thermal contact burns,** which are a result of the skin coming in contact with the hot surfaces of overheated components. They can be caused by contact with objects dispersed as a result of the blast associated with an electric arc.

If a person does suffer a severe shock, it is important to free the victim from the current as quickly as can be done safely. Do not touch the person until the electric power is turned off. You cannot help by becoming a second victim. The victim should be attended to immediately by a person trained in CPR (cardiopulmonary resuscitation).

Figure 1-3 Arc flash.
Photo Courtesy of Honeywell, www.honeywell.com

Arc Flash Hazards

An **arc flash** is the ball of fire that explodes from an electrical **short circuit** between one exposed live conductor and another conductor or to ground. The arc flash creates an enormous amount of energy (Figure 1-3) that can damage equipment and cause severe injury or loss of life.

An arc flash can be caused by dropped tools, unintentional contact with electrical systems, or the buildup of conductive dust, dirt, corrosion, and particles.

Electrical short circuits are either bolted faults or arcing faults. A **bolted fault** is current flowing through bolted bus bars or other electric conductors. An **arcing fault** is current flowing through the air. Because air offers opposition to electric current flow, the arc fault current is always lower than the bolted fault current. An **arc blast** is a flash that causes an explosion of air and metal that produces dangerous pressure waves, sound waves, and molten steel.

In order to understand the hazards associated with an arc flash incident, it is important to understand the difference between an arcing short circuit and a bolted short circuit. A bolted short circuit occurs when the normal circuit current bypasses the load through a very low conductive path, resulting in current flow that can be hundreds or thousands of times the normal load current. In this case, assuming all equipment remains intact, the fault energy is contained within the conductors and equipment, and the power of the fault is dissipated throughout the circuit from the source to the short. All equipment needs to have adequate interrupting ratings to safely contain and clear the high fault currents associated with bolted faults.

In contrast, an arcing fault is the flow of current through a higher-resistance medium, typically the air, between phase conductors or between phase conductors and neutral or ground. Arcing fault currents can be extremely high in current magnitude approaching the bolted short-circuit current but are typically between 38 and 89 percent of the bolted fault. The inverse characteristics of typical overcurrent protective devices generally result in substantially longer clearing times for an arcing fault due to the lower fault values.

Eighty percent of electrical workplace accidents are associated with arc flash and involve burns or injuries caused by intense heat or showers of molten metal or debris. In addition to toxic smoke, shrapnel, and shock waves, the creation of an arc flash produces an intense flash of blinding light. This flash is capable of causing immediate vision damage and can increase a worker's risk of future vision impairment.

An arc flash hazard exists when a person interacts with equipment in a way that could cause an electric arc. Such tasks may include testing or troubleshooting, application of temporary protective grounds, or the opening or closing of power circuit breakers as illustrated in Figure 1-4. *Arcs can produce temperature four times hotter than the surface of the sun.* To address this hazard, safety standards such as National Fire Protection Association (NFPA) 70E have been developed to minimize arc flash hazards. The NFPA standards require that any panel likely to be serviced by a worker be **surveyed** and **labeled**. Injuries can be avoided with training; with proper work practices; and by using protective face shields, hoods, and clothing that are NFPA-compliant.

Figure 1-4 An arc flash hazard exists when a person interacts with equipment.
Chemco Electrical Contractors Ltd.

Figure 1-5 Typical safety signs.

Personal Protective Equipment

Construction and manufacturing worksites, by nature, are potentially hazardous places. For this reason, safety has become an increasingly large factor in the working environment. The electrical industry, in particular, regards **safety** to be unquestionably the most single important priority because of the hazardous nature of the business. A safe operation depends largely upon all personnel being informed and aware of potential hazards. Safety signs and tags indicate areas or tasks that can pose a hazard to personnel and/or equipment. Signs and tags may provide warnings specific to the hazard, or they may provide safety instructions (Figure 1-5).

To perform a job safely, the proper protective clothing must be used. Appropriate attire should be worn for each particular job site and work activity (Figure 1-6). The following points should be observed:

1. Hard hats, safety shoes, and goggles must be worn in areas where they are specified. In addition, hard hats shall be approved for the purpose of the electrical work being performed. *Metal hats are not acceptable!*
2. Safety earmuffs or earplugs must be worn in noisy areas.
3. Clothing should fit snugly to avoid the danger of becoming entangled in moving machinery. Avoid wearing synthetic-fiber clothing such as polyester material as these types of materials may melt or ignite when exposed to high temperatures and may increase the severity of a burn. Instead always wear cotton clothing.
4. Remove all metal jewelry when working on energized circuits; gold and silver are excellent conductors of electricity.
5. Confine long hair or keep hair trimmed when working around machinery.

A wide variety of electrical safety equipment is available to prevent injury from exposure to live electric circuits (Figure 1-7). Electrical workers should be familiar with safety standards such as **NFPA-70E** that pertain to the type of protective equipment required, as well as how such equipment shall be cared for. To make sure electrical protective equipment actually performs as designed, it must be inspected for damage before each day's use and immediately following any incident that can reasonably be suspected of having caused damage. All electrical protection equipment must be listed and may include the following:

Rubber Protective Equipment—Rubber gloves are used to prevent the skin from coming into contact with energized circuits. A separate outer leather cover is used to protect the rubber glove from punctures and other damage. Rubber blankets are used to prevent contact with energized conductors or circuit parts when working near exposed energized circuits. All rubber protective equipment must be marked with the appropriate voltage rating and the last inspection date. It is important that the insulating value of both rubber gloves and blankets have a voltage rating that matches that of the circuit or equipment they are to be used with. Insulating gloves must be given an air test daily, along with inspection. Twirl the glove around quickly or roll it down to trap air inside. Squeeze the palm, fingers, and thumb

Figure 1-6 Appropriate attire should be worn for each particular job site and work activity.
Photo courtesy of Capital Safety, www.capitalsafety.com

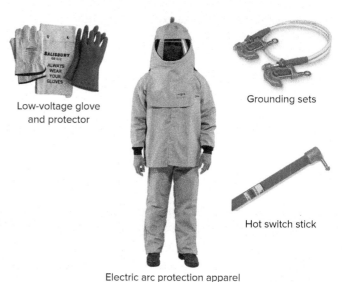

Figure 1-7 Electrical safety equipment.
Photo courtesy of W.W. Grainger, www.grainger.com

to detect any escaping air. If the glove does not pass this inspection, it must be disposed of.

Protection Apparel—Special protective equipment available for high-voltage applications include high-voltage sleeves, high-voltage boots, nonconductive protective helmets, nonconductive eyewear and face protection, switchboard blankets, and flash suits.

Hot Sticks—Hot sticks are insulated tools designed for the manual operation of high-voltage disconnecting switches, high-voltage fuse removal and insertion, as well as the connection and removal of temporary grounds on high-voltage circuits. A hot stick is made up of two parts, the head, or hood, and the insulating rod. The head can be made of metal or hardened plastic, while the insulating section may be wood, plastic, or other effective insulating materials.

Shorting Probes—Shorting probes are used on de-energized circuits to discharge any charged capacitors or built-up static charges that may be present when power to the circuit is disconnected. Also, when working on or near any high-voltage circuits, shorting probes should be connected and left attached as an extra safety precaution in the event of any accidental application of voltage to the circuit. When installing a shorting probe, first connect the test clip to a good ground contact. Next, hold the shorting probe by the handle and hook the probe end over the part or terminal to be grounded. Never touch any metal part of the shorting probe while grounding circuits or components.

Face Shields—Listed face shields should be worn during all switching operations where there is a possibility of injury to the eyes or face from electrical arcs or flashes, or from flying or falling objects that may result from an electrical explosion.

With proper precautions, there is no reason for you to ever receive a serious electrical shock. Receiving an electrical shock is a clear warning that proper safety measures have not been followed. To maintain a high level of electrical safety while you work, there are a number of precautions you should follow. Your individual job will have its own unique safety requirements. However, the following are given as essential basics.

- Never take a shock on purpose.
- Keep material or equipment at least 10 feet away from high-voltage overhead power lines.
- Do not close any switch unless you are familiar with the circuit that it controls and know the reason for its being open.
- When working on any circuit, take steps to ensure that the controlling switch is not operated in your absence. Switches should be padlocked open, and warning notices should be displayed (**lockout/tagout**).
- Avoid working on "live" circuits as much as possible.
- When installing new machinery, ensure that the framework is efficiently and permanently grounded.
- Always treat circuits as "live" until you have proven them to be "dead." Presumption at this point can kill you. It is a good practice to take a meter reading before starting work on a dead circuit.
- Avoid touching any grounded objects while working on electrical equipment.
- Remember that even with a 120 V control system, you may well have a higher voltage in the panel. Always work so that you are clear of any of the higher voltages. (Even though you are testing a 120 V system, you are most certainly in close proximity to 240 V or 480 V power.)
- Don't reach into energized equipment while it is being operated. This is particularly important in high-voltage circuits.
- Use good electrical practices even in temporary wiring for testing. At times you may need to make alternate connections, but make them secure enough so that they are not in themselves an electrical hazard.
- When working on live equipment containing voltages over approximately 30 V, work with only one hand. Keeping one hand out of the way greatly reduces the possibility of passing a current through the chest.
- Safely discharge capacitors before handling them. Capacitors connected in live motor control circuits can store a lethal charge for a considerable time after the voltage to the circuits has been switched off. Although Article 460 of the National Electrical Code (NEC) requires an automatic **discharge** to under 50 volts within 1 minute, never assume that the discharge is working! Always verify that there is no voltage present.

Confined spaces can be found in almost any workplace. Figure 1-8 illustrates examples of typical confined spaces. In general, a "confined space" is an enclosed or partially enclosed space that:

- Is not primarily designed or intended for human occupancy.
- Has a restricted entrance or exit by way of location, size, or means.
- Can represent a risk for the health and safety of anyone who enters, because of its design, construction, location, or atmosphere; the materials or substances in it; work activities being carried out in it; or the mechanical, process, and safety hazards present.

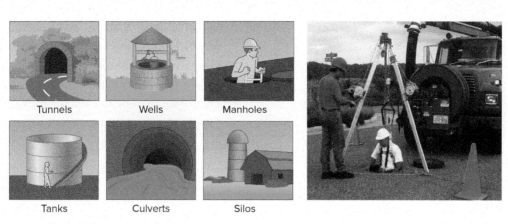

Figure 1-8 Confined spaces.
Photo courtesy of Capital Safety, www.capitalsafety.com

All hazards found in a regular workspace can also be found in a confined space. However, they can be even more hazardous in a confined space than in a regular worksite. Hazards in confined spaces can include poor air quality, fire hazard, noise, moving parts of equipment, temperature extremes, poor visibility, and barrier failure resulting in a flood or release of free-flowing solid. A "permit-required confined space" is a confined space that has specific health and safety hazards associated with it. Permit-required confined spaces require assessment of procedures in compliance with Occupational Safety and Health Administration (OSHA) standards prior to entry.

Machine Safety

Safety Light Curtains

Safety light curtains protect personnel from injury and machines from damage. They create a curtain of photoelectric light beams between an **emitter** and a receiver that then sense whenever an object intrudes into this light field. (Figure 1-9).

- When the light curtain is active, the emitter LEDs emit pulses of infrared light in rapid sequence.
- When the light reaches the corresponding phototransistor in the receiver, it produces an electrical signal.
- When an object such as an operator's hand blocks one of the light beams, the phototransistor that normally detects that beam receives no light.
- As a result, the phototransistor does not produce the signal it normally would at that time in the sequence.
- The light-curtain control circuitry senses this and sends a stop signal to the machinery
- They have an advantage over mechanical guards in that they're relatively small and unobtrusive.

Two general categories of safety light curtains are based on the scale of protection they provide.

Figure 1-9 Safety light curtain.
Banner Engineering

- **Point of Operation Control (POC)** These light curtains are designed to protect on a small scale. Specifically, they help protect hands, fingers and arms that are going to be operating the machinery. These light curtains are generally located very close to the machine, directly where the worker will be interacting with it.
- **Perimeter Access Control (PAC)** These light curtains offer full body protection. They essentially create a fence around machines that don't require up-close usage by workers and are designed to detect people or objects when they intrude into the light barrier.

Safety Interlock switches

Safety interlock switches are a means of safeguarding that monitors the position of a guard or gate. They can be used to shut off power, control personnel access and prevent a machine from starting when the guard is open. Figure 1-10 shows an example of a tongue-operated safety interlock switch. The switching element and tongue actuator are

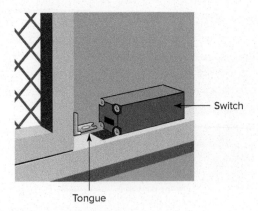

Figure 1-10 Tongue-operated safety interlock switch.

combined or separated during actuation. They fit to the leading edge of sliding of machine guards and provide interlock detection of movement.

Emergency Stop Controls

Emergency stop controls provide workers a means of stopping a device during an emergency in order to prevent injury to personnel and material loss. An emergency stop switch (Figure 1-11) must be highly visible in color and shape and must be easy to operate in emergency situations. Requirements for emergency stop buttons include:

- A direct opening mechanism must be installed on a normally closed (NC) contact. A general-purpose push button switch doesn't have a direct opening mechanism on the NC contact. If the contact welds, conduction will be maintained and the device cannot be stopped in a hazardous situation (load). If this occurs, the device may keep operating in the hazardous state. Therefore, use the NC contact of an emergency stop push button switch for safety applications.
- There must be a self-holding function. This function requires that once emergency stop is activated the control process cannot be started again until the actuating stop switch has been reset to the ON position.
- The button must be a mushroom head design or something equally easy to use.
- The button must be red and the background must be yellow.

Safety Laser Scanners

Safety laser scanners (Figure 1-12) provide a laser safety solution for safeguarding mobile vehicles and stationary applications, such as the interior of robotic work cells that cannot be solved by other safeguarding solutions. Common uses for safety laser scanners include:

- A safety laser scanner can be mounted on an automated guide vehicle or transfer cart (Figure 1-13) to

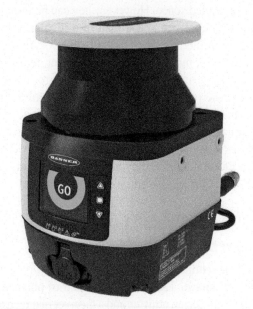

Figure 1-12 Safety laser scanner.
Banner Engineering

Figure 1-11 Emergency stop switch.

Figure 1-13 Safety laser scanner mounted on a transfer cart.
Banner Engineering

eliminate the risk of collisions with objects or people in its path.
- Scanners allow for vertical mounting to detect any undesirable entrances into a hazardous area. This is ideal in locations where it would be too difficult to effectively mount a light curtain.
- Safety scanners prevent hazards from operating when an unintended object or person is in a dangerous area. These safety devices can be unobtrusively mounted to avoid damage or potential impact, while protecting a simple or complex shaped area.

Part 1 Review Questions

1. Does the severity of an electric shock increase or decrease with each of the following changes?
   a. A decrease in the source voltage
   b. An increase in body current flow
   c. An increase in body resistance
   d. A decrease in the length of time of exposure
2. a. Calculate the theoretical body current flow (in amperes and milliamperes) of an electric shock victim who comes in contact with a 120 V energy source. Assume a total resistance of 15,000 Ω (skin, body, and ground contacts).
   b. What effect, if any, would this amount of current likely have on the body?
3. Normally a 6 volt lantern battery capable of delivering 2 A of current is considered safe to handle. Why?
4. Why is AC of a 60 Hz frequency considered to be potentially more dangerous than DC of the same voltage and current value?
5. What circuit fault can result in an arc flash?
6. Define each of the following terms associated with an arc flash:
   a. *Bolted fault*
   b. *Arcing fault*
   c. *Arc blast*
7. Explain why an arc flash is so potentially dangerous.
8. State the piece of electrical safety equipment that should be used to perform each of the following tasks:
   a. A switching operation where there is a risk of injury to the eyes or face from an electric arc.
   b. Using a multimeter to verify the line voltage on a three-phase 480 volt system.
   c. Opening a manually operated high-voltage disconnect switch.
9. Outline the safety procedure to follow when you are connecting shorting probes across de-energized circuits.
10. List three pieces of personal protection equipment required to be worn on most job sites.
11. Explain the way in which safety light curtains operate.
12. a. Describe a typical example of point of operation light curtain control.
    b. Describe a typical example of perimeter access light curtain control.
13. What type of safety switch is used to monitor the position of a guard or gate?
14. Emergency stop controls are required to have a self-holding function. What does this function require?
15. What type of safety device can be mounted on a transfer cart to eliminate the risk of collisions with objects in its path?

PART 2 GROUNDING—LOCKOUT—CODES

Grounding and Bonding

Proper grounding practices protect people from the hazards of electric shock and ensure the correct operation of overcurrent protection devices. Intentional grounding is required for the safe operation of electrical systems and equipment. Unintentional or accidental grounding is considered a fault in electrical wiring systems or circuits.

"Grounding" is the intentional connection of a current-carrying conductor to the earth. For AC premises wiring systems in buildings and similar structures, this ground connection is made on the premise side of the service equipment and the supply source, such as a utility transformer. The prime reasons for grounding are:

- To limit the voltage surges caused by lightning, utility system operations, or accidental contact with higher-voltage lines.
- To provide a ground reference that stabilizes the voltage under normal operating conditions.
- To facilitate the operation of overcurrent devices such as circuit breakers, fuses, and relays under ground-fault conditions.

"Bonding" is the permanent joining together of metal parts that aren't intended to carry current during normal operation, which creates an electrically conductive path that can safely carry current under ground-fault conditions. The prime reasons for bonding are:

- To establish an effective path for fault current that facilitates the operation of overcurrent protective devices.
- To minimize shock hazard to people by providing a low-impedance path to ground. Bonding limits the touch voltage when non-current-carrying metal parts are inadvertently energized by a ground fault.

The Code requires all metal used in the construction of a wiring system to be bonded to, or connected to, the ground system. The intent is to provide a low-impedance path back to the utility transformer in order to quickly clear faults. Figure 1-14 illustrates the ground-fault current path required to ensure that overcurrent devices operate to open the circuit. The earth is not considered an effective ground-fault current path. The resistance of earth is so high that very little fault current returns to the electrical supply source through the earth. For this reason the main bonding jumper is used to provide the connection between the grounded service conductor and the equipment grounding conductor at the service. Bonding jumpers may be located throughout the electrical system, but a main bonding jumper is located only at the service entrance. Grounding is accomplished by connecting the circuit to a metal underground water pipe, the metal frame of a building, a concrete-encased electrode, or a ground ring.

A grounding system has two distinct parts: system grounding and equipment grounding. **System grounding** is the electrical connection of one of the current carrying conductors of the electrical system to the ground. **Equipment grounding** is the electrical connection of all the metal parts that do not carry current to ground. Conductors that form parts of the grounding system include the following:

Equipment grounding conductor (EGC) is an electrical conductor that provides a low-impedance ground path between electrical equipment and enclosures within the distribution system. Figure 1-15 shows the connection for

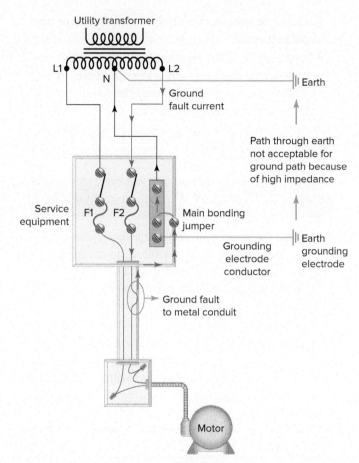

Figure 1-14 Ground-fault current path.

an EGC. Electrical motor windings are normally insulated from all exposed non-current-carrying metal parts of the motor. However, if the insulation system should fail, then the motor frame could become energized at line voltage. Any person contacting a grounded surface and the energized motor frame simultaneously could be

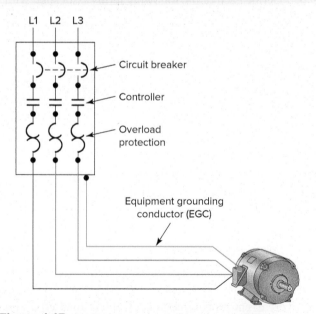

Figure 1-15 Equipment grounding conductor (EGC).

severely injured or killed. Effectively grounding the motor frame forces it to take the same zero potential as the earth, thus preventing this possibility.

Grounded conductor is a conductor that has been intentionally grounded.

Grounding electrode conductor is a conductor used to connect the equipment grounding conductor or the grounded conductor (at the service entrance or at the separately derived system) to the grounding electrode(s). A **separately derived system** is a system that supplies electrical power derived (taken) from a source other than a service, such as the secondary of a distribution transformer.

A **ground fault** is defined as an unintentional, electrically conducting connection between an ungrounded conductor of an electric circuit and the normally non-current-carrying conductors, metallic enclosures, metallic raceways, metallic equipment, or earth. The **ground-fault circuit interrupter (GFCI)** is a device that can sense small ground-fault currents. The GFCI is fast acting; the unit will shut off the current or interrupt the circuit within 1/40 second after its sensor detects a leakage as small as 5 milliamperes (mA). Most circuits are protected against overcurrent by 15 ampere or larger fuses or circuit breakers. This protection is adequate against short circuits. Overloads implies protection for a motor. Leakage currents to ground may be much less than 15 amperes and still be hazardous.

Figure 1-16 shows the simplified circuit of a GFCI receptacle. The device compares the amount of current in the ungrounded (hot) conductor with the amount of current in the grounded (neutral) conductor. Under normal operating conditions, the two will be **equal** in value. If the current in the neutral conductor becomes less than the current in the hot conductor, a ground-fault condition exists. The amount of current that is missing is returned to the source by the ground-fault path. Whenever the ground-fault current exceeds approximately *5 mA,* the device automatically opens the circuit to the receptacle.

GFCIs can be used successfully to reduce electrical hazards on construction sites. The ground-fault protection rules and regulations of OSHA have been determined necessary and appropriate for employee safety and health. According to OSHA, it is the employer's responsibility to provide either (1) ground-fault circuit interrupters on construction sites for receptacle outlets in use and not part of the permanent wiring of the building or structure or (2) a scheduled and recorded assured equipment-grounding conductor program on construction sites, covering all cord sets, receptacles that are not part of the permanent wiring of the building or structure, and equipment connected by cord and plug that are available for use or used by employees.

Lockout and Tagout

Electrical "lockout" is the process of removing the source of electrical power and installing a lock, which prevents the power from being turned ON. Electrical "tagout" is the process of placing a danger tag on the source of electrical power, which indicates that the equipment may not be operated until the danger tag is removed (Figure 1-17). This procedure is necessary for the safety of personnel in that it ensures that no one will inadvertently energize the equipment while it is being worked on. Electrical lockout and tagout is used when servicing electrical equipment that does not require power to be on to perform the service as in the case of motor alignment or replacement of a motor or motor control component.

Lockout means achieving a zero state of energy while equipment is being serviced. Just pressing a stop button to shut down machinery won't provide you with security. Someone else working in the area can simply reset it. Even a separate automated control could be activated to override the manual controls. It's essential that all interlocking or dependent systems also be deactivated. These could

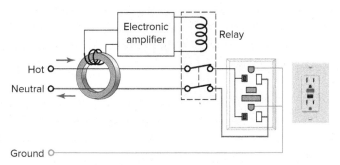

Zero current flows in this conductor under normal operating conditions.

Figure 1-16 GFCI receptacle.
Steve Wisbauer/Stockbyte/Getty Images

Figure 1-17 Lockout/tagout devices.
Photo courtesy of Panduit Corp, www.panduit.com

feed into the system being isolated, either mechanically or electrically. It's important to test the start button before resuming any work in order to verify that all possible energy sources have been isolated.

The "danger tag" has the same importance and purpose as a lock and is used alone only when a lock does not fit the disconnect means. Danger tags are required to be securely attached at the disconnect device with space provided for the worker's name, craft, and procedure that is taking place.

The following are the basic steps in a lockout procedure:

- **Prepare for machinery shutdown:** Document all lockout procedures in a plant safety manual. This manual should be available to all employees and outside contractors working on the premises. Management should have policies and procedures for safe lockout and should also educate and train everyone involved in locking out electrical or mechanical equipment. Identify the location of all switches, power sources, controls, interlocks, and other devices that need to be locked out in order to isolate the system.

- **Machinery or equipment shutdown:** Stop all running equipment by using the controls at or near the machine.

- **Machinery or equipment isolation:** Disconnect the switch (do not operate if the switch is still under load). Stand clear of the box and face away while operating the switch with the left hand (if the switch is on the right side of the box).

- **Lockout and tagout application:** Lock the disconnect switch in the OFF position. If the switch box is the breaker type, make sure the locking bar goes right through the switch itself and not just the box cover. Some switch boxes contain fuses, and these should be removed as part of the lockout process. If this is the case, use a fuse puller to remove them. Use a tamper-proof lock with one key, which is kept by the individual who owns the lock. Combination locks, locks with master keys, and locks with duplicate keys are not recommended.

  Tag the lock with the signature of the individual performing the repair and the date and time of the repair. There may be several locks and tags on the disconnect switch if more than one person is working on the machinery. The machine operator's (and/or the maintenance operator's) lock and tag will be present as well as the supervisor's.

- **Release of stored energy:** All sources of energy that have the potential to unexpectedly start up, energize, or release must be identified and locked, blocked, or released.

  Capacitors retain their charge for a considerable period of time after having been disconnected from

Figure 1-18 Testing for the presence of voltage.
Photo courtesy of Fluke, www.fluke.com. Reproduced with Permission.

the power source. Always assume there is a voltage present when working with circuits having high capacitance, even when the circuit has been disconnected from its power source.

- **Verification of isolation:** Use a voltage test to determine that voltage is present at the line side of the switch or breaker. When all phases of outlet are dead with the line side live, you can verify the isolation. Ensure that your voltmeter is working properly by performing the "live-dead-live" check before each use: First check your voltmeter on a known live voltage source of the same voltage range as the circuit you will be working on. Next check for the presence of voltage on the equipment you have locked out (Figure 1-18). Finally, to ensure that your voltmeter did not malfunction, check it again on the known live source.

- **Lockout/tagout removal:** Remove tags and locks when the work is completed. Each individual must remove his or her own lock and tag. If there is more than one lock present, the person in charge of the work is the last to remove his or her lock. Before reconnecting the power, check that all guards are in place and that all tools, blocks, and braces used in the repair are removed. Make sure that all employees stand clear of the machinery.

Electrical Codes and Standards

Occupational Safety and Health Administration (OSHA) In 1970, Congress created a regulatory agency known as the Occupational Safety and Health Administration

(OSHA). The purpose of OSHA is to assure safe and healthful working conditions for working men and women by authorizing enforcement of standards developed under the Act, by encouraging and assisting state governments to improve and expand their own occupational safety and health programs, and by providing for research, information, education, and training in the field of occupational health and safety.

OSHA inspectors check on companies to make sure they are following prescribed safety regulations. OSHA also inspects and approves safety products. OSHA's electrical standards are designed to protect employees exposed to dangers such as electric shock, electrocution, fires, and explosions.

National Electrical Code (NEC) The National Electrical Code (NEC) comprises a set of rules that, when properly applied, are intended to provide a safe installation of electrical wiring and equipment. This widely adopted minimum electrical safety standard has as its primary purpose "the practical safeguarding of persons and property from hazards arising from the use of electricity." Standards contained in the NEC are enforced by being incorporated into the different city and community ordinances that deal with electrical installations in residences, industrial plants, and commercial buildings. The NEC is the most widely adopted code in the world and many jurisdictions adopt it in its entirety without exception or local amendments or supplements.

An "Article" of the Code covers a specific subject. For example, Article 430 of the NEC covers motors and all associated branch circuits, overcurrent protection, overload, and so on. The installation of motor-control centers is covered in Article 409, and air-conditioning equipment is covered in Article 440. Each Code rule is called a "Code Section." A Code Section may be broken down into subsections. For example, the rule that requires a motor disconnecting means be mounted within sight of the motor and driven machinery is contained in Section 430.102 (B). "In sight" is defined by the Code as visible and not more than 50 feet in distance (Article 100—definitions).

Article 430 on motors is the longest article in the Code. One of the reasons for this is that the characteristics of a motor load are quite different from heating or incandescent lighting loads and so the method of protecting branch circuit conductors against excessive current is slightly different. Non-motor branch circuits are protected against overcurrent, whereas motor branch circuits are protected against overload conditions as well as groundfaults and short circuits. The single-line diagram of Figure 1-19 illustrates some of the motor terminology used throughout the Code and by motor control equipment manufacturers.

The use of electrical equipment in hazardous locations increases the risk of fire or explosion. Hazardous locations can contain gas, dust (e.g., grain, metal, wood, or

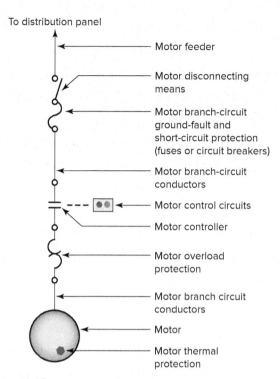

Figure 1-19 Motor terminology.

coal), or flying fibers (textiles or wood products). A substantial part of the NEC is devoted to the discussion of hazardous locations, because electrical equipment can become a source of ignition in these volatile areas. Articles 500 through 504 and 510 through 517 provide classification and installation standards for the use of electrical equipment in these locations. Explosion-proof apparatus, dust-ignition-proof equipment, and purged and pressurized equipment are examples of protection techniques that can be used in certain hazardous (classified) locations. Figure 1-20 shows a motor start/stop station designed to meet hazardous location requirements.

Figure 1-20 Pushbutton station designed for hazardous locations.
Photo courtesy of Rockwell Automation, www.rockwellautomation.com

National Fire Protection Association (NFPA) The National Fire Protection Association (NFPA) develops codes governing construction practices in the building and electrical trades. It is the world's largest and most influential fire safety organization. NFPA has published almost 300 codes and standards, including the National Electrical Code, with the mission of preventing the loss of life and property. Fire prevention is a very important part of any safety program. Figure 1-21 illustrates some of the common types of fire extinguishers and their applications. Icons found on the fire extinguisher indicate the types of fire the unit is intended to be used on.

It is important to know where your fire extinguishers are located and how to use them. In case of an electrical fire, the following procedures should be followed:

1. Trigger the nearest fire alarm to alert all personnel in the workplace as well as the fire department.
2. If possible, disconnect the electric power source.
3. Use a carbon dioxide or dry-powder fire extinguisher to put out the fire. *Under no circumstances use water*, as the stream of water may conduct electricity through your body and give you a severe shock.
4. Ensure that all persons leave the danger area in an orderly fashion.
5. Do not reenter the premises unless advised to do so.

There are four classes of fires, categorized according to the kind of material that is burning (see Figure 1-21):

- **Class A** fires are those fueled by materials that, when they burn, leave a residue in the form of ash, such as paper, wood, cloth, rubber, and certain plastics.
- **Class B** fires involve flammable liquids and gases, such as gasoline, paint thinner, kitchen grease, propane, and acetylene.
- **Class C** fires involve energized electrical wiring or equipment such as motors and panel boxes.
- **Class D** fires involve combustible metals such as magnesium, titanium, zirconium, sodium, and potassium.

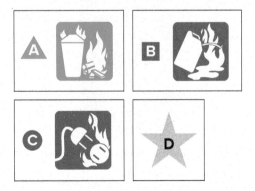

Figure 1-21 Types of fire extinguishers and their applications.

Nationally Recognized Testing Laboratory (NRTL) Article 100 of the NEC defines the terms "labeled" and "listed," which are both related with product evaluation. Labeled or listed indicates the piece of electrical equipment or material has been tested and evaluated for the purpose for which it is intended to be used. Products that are big enough to carry a label are usually labeled. The smaller products are usually listed. Any modification of a piece of electrical equipment in the field may void the label or listing.

In accordance with OSHA Safety Standards, a Nationally Recognized Testing Laboratory (NRTL) must test electrical products for conformity to national codes and standards before they can be listed or labeled. The biggest and best-known testing laboratory is the Underwriters' Laboratories, identified with the **UL logo** shown in Figure 1-22. The purpose of the Underwriters' Laboratories is to establish, maintain, and operate laboratories for the investigation of materials, devices, products, equipment, construction, methods, and systems with regard to hazards affecting life and property.

National Electrical Manufacturers Association (NEMA) The National Electrical Manufacturers Association (NEMA) is a group that defines and recommends safety standards for electrical equipment. Standards established by NEMA assist users in proper selection of industrial control equipment. As an example, NEMA standards provide practical information concerning the rating, testing, performance, and manufacture of motor control devices such as enclosures, contactors, and starters.

International Electrotechnical Commission (IEC) The International Electrotechnical Commission (IEC) is a Europe-based organization made up of national committees from more than 60 countries. There are basically two major mechanical and electrical standards for motors: NEMA in North America and IEC in most of the rest of the world. Dimensionally, IEC standards are expressed in metric units. Though NEMA and IEC standards use different units of measurements and terms, they are essentially analogous in ratings and, for most common applications, are largely interchangeable. NEMA standards tend to be more conservative—allowing more room for "design interpretation," as has been U.S. practice. Conversely, IEC standards tend to be more specific, more categorized—some say more precise—and designed with less over

Figure 1-22 Underwriters' Laboratories logo.
Source: Logo of Underwriters' Laboratories.

current tolerance. As an example, a NEMA-rated motor starter will typically be larger than its IEC counterpart.

Institute of Electrical and Electronics Engineers (IEEE) The Institute of Electrical and Electronics Engineers (IEEE) is a technical professional association whose primary goal is to foster and establish technical developments and advancements in electrical and electronic standards. IEEE is a leading authority in technical areas. Through its technical publishing, conferences, and consensus-based standards activities, the IEEE produces more than 30 percent of the world's published literature in electrical and electronic engineering. For example, IEEE Standard 142 provides all the information you need for a good grounding design.

Part 2 Review Questions

1. Explain how grounding the frame of a motor can prevent someone from receiving an electric shock.
2. Compare the terms *grounding* and *bonding*.
3. What is the minimum amount of leakage ground current required to trip a ground-fault circuit interrupter?
4. List the seven steps involved in a lockout/tagout procedure.
5. A disconnect switch is to be pulled open as part of a lockout procedure. Explain the safe way to proceed.
6. What is the prime objective of the National Electrical Code?
7. How are the standards contained in the NEC enforced?
8. Explain the difference between a Code Article and a Section.
9. What do the icons found on most fire extinguishers indicate?
10. What does a UL-labeled or -listed electrical device signify?
11. List three motor control devices that are rated by NEMA.
12. Compare NEMA and IEC motor standards.

Troubleshooting Scenarios

1. The voltage between the frame of a three-phase 208 V motor and a grounded metal pipe is found to be 120 V. What does this indicate? Why?
2. A ground-fault circuit interrupter does not provide overload protection. Why?
3. A listed piece of electrical equipment is not installed according to the manufacturer's instructions. Discuss why this will void the listing.
4. A hot stick is to be used to open a manually operated high-voltage disconnect switch. Why is it important to make certain that no loads are connected to the circuit when the switch is opened?
5. An employee is contemplating using his lockout lock to secure his personal tool crib. Why is this not acceptable?

Discussion Topics and Critical Thinking Questions

1. Worker A makes contact with a live wire and receives a mild shock. Worker B makes contact with the same live wire and receives a fatal shock. Discuss some of the reasons why this might occur.
2. The victim of death by electrocution is found with his fist still clenched firmly around the live conductor he made contact with. What does this indicate?
3. Why can birds safely rest on high-voltage power lines without getting shocked?
4. You have been assigned the task of explaining the company lockout procedure to new employees. Outline what you would consider the most effective way of doing this.
5. Visit the website of one of the groups involved with electrical codes and standards. Report on the service it provides.

CHAPTER TWO

Understanding Electrical Drawings

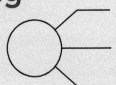

Three-phase motor

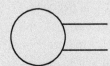

Single-phase motor

Photo courtesy of Ideal Industries, www.idealindustries.com

CHAPTER OBJECTIVES

This chapter will help you:

- Recognize symbols frequently used on motor and control diagrams.
- Differentiate between NEMA and IEC motor control symbols.
- Interpret and construct ladder diagrams.
- Interpret wiring, single-line, and block diagrams.
- Explain the terminal connections for different types of motors.
- Interpret connection schemes used for dual-voltage three-phase motors.
- Interpret information found on motor nameplates.
- Explain the terminology used in motor circuits.
- Understand the operation of manual and magnetic motor starters.
- Follow IEC type 2-wire and 3-wire circuit schematics.

Different types of electrical drawings are used in working with motors and their control circuits. In order to facilitate making and reading electrical drawings, certain standard symbols are used. To read electrical motor drawings, it is necessary to know both the meaning of the symbols and how the equipment operates. This chapter will help you understand the use of symbols in electrical drawings. The chapter also explains motor terminology and illustrates it with practical applications.

PART 1 SYMBOLS— ABBREVIATIONS—LADDER DIAGRAMS

Motor Symbols

A motor **control circuit** can be defined as a means of supplying power to and removing power from a motor. The symbols used to represent the different components of a motor control system can be considered a type of technical shorthand. The use of these symbols tends to make circuit diagrams less complicated and easier to read and understand.

In motor control systems, symbols and related lines show how the parts of a circuit are connected to one another. Unfortunately, not all electrical and electronic symbols are standardized. You will find slightly different symbols used by different manufacturers. Also, symbols sometimes look nothing like the real thing, so you have to learn what the symbols mean. Figure 2-1 shows some of the typical NEMA symbols used in motor circuit diagrams.

There are two standards for electric motor control symbols: NEMA and IEC. Figure 2-2 shows a comparison of common motor NEMA and IEC symbols.

Abbreviations for Motor Terms

An abbreviation is the shortened form of a word or phase. Uppercase letters are used for most abbreviations. The following is a list of some of the abbreviations commonly used in motor circuit diagrams.

AC	alternating current
ARM	armature
AUTO	automatic
BKR	breaker
COM	common
CR	control relay
CT	current transformer
DC	direct current
DB	dynamic braking
FLD	field
FWD	forward
GND	ground
HP	horsepower
L1, L2, L3	power line connections
LS	limit switch
MAN	manual
MTR	motor
M	motor starter
NEG	negative
NC	normally closed
NO	normally open
OL	overload relay
PH	phase
PL	pilot light
POS	positive
PWR	power
PRI	primary
PB	push button
REC	rectifier
REV	reverse
RH	rheostat
SSW	safety switch
SEC	secondary
1PH	single-phase
SOL	solenoid
SW	switch
T1, T2, T3	motor terminal connections
3PH	three-phase
TD	time delay
TRANS	transformer

Motor Ladder Diagrams

Motor control drawings provide information on circuit operation, device and equipment location, and wiring instructions. Symbols used to represent switches consist of node points (places where circuit devices attach to each other), contact bars, and the specific symbol that identifies that particular type of switch, as illustrated in Figure 2-3. Although a control device may have more than one set of contacts, only the contacts used in the circuit are represented on control drawings.

A variety of control diagrams and drawings are used to install, maintain, and troubleshoot motor control systems. These include ladder diagrams, wiring diagrams, line diagrams, and block diagrams. A "ladder diagram" (considered by some as a form of a schematic diagram) focuses on the electrical operation of a circuit, not the physical location of a device. For example, two stop push buttons may be physically at opposite ends of a long conveyor, but electrically side by side in the ladder diagram.

Ladder diagrams, such as the one shown in Figure 2-4, are drawn with two vertical lines and any number of horizontal lines.

- The vertical lines (called **rails**) connect to the power source and are identified as line 1 (L1) and line 2 (L2).
- The horizontal lines (called **rungs**) are connected across L1 and L2 and contain the control circuitry.
- Ladder diagrams are designed to be *read* like a book, starting at the top left and reading from left to right and top to bottom.

Because ladder diagrams are easier to read, they are often used in tracing through the operation of a circuit. Most programmable logic controllers (PLCs) use the ladder-diagramming concept as the basis for their programming language.

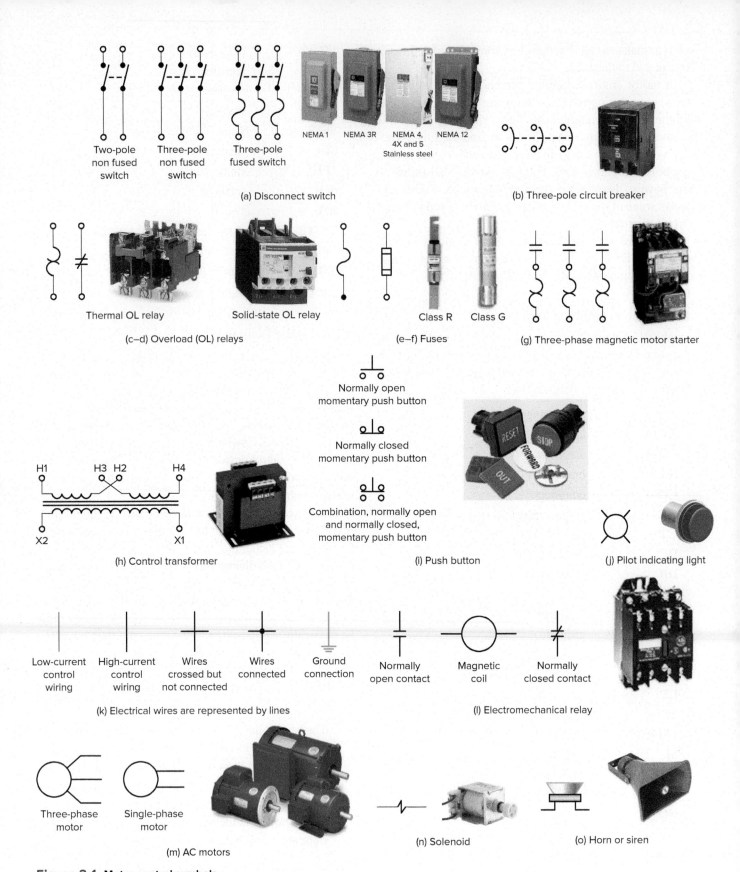

Figure 2-1 Motor control symbols.
Photos *a–d, g:* This material and associated copyrights are proprietary to, and used with the permission of, Schneider Electric; *e–f:* Courtesy of Cooper Industries, Copyright © 2011 Cooper Industries; *h–j, l:* Photo courtesy of Rockwell Automation, www.rockwellautomation.com; *m:* Photo Courtesy of Leeson, www.leeson.com; *n:* Photo courtesy of Ledex, www.ledex.com; *o:* Ingram Publishing/SuperStock

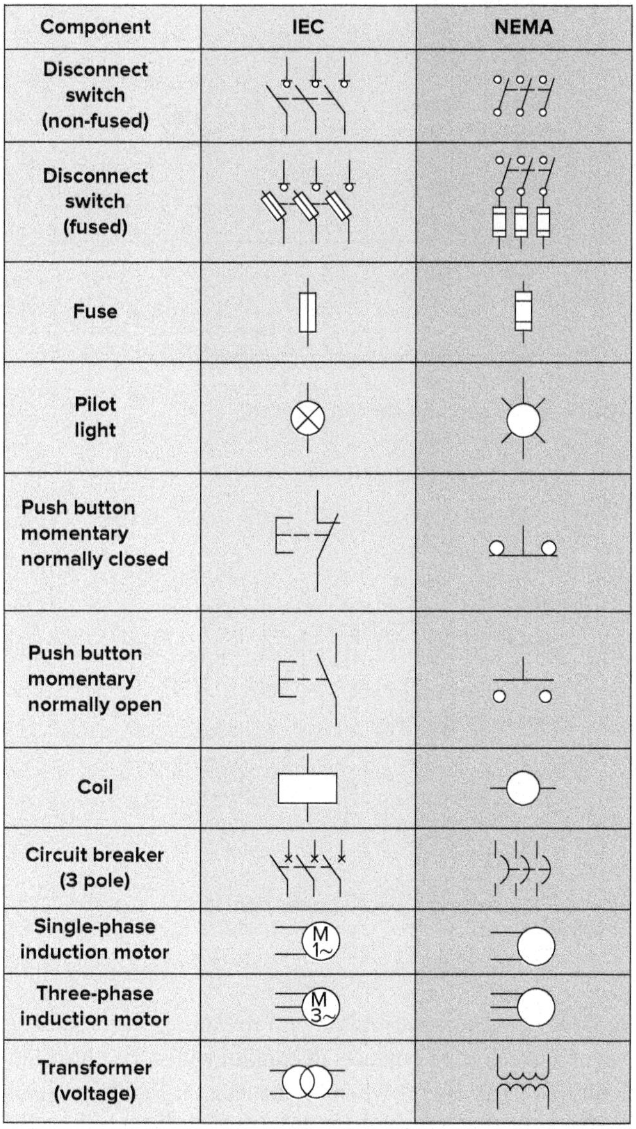

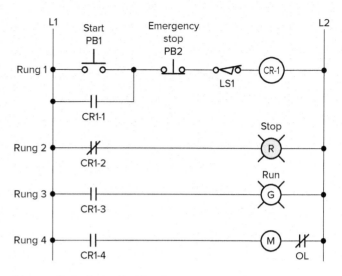

Figure 2-4 Typical ladder diagram.

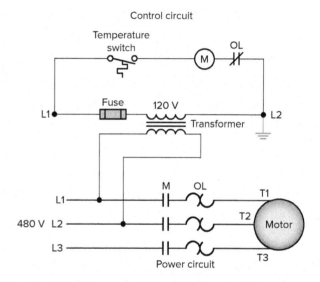

Figure 2-5 Motor power and control circuit wiring.

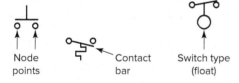

Figure 2-2 Comparison of common motor NEMA and IEC symbols.

Figure 2-3 Switch symbol component parts.

Some motor ladder diagrams illustrate only the single-phase control circuit connected to L1 and L2, and not the three-phase power circuit supplying the motor. Figure 2-5 shows both the power circuit and control circuit wiring.

- On diagrams that include power and control circuit wiring, you may see both heavy and light conductor lines. The **heavy** lines are used for the **higher-current** power circuit and the **lighter** lines for the **lower-current** control circuit.

- Conductors that cross each other but make **no electrical contact** are represented by intersecting lines with **no dot.**
- Conductors that **make contact** are represented by a **dot** at the junction.
- In most instances, the control voltage is obtained directly from the power circuit or from a **step-down control transformer** connected to the power circuit. Using a transformer allows a lower voltage (120 V AC) for the control circuit while supplying the three-phase motor power circuit with a higher voltage (480 V AC) for more efficient motor operation.

A ladder diagram gives the necessary information for easily following the sequence of operation of the circuit. It is a great aid in troubleshooting as it shows, in a simple way,

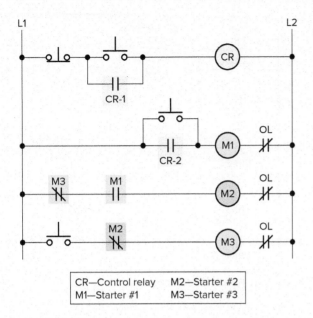

Figure 2-6 Identification of coils and associated contacts.

| CR—Control relay | M2—Starter #2 |
| M1—Starter #1 | M3—Starter #3 |

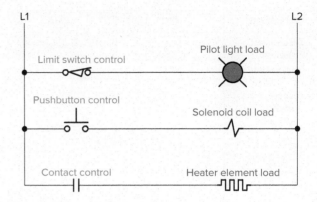

Figure 2-7 Load and control devices.

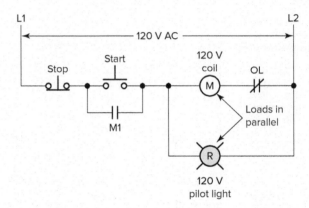

Figure 2-8 Loads are placed on the right and contacts on the left.

the effect that opening or closing various contacts has on other devices in the circuit. All switches and relay contacts are classified as normally open (NO) or normally closed (NC). The positions drawn on diagrams are the electrical characteristics of each device as would be found when it is purchased and not connected in any circuit. This is sometimes referred to as the "off-the-shelf" or **de-energized** state. It is important to understand this because it may also represent the de-energized position in a circuit. The de-energized position refers to the component position when the circuit is de-energized, or no power is present on the circuit. This point of reference is often used as the starting point in the analysis of the operation of the circuit.

A common method used to identify the relay coil and the contacts operated by it is to place a letter or letters in the circle that represents the coil (Figure 2-6). Each contact that is operated by this coil will have the coil letter or letters written next to the symbol for the contact. Sometimes, when there are several contacts operated by one coil, a number is added to the letter to indicate the contact number "separated by dash" or other text to be consistent. Although there are standard meanings of these letters, most diagrams provide a key list to show what the letters mean; generally they are taken from the name of the device.

A **load** is a circuit component that has resistance and consumes electric power supplied from L1 to L2, as illustrated in Figure 2-7. Control coils, solenoids, horns, and pilot lights are examples of loads. At least one load device must be included in each rung of the ladder diagram. Without a load device, the control devices would be switching an open circuit to a short circuit between L1 and L2. **Contacts** from control devices such as switches, push buttons, and relays are considered to have little or no resistance in the closed state. Connection of contacts in parallel with a load also can result in a short circuit when the contact closes. The circuit current will take the path of least resistance through the closed contact, shorting out the energized load.

Normally loads are placed on the right side of the ladder diagram next to L2 and contacts on the left side next to L1. One exception to this rule is the placement of the normally closed contacts controlled by the motor overload protection device. These contacts are drawn on the right side of the motor starter coil as shown in Figure 2-8. When two or more loads are required to be energized simultaneously, they must be connected in parallel. This will ensure that the full line voltage from L1 and L2 will appear across each load. If the loads are connected in series, neither will receive the entire line voltage necessary for proper operation. Recall that in a **series** connection of loads, the applied voltage is divided between each of the loads. In a **parallel** connection of loads, the voltage across each load is the same and is equal in value to the applied voltage.

Control devices such as switches, push buttons, limit switches, and pressure switches operate loads. Devices that start a load are usually connected in parallel, while devices

20 Chapter 2 Understanding Electrical Drawings

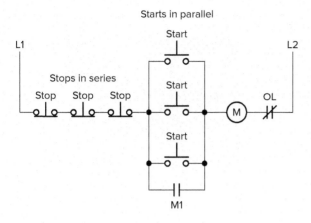

Figure 2-9 Stop devices connect in series and start devices connect in parallel.

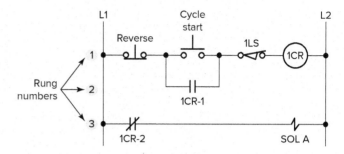

Figure 2-11 Ladder diagram with rung numbers detailed.

that stop a load are connected in series. For example, multiple start push buttons controlling the same motor starter coil would be connected in parallel, while multiple stop push buttons would be connected in series (Figure 2-9). All control devices are identified with the appropriate nomenclature for the device (e.g., stop, start). Similarly, all loads are required to have abbreviations to indicate the type of load (e.g., M for starter coil). Often an additional numerical suffix is used to differentiate multiple devices of the same type. For example, a control circuit with two motor starters might identify the coils as M1 (contacts 1-M1, 2-M1, etc.) and M2 (contacts 1-M2, 2-M2, etc.), as illustrated in Figure 2-10.

As the complexity of a control circuit increases, its ladder diagram increases in size, making it more difficult to read and locate which contacts are controlled by which coil. "Rung numbering" is used to assist in reading and understanding larger ladder diagrams. Each rung of the ladder diagram is marked (rung 1, 2, 3, etc.), starting with the top rung and reading down. A rung can be defined as a complete path from L1 to L2 that contains a load. Figure 2-11 illustrates the marking of each rung in a line diagram with three separate rungs:

- The path for **rung 1** is completed through the reverse push button, cycle start push button, limit switch 1LS, and coil 1CR.
- The path for **rung 2** is completed through the reverse push button, relay contact 1CR-1, limit switch 1LS, and coil 1CR. Note that rung 1 and rung 2 are identified as two separate rungs even though they control the same load. The reason for this is that either the cycle start push button or the 1CR-1 relay contact completes the path from L1 to L2.
- The path for **rung 3** is completed through relay contact 1CR-2 and solenoid SOL A.

"Numerical cross-referencing" is used in conjunction with the rung numbering to locate auxiliary contacts controlled by coils in the control circuit. At times auxiliary contacts are not in close proximity on the ladder diagram to the coil controlling their operation. To locate these contacts, rung numbers are listed to the right of L2 in parentheses on the rung of the coil controlling their operation. In the example shown in Figure 2-12:

- The contacts of coil 1CR appear at two different locations in the line diagram.

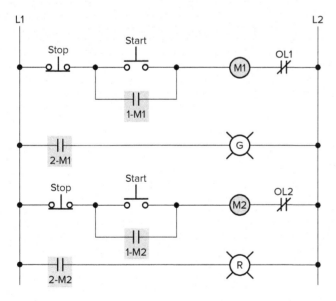

Figure 2-10 Differentiating between multiple devices of the same type.

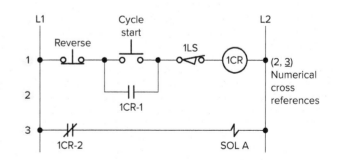

Figure 2-12 Numerical cross-reference system.

Part 1 Symbols—Abbreviations—Ladder Diagrams

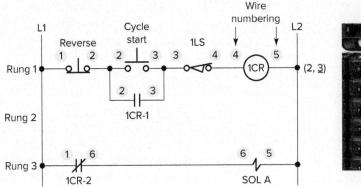

Figure 2-13 Wire numbering.
Photo courtesy of Ideal Industries, www.idealindustries.com

- The numbers in parentheses to the right of the line diagram identify the line location and type of contacts controlled by the coil.
- Numbers appearing in the parentheses for normally open contacts have no special markings.
- Numbers used for normally closed contacts are identified by underlining or overscoring the number to distinguish them from normally open contacts.
- In this circuit, control relay coil 1CR controls two sets of contacts: 1CR-1 and 1CR-2. This is shown by the numerical code 2, 3.

Some type of "wire identification" is required to correctly connect the control circuit conductors to their components in the circuit. The method used for wire identification varies for each manufacturer. Figure 2-13 illustrates one method where each common point in the circuit is assigned a reference number:

- Numbering starts with all wires that are connected to the L1 side of the power supply identified with the number 1.
- Continuing at the top left of the diagram with rung 1, a new number is designated sequentially for each wire that crosses a component.
- Wires that are electrically common are marked with the same numbers.
- Once the first wire directly connected to L2 has been designated (in this case 5), all other wires directly connected to L2 will be marked with the same number.
- The number of components in the first line of the ladder diagram determines the wire number for conductors directly connected to L2.

Figure 2-14 illustrates an alternative method of assigning wire numbers. With this method all wires directly connected to L1 are designated 1 while all those connected to L2 are designated 2. After all the wires with 1 and 2 are marked, the remaining numbers are assigned in a sequential order starting at the top left of the diagram. This method has as its advantage the fact that all wires directly connected to L2 are always designated as 2. Ladder diagrams may also contain a series of descriptions located to the right of L2, which are used to document the function of the circuit controlled by the output device.

A broken line normally indicates a **mechanical** connection. Do not make the mistake of reading a broken

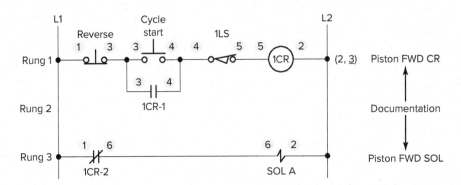

Figure 2-14 Alternative wiring identification with documentation.

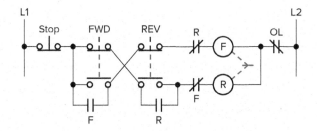

Figure 2-15 Representing mechanical functions.

line as a part of the electrical circuit. In Figure 2-15, the vertical broken lines on the forward and reverse push buttons indicate that their normally closed and normally open contacts are mechanically connected. Thus, pressing the button will open the one set of contacts and close the other. The broken line between the F and R coils indicates that the two are mechanically **interlocked.** Therefore, coils F and R cannot close contacts simultaneously because of the mechanical interlocking action of the device.

When a control transformer is required to have one of its secondary lines grounded, the ground connection must be made so that an accidental ground in the control circuit will not start the motor or make the stop button or control inoperative. Figure 2-16a illustrates the secondary of a control transformer properly grounded to the L2 side of the circuit. When the circuit is operational, the entire circuit to the left of coil M is the ungrounded circuit (it is the "hot" leg). A fault path to ground in the ungrounded circuit will create a short-circuit condition causing the control transformer fuse to open.

Figure 2-16b shows the same circuit improperly grounded at L1. In this case, a short to ground to the left of coil M would *energize* the coil, starting the motor unexpectedly. The fuse would not operate to open the circuit and pressing the stop button would not de-energize the M coil. Equipment damage and personnel injuries would be very likely. Clearly, output devices (loads) must be directly connected to the **grounded side** of the circuit.

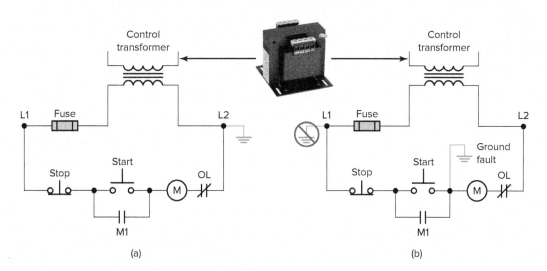

Figure 2-16 Control transformer ground connection: (*a*) control transformer properly grounded to the L2 side of the circuit; (*b*) control transformer improperly grounded at the L1 side of the circuit.
Photo courtesy of Rockwell Automation, www.rockwellautomation.com

Part 1 Review Questions

1. Define the term *motor control circuit.*
2. Why are symbols used to represent components on electrical diagrams?
3. An electrical circuit contains three pilot lights. What acceptable symbol could be used to designate each light?
4. Describe the basic structure of an electrical ladder diagram schematic.
5. Lines are used to represent electrical wires on diagrams.
   a. How are wires that carry high current differentiated from those that carry low current?
   b. How are wires that cross but do not electrically connect differentiated from those that connect electrically?

6. The contacts of a pushbutton switch open when the button is pressed. What type of push button would this be classified as? Why?

7. A relay coil labeled TR contains three contacts. What acceptable coding could be used to identify each of the contacts?

8. A rung on a ladder diagram requires that two loads, each rated for the full line voltage, be energized when a switch is closed. What connection of loads must be used? Why?

9. One requirement for a particular motor application is that six pressure switches be closed before the motor is allowed to operate. What connections of switches should be used?

10. The wire identification labels on several wires of an electrical panel are examined and found to have the same number. What does this mean?

11. A broken line representing a mechanical function on an electrical diagram is mistaken for a conductor and wired as such. What two types of problems could this result in?

12. Compare the shape of the symbol used to represent an electromagnetic coil on a NEMA and IEC motor schematic.

PART 2 WIRING—SINGLE LINE—BLOCK DIAGRAMS

Wiring Diagrams

Wiring diagrams are used to show the **point-to-point** wiring between components of an electric system and sometimes their physical relation to each other. They may include wire identification numbers assigned to conductors in the ladder diagram and/or color coding. Coils, contacts, motors, and the like are shown in the actual position that would be found on an installation. These diagrams are helpful in wiring up systems, because connections can be made exactly as they are shown on the diagram. A wiring diagram gives the necessary information for actually wiring up a device or group of devices or for physically tracing wires in troubleshooting. However, it is difficult to determine circuit operation from this type of drawing.

Wiring diagrams are provided for most electrical devices. Figure 2-17 illustrates a typical wiring diagram provided for a motor starter. The diagram shows, as closely as possible, the actual location of all of the component parts of the device. The open terminals (marked by an open circle) and arrows represent connections made by the user. Note that bold black lines denote the power circuit, and thinner red lines are used to show the control circuit.

The routing of wires in cables and conduits, as illustrated in Figure 2-18, is an important part of a wiring diagram. A **conduit** layout diagram indicates the start and the finish of the electrical conduits and shows the approximate path taken by any conduit in progressing from one point to another. Integrated with a drawing of this nature is the conduit and cable schedule, which tabulates each conduit as to number, size, function, and service and also includes the number and size of wires to be run in the conduit.

Wiring diagrams show the details of actual connections. Rarely do they attempt to show complete details of panel board or equipment wiring. The wiring diagram of Figure 2-18, reduced to a simpler form, is shown in Figure 2-19 with the internal connections of the magnetic starter omitted. Wires encased in conduit C1 are part of the power circuit and sized for the current requirement of the motor. Wires encased in conduit C2 are part of the lower-voltage control circuit and sized to the current requirements of the control transformer.

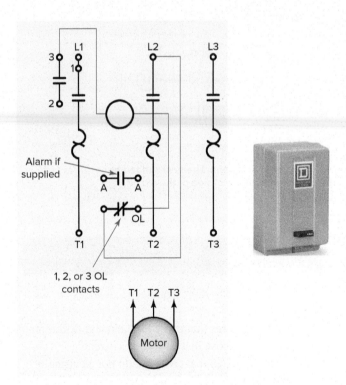

Figure 2-17 Typical motor starter wiring diagram.
This material and associated copyrights are proprietary to, and used with the permission of, Schneider Electric.

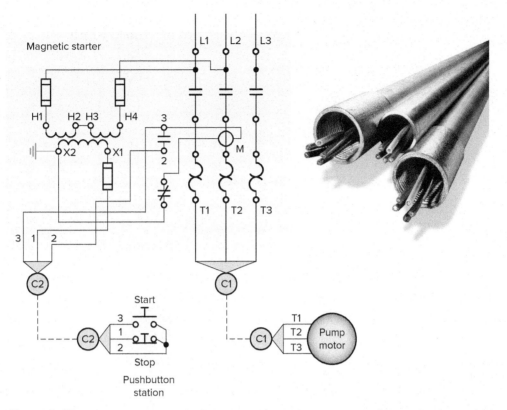

Figure 2-18 Routing of wires in cables and conduits.
Photo courtesy of JMC Steel Group

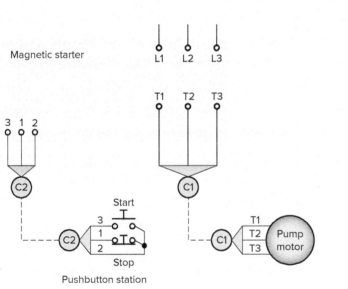

Figure 2-19 Wiring with the internal connections of the magnetic starter omitted.

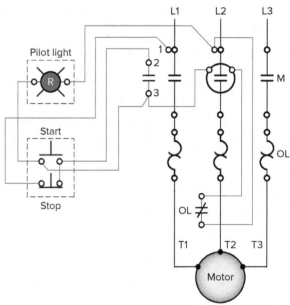

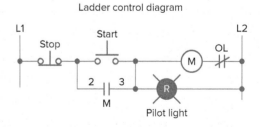

Figure 2-20 Combination wiring and ladder diagram.

Wiring diagrams are often used in conjunction with ladder diagrams to simplify understanding of the control process. An example of this is illustrated in Figure 2-20. The wiring diagram shows both the power and control circuits. A separate ladder diagram of the control circuit is included to give a clearer understanding of its operation. By following

Part 2 Wiring—Single Line—Block Diagrams 25

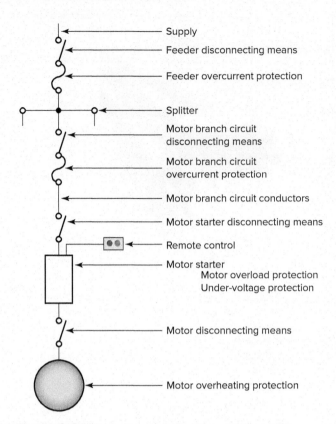

Figure 2-21 Single-line diagram of a motor installation.

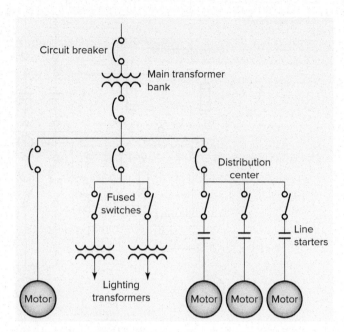

Figure 2-22 Single-line diagram of a power distribution system.

the ladder diagram, it can be seen that the pilot light is wired so that it will be on whenever the starter is energized. The power circuit has been omitted for clarity, since it can be traced readily on the wiring diagram (heavy black lines).

Single-Line Diagrams

A **single-line** (also called a one-line) diagram uses symbols along with a single line to show all major components of an electric circuit. Some motor control equipment manufacturers use a single-line drawing, like the one shown in Figure 2-21, as a road map in the study of motor control installations. The installation is reduced to the simplest possible form, yet it still shows the essential requirements and equipment in the circuit.

Power systems are extremely complicated electrical networks that may be geographically spread over very large areas. For the most part, they are also three-phase networks—each power circuit consists of three conductors and all devices such as generators, transformers, breakers, and disconnects installed in all three phases. These systems can be so complex that a complete conventional diagram showing all the connections is impractical. When this is the case, use of a single-line diagram is a concise way of communicating the basic arrangement of the power system's component. Figure 2-22 shows a single-line diagram of a small power distribution system. These types of diagrams are also called "power riser" diagrams.

Block Diagrams

A block diagram represents the major functional parts of complex electrical/electronic systems by **blocks** rather than symbols. Individual components and wires are not shown. Instead, each block represents electrical circuits that perform specific functions in the system. The functions the circuits perform are written in each block. Arrows connecting the blocks indicate the general direction of current paths.

Figure 2-23 shows a block diagram of a variable-frequency AC motor drive. A **variable-frequency drive** controls the speed of an AC motor by varying the frequency supplied to the motor. The drive also regulates the output voltage in proportion to the output frequency to provide a relatively constant ratio (volts per hertz; V/Hz) of voltage

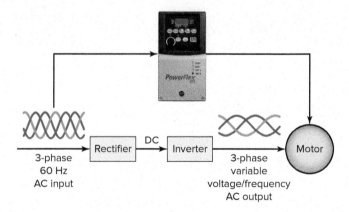

Figure 2-23 Block diagram of a variable-frequency AC drive.
Photo courtesy of Rockwell Automation, www.rockwellautomation.com

to frequency, as required by the characteristics of the AC motor to produce adequate torque. The function of each block is summarized as follows:

- 60 Hz three-phase power is supplied to the rectifier block.
- The **rectifier block** is a circuit that converts or rectifies its three-phase AC voltage into a DC voltage.
- The **inverter block** is a circuit that inverts, or converts, its DC input voltage back into an AC voltage. The inverter is made up of electronic switches, which switch the DC voltage on and off to produce a controllable AC power output at the desired frequency and voltage.

Riser Diagrams

A **riser diagram** is similar to line and block diagrams in that it shows the relationship of components within a circuit and the single line connections between them. The main difference is that a riser diagram shows the circuit as an **elevation** and is more physical than electrical oriented. Riser diagrams are often part of construction installation drawings. Figure 2-24 shows an example of a **power riser diagram** that shows at a glance how electrical power is distributed throughout the floors of a building.

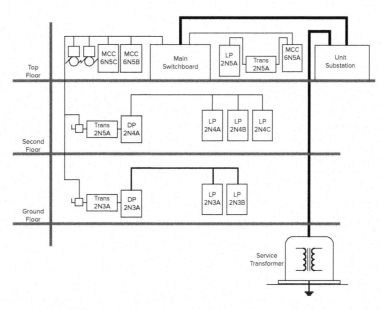

Figure 2-24 Electrical power riser diagram.

Part 2 Review Questions

1. What is the main purpose of a wiring diagram?
2. In addition to numbers, what other method can be used to identify wires on a wiring diagram?
3. What role can a wiring diagram play in the troubleshooting of a motor control circuit?
4. List the pieces of information most likely to be found in the conduit and cable schedule for a motor installation.
5. Explain the purpose of using a motor wiring diagram in conjunction with a ladder diagram of the control circuit.
6. What is the main purpose of a single-line diagram?
7. What is the main purpose of a block diagram?
8. Explain the function of the rectifier and inverter blocks of a variable-frequency AC drive.
9. In what way do electrical risers diagrams differ from line and block diagram types?

PART 3 MOTOR TERMINAL CONNECTIONS

Motor Classification

Electric motors have been an important element of our industrial and commercial economy for over a century. Most of the industrial machines in use today are driven by electric motors. Industries would cease to function without properly designed, installed, and maintained motor control systems. In general, motors are classified according to the type of power used (AC or DC) and the motor's principle of operation. The "family tree" of motor types is illustrated in Figure 2-25.

In the United States the Institute of Electrical and Electronics Engineers (IEEE) establishes the standards for motor testing and test methodologies, while the National Electrical Manufacturers Association (NEMA) prepares the standards for motor performance and classifications. Additionally, motors shall be installed in accordance with Article 430 of the National Electrical Code (NEC).

DC Motor Connections

Industrial applications use DC motor because the speed–torque relationship can be easily varied. DC motor applications include cranes (Figure 2-26), conveyors, elevators, and material-handling processes.

- DC motors feature a speed, which can be controlled smoothly down to zero, immediately followed by acceleration in the opposite direction.
- In emergency situations, DC motors can supply over **five times rated torque** without stalling.

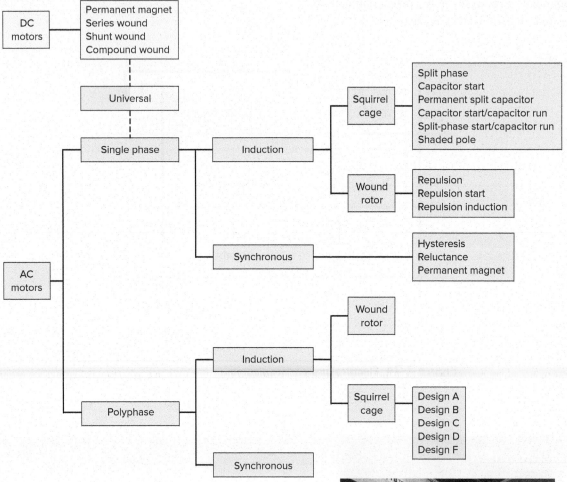

Figure 2-25 Family tree of motor types.

Figure 2-26 DC motor used on cranes.
Photo courtesy of Wilcox Door Service Inc., www.wilcoxdoor.com

28   Chapter 2   Understanding Electrical Drawings

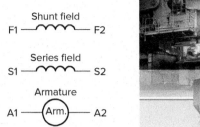

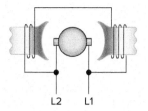

Figure 2-27 Parts of a DC compound motor.
Photo courtesy of Siemens, www.siemens.com

- **Dynamic braking** (DC motor-generated energy is fed to a resistor grid) or **regenerative braking** (DC motor-generated energy is fed back into the DC motor supply) can be obtained with DC motors on applications requiring quick stops, thus eliminating the need for, or reducing the size of, a mechanical brake.

Figure 2-27 shows the symbols used to identify the basic parts of a direct current (DC) compound motor. The rotating part of the motor is referred to as the armature; the stationary part of the motor is referred to as the stator, which contains the series field winding and the shunt field winding. In DC machines **A1 and A2** always indicate the armature leads, **S1 and S2** indicate the series field leads, and **F1 and F2** indicate the shunt field leads.

It is the kind of field excitation provided by the field that distinguishes one type of DC motor from another; the construction of the armature has nothing to do with the motor classification. There are three general types of DC motors, classified according to the method of field excitation as follows:

- A shunt DC motor (Figure 2-28) uses a comparatively high resistance shunt field winding, made up of many turns of fine wire, connected in parallel (shunt) with the armature.
- A series DC motor (Figure 2-29) uses a very low resistance series field winding, made up of very few turns of heavy wire, connected in series with the armature.
- A compound DC motor (Figure 2-30) uses a combination of a shunt field (many turns of fine wire) in parallel with the armature, and series field (few turns of heavy wire) in series with the armature.

Figure 2-28 Standard DC shunt motor connections for counterclockwise and clockwise rotation.

All connections shown in Figures 2-28, 2-29, and 2-30 are for counterclockwise and clockwise rotation facing the end opposite the drive (commutator end). One purpose of applying markings to the terminals of motors according to a standard is to aid in making connections when a predictable rotation direction is required. This may be the case when improper rotation could result in unsafe operation or damage. Terminal markings are normally used to tag only

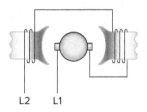

Figure 2-29 Standard DC series motor connections for counterclockwise and clockwise rotation.

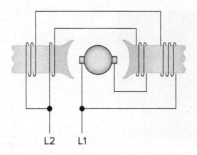

Counterclockwise			Clockwise		
Line 1	Tie	Line 2	Line 1	Tie	Line 2
F1-A1	A2-S1	F2-S2	F1-A2	A1-S1	S2-F2

Figure 2-30 Standard DC compound (cumulative) motor connections for counterclockwise and clockwise rotation. For differential compound connection, reverse S1 and S2.

those terminals to which connections must be made from outside circuits.

The direction of rotation of a DC motor depends on the direction of the magnetic field and the direction of current flow in the armature. If **either** the direction of the field or the direction of current flow through the armature is reversed, the rotation of the motor will reverse. However, if **both** of these factors are reversed at the same time, the motor will continue rotating in the same direction.

AC Motor Connections

The **AC induction motor** is the dominant motor technology in use today, representing more than 90 percent of installed motor capacity. Induction motors are available in single-phase (1ϕ) and three-phase (3ϕ) configurations, in sizes ranging from fractions of a horsepower to tens of thousands of horsepower. They may run at fixed speeds—most commonly 900, 1200, 1800, or 3600 rpm—or be equipped with an adjustable-speed drive.

The most commonly used AC motors by far have a **squirrel-cage** configuration (Figure 2-31), so named because of the aluminum or copper squirrel cage imbedded within the iron laminates of the rotor. There is no physical electrical connection to the squirrel cage. Current in the rotor is induced by the rotating magnetic field of the stator. **Wound-rotor** models, in which coils of wire in the rotor windings, are also available. These are expensive but offer greater control of the motor's performance characteristics, so they are most often used for special

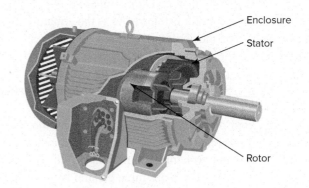

Figure 2-31 Three-phase squirrel-cage AC induction motor.
Photo courtesy of Siemens, www.siemens.com

torque and acceleration applications and for adjustable-speed applications.

Single-Phase Motor Connections The majority of single-phase AC induction motors are constructed in fractional horsepower sizes for 120 to 240 V, 60 Hz power sources. Although there are several types of single-phase motors, they are basically identical except for the means of starting. The "split-phase motor" is most widely used for medium starting applications (Figure 2-32). The operation of the split-phase motor is summarized as follows:

- The motor has a start and a main winding, which are both energized when the motor is started.
- The start-winding produces a phase difference to start the motor and is switched out by a centrifugal switch as running speed is approached. When the

30 Chapter 2 Understanding Electrical Drawings

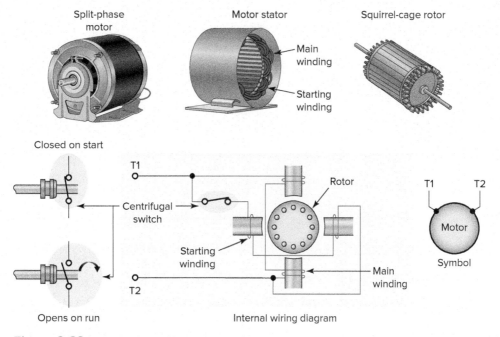

Figure 2-32 AC split-phase induction motor.

motor reaches about 75 percent of its rated full load speed, the starting winding is disconnected from the circuit.

- Split-phase motor sizes range up to about ½ horsepower. Popular applications include fans, blowers, home appliances such as washers and dryers, and tools such as small saws or drill presses where the load is applied after the motor has obtained its operating speed.
- The motor can be reversed by reversing the leads to the start-winding or main winding, but not to both. Generally, the industry standard is to reverse the start winding leads

In a **dual-voltage** split-phase motor (Figure 2-33), the running winding is split into two sections and can be connected to operate from a **120 V or 240 V** source. The two run windings are connected in series when operated from a 240 V source, and in parallel for 120 V operation. The start winding is connected across the supply lines for low voltage and at one line to the midpoint of the run windings for high voltage. This ensures that all windings receive the 120 V they are designed to operate at. To reverse the direction of rotation of a dual-voltage split-phase motor, interchange the two start winding leads. Dual-voltage motors are connected for the desired voltage by following the connection diagram on the nameplate.

The nominal dual-voltage split-phase motor rating is 120/240 V. With any type of dual-voltage motor, the **higher** voltage is preferred when a choice between voltages is available. The motor uses the same amount of power and produces the same amount of horsepower when operating from a 120 V or 240 V supply. However, as the voltage is doubled from 120 V to 240 V, the current is cut in half. Operating the motor at this reduced current level allows you to use smaller circuit conductors and reduces line power losses.

Many single-phase motors use a capacitor in series with one of the stator windings to optimize the phase difference between the start and run windings for starting. The result is a higher starting **torque** than a split-phase motor can produce. There are three types of capacitor motors: **capacitor start,** in which the capacitor phase is in the circuit only during starting; **permanent-split capacitor,** in which the capacitor phase is in the circuit for both starting and running; and **two-value capacitor,** in which there are different values of capacitance for starting and running. The permanent-split capacitor motor, illustrated in Figure 2-34, uses a capacitor permanently connected in series with one of the stator windings. This design is lower in cost than the capacitor-start motors that incorporate capacitor switching systems. Installations include compressors, pumps, machine tools, air conditioners, conveyors, blowers, fans, and other hard-to-start applications.

Three-Phase Motor Connections The three-phase AC induction motor is the most common motor used in commercial and industrial applications. Single-phase larger horsepower motors are not normally used because they are

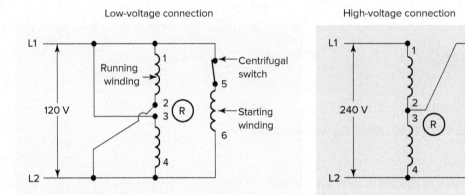

Figure 2-33 Dual-voltage split-phase motor stator connections.

Figure 2-34 Permanent-split capacitor motor.
Photo courtesy of Leeson, www.leeson.com

inefficient compared to three-phase motors. In addition, single-phase motors are not self-starting on their running windings, as are three-phase motors.

Large horsepower AC motors are usually three-phase. All three-phase motors are constructed internally with a number of individually wound coils. Regardless of how many individual coils there are, the individual coils will always be wired together (series or parallel) to produce three distinct windings, which are referred to as phase A, phase B, and phase C. All three-phase motors are wired so that the phases are connected in either **wye (Y)** or **delta (Δ)** configuration, as illustrated in Figure 2-35.

Dual-Voltage Motor Connections Manufacturers deploy various external connection schemes to produce

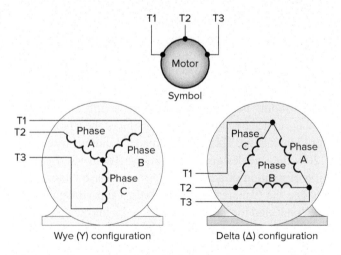

Figure 2-35 Three-phase wye and delta motor connections.
Photo courtesy of Leeson, www.leeson.com

32 Chapter 2 Understanding Electrical Drawings

three-phase induction motors for multiple voltages and/or starting methods. The following apply to common three-phase motor connections.

- **Three-lead** single voltage connections are the most common and simple. Supply line leads L1, L2, L3 are directly connected to motor leads T1, T2, T3. If there's any doubt about the connection, it's a good idea to run the machine unloaded to determine the direction of rotation.
- **Six-lead** dual voltage connections are numbered 1 through 6. The winding can be connected wye or delta. On machines rated for two voltages, the wye connection is for the high voltage; the delta connection is for the low voltage. The voltage ratio is 1.73, for example, 220 volts low voltage and 380 volts high voltage.
- **Nine-lead** dual voltage connections are numbered 1 through 9. The motor is typically rated for two voltages and could be designed with either a wye connection or a delta connection. The two connections have a 2-to-1 voltage difference. The nine-lead wye connected motor is the most common type of three-phase motor used in commercial and industrial installations.
- **Twelve-lead** dual voltage connections are numbered 1 through 12. The motor is typically rated for standardized voltages of 230/460 volts that requires 12 leads to accommodate.

It is common practice to manufacture three-phase motors that can be connected to operate at different voltage levels. The most common multiple-voltage rating for three-phase motors is 208/230/460 V. Always check the motor specifications or nameplate for the proper voltage rating and wiring diagram for method of connection to the voltage source.

Figure 2-36 illustrates the typical terminal identification and connection table for a nine-lead dual-voltage wye-connected three-phase motor. One end of each phase is internally permanently connected to the other phases. Each phase coil (A, B, C) is divided into two equal parts and connected in either series for high-voltage operation or parallel for low-voltage operation. According to NEMA nomenclature, these leads are marked T1 through T9. High-voltage and low-voltage connections are given in the accompanying connection table and motor terminal board. The same principle of series (high-voltage) and parallel (low-voltage) coil connections is applied for dual-voltage wye–delta connected

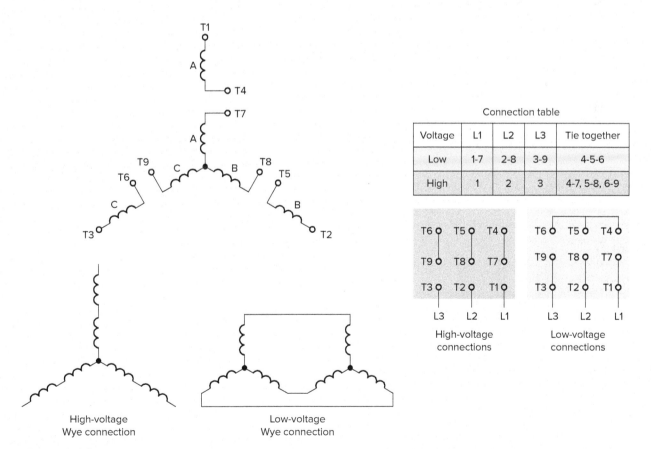

Figure 2-36 **Dual-voltage wye connections.**

three-phase motors. In all cases refer to the wiring diagram supplied with the motor to ensure proper connection for the desired voltage level.

Figure 2-37 illustrates the IEC nomenclature for an equivalent IEC nine-lead dual-voltage wye connected three-phase motor. For the **high-voltage** connection:

- Nine leads are brought out of the motor.
- These leads are labeled U1, U2, U5 – V1, V2, V5 – W1, W2, W5.
- They are externally connected for either of the two voltages.
- To connect the wye configuration for high voltage, connect L1 to U1, L2 to V1, L3 to W1, and tie U2 to U5, V2 to V5, and W2 to W5.
- The internally connected *Y points*, which otherwise would have been U6, V6, and W6, are not brought out.
- This connects the two individual parts of the phase coils A, B, and C in **series,** with each coil receiving 50 percent of the line-to-neutral point voltage.

In a similar fashion, for the **low-voltage** wye configuration:

- Connect L1 to U1 and U5, L2 to V1 and V5, L3 to W1 and W5 and tie U2, V2, and W2 together.
- This connects the two individual parts of the phase coils in **parallel** with each coil receiving 100 percent of the line-to-neutral point voltage.

Multispeed Motor Connections Some three-phase motors, referred to as **multispeed motors,** are designed to provide two separate speed ranges. The speed of an induction motor depends on the number of poles built into the motor and the frequency of the electrical power supply. Changing the number of poles provides specific speeds that correspond to the number of poles selected. The more poles per phase, the slower the operating rpm of the motor.

$$\text{RPM} = 120 \times \frac{\text{Frequency}}{\text{Number of poles}}$$

Two-speed motors with single windings can be reconnected, using a controller, to obtain different speeds. The controller circuitry serves to change the connections of the stator windings. These motors are wound for one speed, but when the winding is reconnected, the number of magnetic poles within the stator is doubled and the motor speed is reduced to one-half the original speed. This type of reconnection should not be confused

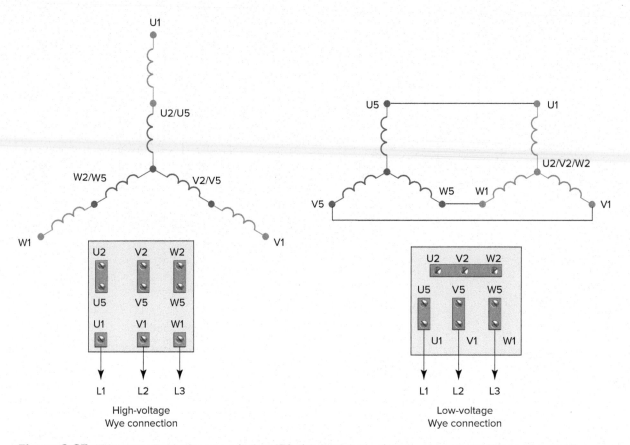

Figure 2-37 IEC nomenclature for an equivalent IEC nine-lead dual-voltage wye connected three-phase motor.

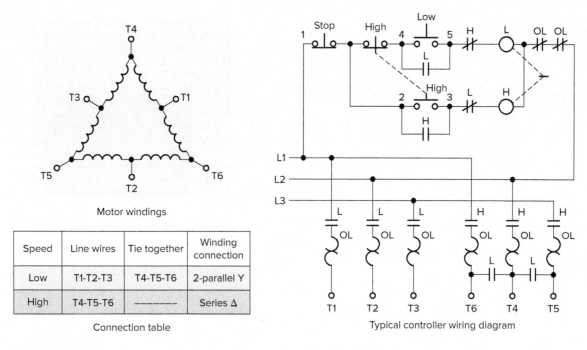

Figure 2-38 Constant-horsepower two-speed, three-phase motor and controller.

with the reconnection of dual-voltage three-phase motors. In the case of multispeed motors, the reconnection results in a motor with a different number of magnetic poles. Three types of single-winding two-speed motors are available: **constant horsepower, constant torque,** and **variable torque.** Figure 2-38 shows the connections for a constant-horsepower two-speed, three-phase motor and controller.

To reverse the direction of rotation of any three-phase wye- or delta-connected motor, simply **reverse or interchange** any two of the three main power leads to the motor. Standard practice is to interchange L1 and L3 as illustrated in Figure 2-39. When you are connecting a motor, the direction of rotation is usually not known until the motor is started. In this case, the motor may be temporarily connected to determine the direction of rotation before making permanent connections.

In certain applications, unintentional reversal of motor rotation can result in serious damage. When this is the case, phase failure and phase reversal relays are used to protect motors, machines, and personnel from the hazards of open-phase or reversed phase conditions. The schematic diagram of Figure 2-40 shows a typical application for a **reverse-phase relay.** The operation of this circuit can be summarized as follows:

- The relay is designed to continuously **monitor phase rotation** of the three-phase lines.
- A solid-state sensing circuit within the relay controls an electromechanical relay coil, the **normally open (NO) contact,** that **closes** when power with correct phase rotation is applied.
- The relay coil **will not energize** if the applied phases are reversed and will **de-energize** if phase rotation is reversed while the motor is running.

The speed of an AC induction motor depends on two factors: the number of motor poles and the frequency of the applied power. In variable-frequency motor drives, variable speed of an induction motor is achieved by varying the frequency of the voltage applied to the motor. The lower the frequency, the slower the operating rpm of the motor.

Standard induction motors can be detrimentally affected when operated by variable-frequency drives. **"Inverter duty"** and **"vector duty"** describe a class of motors that are capable of operation from a variable-frequency drive. Low temperature rise in this class of motor is accomplished with better insulation systems, additional active material (iron and copper), and/or external fans for better cooling at low-speed operation.

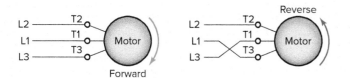

Figure 2-39 Reversing the direction of rotation of a three-phase motor.

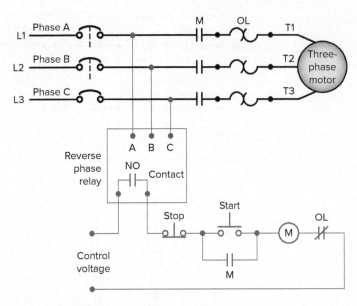

Figure 2-40 Reverse-phase relay circuit.
Time Mark Corporation, 2014.

Part 3 Review Questions

1. In what two general ways are motors classified?
2. List three major organizations involved with motor standards and installation requirements in the United States.
3. What two DC motor operating features make them useful for industrial applications?
4. What part of a DC motor is identified by each of the following load designations?
   a. A1 and A2
   b. S1 and S2
   c. F1 and F2
5. List the three general types of DC motors.
6. What two factors determine the direction of rotation of a DC motor?
7. In what phase configurations are AC induction motors available?
8. What terms are used to identify the rotating and stationary parts of an AC induction motor?
9. Describe the construction of and external electrical connection to the squirrel-cage rotor used in AC induction motors.
10. Outline the starting sequence of a split-phase motor.
11. Assume the direction of rotation of a split-phase motor needs to be reversed. How is this done?
12. A dual-voltage split-phase motor is to be connected for the lower voltage. What connection of the two run windings would be used?
13. You have the option of operating a dual-voltage motor at either the high- or low-voltage level. What are the advantages of operating it at the high-voltage level?
14. What is the main advantage of the capacitor motor over the standard split-phase type?
15. How are the three distinct windings of a three-phase motor identified?
16. Large horsepower AC motors are usually three-phase. Why?
17. What two basic configurations are used for the connection of all three-phase motors?
18. According to NEMA nomenclature, how are the leads of a nine-lead dual-voltage three-phase motor labeled?
19. State the relationship between the speed of a three-phase induction motor and the number of poles per phase.
20. Assume the direction of rotation of a three-phase motor needs to be reversed. How is this done?
21. State the relationship between the speed of a three-phase induction motor and the frequency of the power source.

22. Why should inverter-duty AC induction motors be used in conjunction with variable-frequency motor drives?
23. According to NEMA nomenclature the output leads of a three-phase motor are identified as T1, T2, T3. What is the equivalent IEC nomenclature?
24. For a dual-voltage nine-lead three-phase motor each phase coil (A, B, C) is divided into two equal parts. What electrical connection of the phase coils is used for high-voltage operation and low-voltage operation?
25. For six-lead dual-voltage three-phase motors what three-phase configuration is used for the low- and high-voltage connections?

PART 4 MOTOR NAMEPLATE AND TERMINOLOGY

The motor nameplate (Figure 2-41) contains important information about the connection and use of the motor. An important part of making motors interchangeable is ensuring that nameplate information is common among manufacturers.

NEC Required Nameplate Information

Motor Manufacture This will include the name and logo of the manufacturer along with catalog numbers, parts numbers, and model numbers used to identify a motor. Each manufacturer uses a unique coding system.

Voltage Rating Voltage rating is abbreviated **V** on the nameplate of a motor. It indicates the voltage at which the motor is designed to operate. The voltage of a motor is usually determined by the supply to which it is being attached. NEMA requires that the motor be able to carry its rated horsepower at nameplate voltage ±10 percent although not necessarily at the rated temperature rise. Thus, a motor with a rated nameplate voltage of 460 V should be expected to operate successfully between 414 V and 506 V.

The voltage may be a single rating such as 115 V or, for dual-voltage motors, a dual rating such as 115 V/230 V. Most 115/230 V motors are shipped from the factory connected for 230 V. A motor connected for 115 V that has 230 V applied will burn up immediately. A motor connected for 230 V that has 115 V applied will be a slow-running motor that overheats and trips out.

NEMA standard motor voltages are:

Single-phase motors—115, 230, 115/230, 277, 460, and 230/460 V

Three-phase motors up to 125 hp—208, 230, 460, 230/460, 575, 2,300, and 4,000 V

Three-phase motors above 125 Hp—460, 575, 2,300, and 4,000 V

When dealing with motors, it is important to distinguish between **nominal** system and **nameplate** voltages. Examples of the differences between the two are as follows:

Nominal system voltage	Nameplate voltage
120 V	115 V
220 V	208 V
240 V	230 V
480 V	460 V
600 V	575 V
2,400 V	2,300 V
4,160 V	4,000 V
6,900 V	6,600 V

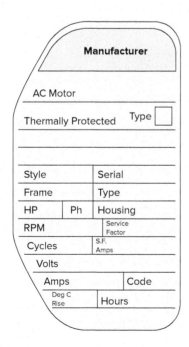

Figure 2-41 Typical motor nameplate.

Current Rating The nameplate current rating of a motor is abbreviated **A or AMPS.** The nameplate current rating is the full-load current (also known as FLA) at rated load, rated voltage, and rated frequency. Motors that are not fully loaded draw less than the rated nameplate current. Similarly, motors that are overloaded draw more than the rated nameplate current.

Motors that have dual voltage ratings also have dual current ratings. A dual-voltage motor operated at the higher voltage rating will have the lower current rating. For example, a ½ hp motor rated 115/230 V and 7.4/3.7 A will have a rated current of 3.7 A when operating from a 230 V supply.

Line Frequency The line frequency rating of a motor is abbreviated on the nameplate as **CY or CYC (cycle), or Hz (hertz).** A cycle is one complete wave of alternating voltage or current. Hertz is the unit of frequency and equals the number of cycles per second. In the United States, 60 cycles/second (Hz) is the standard, while in other countries 50 Hz (cycles) is more common.

Phase Rating The phase rating of a motor is abbreviated on the nameplate as **PH.** The phase rating is listed as direct current (DC), single-phase alternating current (1ϕ AC), or three-phase alternating current (3ϕ AC).

Motor Speed The rated speed of a motor is indicated on the nameplate in revolutions per minute **(rpm).** This rated motor speed is not the exact operating speed, but the approximate speed at which a motor rotates when delivering rated horsepower to a load.

The number of poles in the motor and the frequency of the supply voltage determine the speed of an AC motor. The speed of a DC motor is determined by the amount of supply voltage and/or the amount of field current.

Ambient Temperature The ambient temperature rating of a motor is abbreviated **AMD or DEG** on the nameplate of a motor. Ambient temperature is the temperature of the air surrounding the motor. In general, maximum ambient temperature for motors is **40°C or 104°F** unless the motor is specifically designed for a different temperature and indicates so on its nameplate.

Motors operating at or near rated full load will have reduced life if operated at ambient temperatures above their ratings. If the ambient temperature is over 104°F, a higher-horsepower motor or a special motor designed for operation at higher ambient temperatures must be used.

Temperature Rise A motor's permissible temperature rise is abbreviated **Deg.C/Rise** on the nameplate of the motor. This indicates the amount the motor winding temperature will increase above the ambient temperature because of the heat from the current drawn by the motor at full load. It can also be thought of as the amount by which a motor operating under rated conditions is hotter than its surrounding temperature.

Thermal imagers (also known as infrared cameras or infrared imagers) capture images of infrared energy or temperature. Thermal images of electric motors (Figure 2-42) reveal their temperature operating conditions as reflected by their surface temperature. While the infrared camera cannot see the inside of the motor, the exterior surface temperature is an indicator of the internal temperature. As the motor gets hotter inside, it also gets hotter outside. Such condition monitoring is one way to avert many unexpected motor malfunctions in systems.

Insulation Class Motor insulation prevents windings from shorting to each other or to the frame of the motor. The type of insulation used in a motor depends on the operating temperature the motor will experience. As the heat in a motor increases beyond the temperature rating of the insulation, the life of the insulation and of the motor is shortened.

Standard NEMA insulation classes are given by **alphabetic classifications** according to their maximum temperature rating. A replacement motor must have the same

Figure 2-42 Motor thermal imaging inspection.
Photo courtesy of Fluke, www.fluke.com. Reproduced with Permission.

insulation class or a higher temperature rating than the motor it is replacing. The four major NEMA classifications of motor insulation are as follows:

NEMA classification	Maximum operating temperatures
A	221°F (105°C)
B	266°F (130°C)
F	311°F (155°C)
H	356°F (180°C)

Duty Cycle The duty cycle is listed on the motor nameplate as DUTY, DUTY CYCLE, or TIME RATING. Motors are classified according to the length of time they are expected to operate under full load as either continuous duty or intermittent duty. Continuous duty cycle–rated motors are identified as **CONT** on the nameplate, while intermittent-duty cycle motors are identified as **INTER** on the nameplate.

Continuous-duty motors are rated to operate continuously without any damage or reduction in the life of the motor. General-purpose motors will normally be rated for continuous duty. Intermittent-duty motors are rated to operate continuously only for short time periods and then must be allowed to stop and cool before restarting.

Horsepower Rating The horsepower rating of the motor is abbreviated on the nameplate as **HP.** Motors below 1 horsepower are referred to as fractional-horsepower motors and motors 1 horsepower and above are called integral-horsepower motors. The HP rating is a measure of the full load output power the shaft of the motor can produce without reducing the motor's operating life. NEMA has established standard motor horsepower ratings from 1 hp to 450 hp.

Some small fractional-horsepower motors are rated in watts (1 hp = 746 W). Motors rated by the International Electrotechnical Commission (IEC) are rated in kilowatts (kW). When an application calls for a horsepower falling between two sizes, the larger size is chosen to provide the appropriate power to operate the load.

Code Letter An alphabetic letter is used to indicate the National Electric Code Design Code letter for the motor. When AC motors are started with full voltage applied, they draw a starting or "locked-rotor" line current substantially greater than their full-load running current rating. The value of this high current is used to determine circuit breaker and fuse sizes in accordance with NEC requirements. In addition, the starting current can be important on some installations where high starting currents can cause a voltage dip that might affect other equipment.

Motors are furnished with a code letter on the nameplate that designates the locked-rotor rating of the motor in kilovolt-amperes (kVA) per nameplate horsepower. Code letters from A to V are listed in Article 430 of the National Electrical Code. As an example, an M rating allows for 10.0 to 11.19 kVA per horsepower.

Code	kVA/hp	Code	kVA/hp
A	0–3.14	L	9.0–9.99
B	3.15–3.54	M	10.0–11.19
C	3.55–3.99	N	11.2–12.49
D	4.0–4.49	P	12.5–13.99
E	4.5–4.99	R	14.0–15.99
F	5.0–5.59	S	16.0–17.99
G	5.6–6.29	T	18.0–19.99
H	6.3–7.09	U	20.0–22.39
J	7.1–7.99	V	22.4 & Up
K	8.0–8.99		

Code Letters

Design Letter The design letter is an indication of the shape of the motor's torque–speed curve. The most common design letters are A, B, C, D, and E.

Design B is the standard industrial-duty motor, which has reasonable starting torque with moderate starting current and good overall performance for most industrial applications.

Optional Nameplate Information

Service Factor Service factor (abbreviated **SF** on the nameplate) is a multiplier that is applied to the motor's normal horsepower rating to indicate an increase in power output (or overload capacity) that the motor is capable of providing under certain conditions. For example, a 10 hp motor with a service factor of 1.25 safely develops 125 percent of rated power, or 12.5 hp. Generally, electric motor service factors indicate that a motor can:

- Handle a known overload that is occasional.
- Provide a factor of safety where the environment or service condition is not well defined, especially for general-purpose electric motors.
- Operate at a cooler-than-normal temperature at rated load, thus lengthening insulation life.

Common values of service factor are 1.0, 1.15, and 1.25. When the nameplate does not list a service factor, a service factor of 1.00 is assumed. In some cases, the

Figure 2-43 Totally enclosed motor.
Photo courtesy of Siemens, www.siemens.com

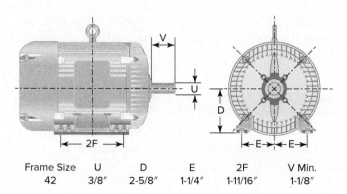

| Frame Size | U | D | E | 2F | V Min. |
| 42 | 3/8" | 2-5/8" | 1-1/4" | 1-11/16" | 1-1/8" |

Figure 2-44 Typical NEMA frame size dimensions.

running current at service factor loading is also indicated on the nameplate as service factor amperes (SFA).

Motor Enclosure The selection of a motor enclosure depends on the ambient temperature and surrounding conditions. The two general classifications of motor enclosures are open and totally enclosed. An **open** motor has ventilating openings, which permit passage of external air over and around the motor windings. A **totally enclosed** motor is constructed to prevent the free exchange of air between the inside and outside of the frame, but not sufficiently enclosed to be termed airtight (Figure 2-43).

Frame Size Refers to a set of **physical dimensions** of motors as established by NEMA and IEC. Frame sizes include physical size, construction, dimensions, and certain other physical characteristics of a motor. When you are changing a motor, selecting the same frame size regardless of manufacturer ensures the mounting mechanism and hole positions will match.

Dimensionally, NEMA standards are expressed in English units (Figure 2-44) and IEC standards are expressed in metric units. NEMA and IEC standards both use letter codes to indicate specific mechanical dimensions, plus number codes for general frame size.

Efficiency Efficiency is included on the nameplate of many motors. The efficiency of a motor is a measure of the effectiveness with which the motor converts electrical energy into mechanical energy. Motor efficiency varies from the nameplate value depending on the percentage of the rated load applied to the motor. Most motors operate near their maximum efficiency at rated load.

Energy-efficient motors, also called premium or high-efficiency motors, are 2 to 8 percent more efficient than standard motors. Motors qualify as "energy efficient" if they meet or exceed the efficiency levels listed in the NEMA's MG1 publication. Energy-efficient motors owe their higher performance to key design improvements and more accurate manufacturing tolerances.

Power Factor The letters **P.F.** when marked on the nameplate of motors stand for power factor. The power factor rating of a motor represents the motor's power factor at rated load and voltage. Motors are inductive loads and have power factors less than 1.0, usually between 0.5 and 0.95, depending on their rated size. A motor with a low power factor will draw more current for the same horsepower than a motor with a high power factor. The power factor of induction motors varies with load and drops significantly when the motor is operated at below 75 percent of full load.

Thermal Protection Thermal protection, when marked on the motor nameplate, indicates that the motor was designed and manufactured with its own **built-in** thermal protection device. There are several types of protective devices that can be built into the motor and used to sense excessive (overload) temperature rise and/or current flow. These devices disconnect the motor from its power source if they sense the overload to prevent damage to the insulation of the motor windings.

The primary types of thermal overload protectors include automatic and manual reset devices that sense either current or temperature. With **automatic-reset** devices, after the motor cools, this electrical circuit–interrupting device automatically restores power to the motor. With **manual reset** devices, the electrical circuit–interrupting device has an external button located on the motor enclosure that must be manually pressed to restore power to the motor. Manual reset protection should be provided where automatic restart of the motor after it cools down could cause personal injury should the motor start unexpectedly. Some low-cost motors have no internal thermal protection

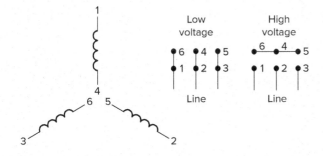

Figure 2-45 Typical dual-voltage motor connection diagram.

and rely on external protection between the motor and the electrical power supply for safety.

Connection Diagrams Connection diagrams can be found on the nameplate of some motors, or the diagram may be located inside the motor conduit box or on a special connection plate. The diagram will indicate the specific connections for dual-voltage motors (Figure 2-45). Some motors can operate in either direction, depending on how the connections to the motor are made, and this information may also be given on the nameplate.

Guide to Motor Terminology

Terminology is of the utmost importance in understanding electrical motor control. Common motor control terms are listed below. Each of these terms will be discussed in detail as they are encountered in the text.

Across-the-line A method of motor starting. Connects the motor directly to the supply line on starting or running. (Also called full voltage.)

Automatic starter A self-acting starter. Completely controlled by the master or pilot switch or some other sensing device.

Auxiliary contact The contact of a switching device in addition to the main circuit contacts (Figure 2-46). Operated by the contactor or starter.

Contactor A type of relay used for power switching.

Jog Momentary operation. Small movement of a driven machine.

Locked-rotor current Measured current with the rotor locked and with rated voltage and frequency applied to the motor.

Low-voltage protection (LVP) Magnetic control only; not automatic restarting. A three-wire control. A power failure disconnects service; when power is restored, manual restarting is required.

Low-voltage release (LVR) Magnetic control only; automatic restarting. A two-wire control. A power failure disconnects service; when power is restored, the controller automatically restarts.

Magnetic contactor A contactor that is operated electromechanically.

Multispeed starter An electric controller with two or more speeds (reversing or nonreversing) and full or reduced voltage starting.

Overload relay Running overcurrent protection. Operates on excessive current. It does not necessarily provide protection against a short circuit. It causes and maintains interruption of the motor from a power supply.

Plugging Braking by reverse rotation. The motor develops retarding force.

Push button A switch, Figure 2-47, that is a manually operable plunger or button for actuating a device, assembled into pushbutton stations.

Reduced voltage starter Applies a reduced supply voltage to the motor during starting.

Relay Used in control circuits and operated by a change in one electrical circuit to control a device in the same circuit or another circuit. Ampere rated.

Remote control Controls the function initiation or change of electrical device from some remote point.

Selector switch A manually operated switch that has the same construction as push buttons, except that rotating a handle actuates the contacts. The rotating cam may be arranged with incremental indices so that multiple positions can be used to select exclusive operations.

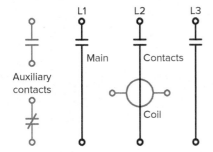

Figure 2-46 Main and auxiliary contacts.

Figure 2-47 Pushbutton switches.

Slip The difference between the actual speed (motor rpm) and the synchronous speed (rotation of the magnetic field).

Starter An electric controller used to start, stop, and protect a connected motor.

Timer A pilot device, also considered a timing relay, that provides an adjustable time period to perform its function. It can be motor driven, solenoid actuated, or electronically operated.

Torque The twisting or turning force that causes an object to rotate. There are two types of torque that are considered when looking at motors: starting torque and running torque.

Part 4 Review Questions

1. Interpret what each of the following pieces of nameplate information specifies:
   a. Voltage rating
   b. Current rating
   c. Phase rating
   d. Motor speed
   e. Ambient temperature
   f. Temperature rise
   g. Insulation class
   h. Duty cycle
   i. Horsepower rating
   j. Code letter
   k. Design letter

2. List three applications where a motor service factor greater than 1.0 may be desirable.

3. What factors enter into the selection of a proper motor enclosure?

4. Why is it important to consider the frame size when you are replacing a motor?

5. To what do energy-efficient motors owe their higher efficiency levels?

6. In what way does the power factor rating of a given horsepower motor affect its operating current?

7. A motor nameplate indicates that the motor has thermal protection. What exactly does this mean?

8. State the correct motor terminology used to describe each of the following:
   a. The current drawn by a motor at standstill with rated voltage and frequency applied.
   b. The twisting or turning force of a motor.
   c. The difference in speed between the rotation of the magnetic field of a motor and the rotation of the rotor shaft.
   d. A device that provides an adjustable time period to perform a function.
   e. Used in control circuits and operated by a change in one electrical circuit to control a device in the same circuit or another circuit.
   f. Running overcurrent motor protection.
   g. Braking a motor by reversing its direction of rotation.
   h. Applies a reduced voltage to the motor during starting.

PART 5 MANUAL AND MAGNETIC MOTOR STARTERS

Manual Starter

Manual motor starters are a very basic way to supply power to a motor. A manual control circuit is a circuit that requires the operator to control the motor directly at the location of the starter. Figure 2-48 shows an example of a three-phase **manual-start** motor control circuit. The dotted line across the contacts designates a manual starter (as opposed to a magnetic starter). Incoming power supply wires (L1, L2, and L3) connect to the top of the contacts, and the opposite sides of the contacts are connected to the overload heater elements. The motor terminal connections (T1, T2, and T3) connect to the 3ϕ motor.

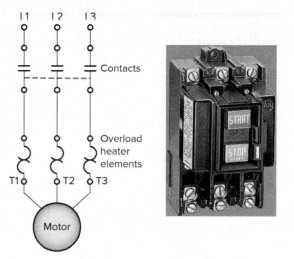

Figure 2-48 Manual motor starter.
Photo courtesy of Rockwell Automation, www.rockwellautomation.com

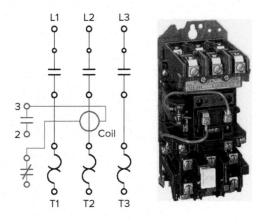

Figure 2-49 Typical three-phase, across-the-line (full-voltage) magnetic starter.
Photo courtesy of Rockwell Automation, www.rockwellautomation.com

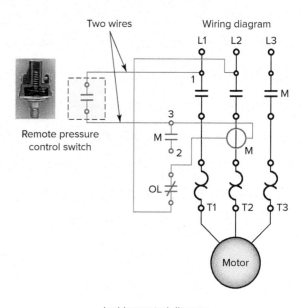

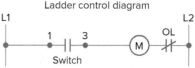

Figure 2-50 Two-wire control circuit.
Photo courtesy of Honeywell, www.honeywell.com

Manual starters are operated by the manual start/stop mechanism located on the front of the starter enclosure. The start/stop mechanism moves all three contacts at once to close (start) or open (stop) the circuit to the motor. The National Electrical Code requires that a starter not only turn a motor on and off but also protect it from overloads. The three thermal overload protective devices are installed to mechanically trip open the starter contacts when an overload condition is sensed. Manual three-phase starters are used in low horsepower applications such as drill presses and table saws where remote pushbutton control is not required.

Magnetic Starter

Magnetic motor starters allow a motor to be controlled from any location. Figure 2-49 shows a typical three-phase across-the-line (**full-voltage**) magnetic starter. The line terminals, load terminals, motor starter coil, overload relays, and auxiliary holding contact are shown. When the starter coil is energized, the three main contacts as well as the holding contact close. Should an overload condition occur, the normally closed OL relay contact would open. In addition to the power circuit, the manufacturer provides some control circuit wiring. In this case the prewired control circuit wiring consists of two connections to the starter coil. One side of the starter coil is factory wired to the overload relay contact and the other side to the holding contact.

Magnetic motor control circuits are divided into two basic types: the two-wire control circuit, and the three-wire control circuit. **Two-wire control** circuits are designed to start or stop a motor when a remote control device such as a thermostat or pressure switch is activated or deactivated. Figure 2-50 shows a typical two-wire control circuit. The operation of the circuit can be summarized as follows:

- The circuit has only **two wires** leading from the control device to the magnetic starter.
- The starter operates automatically in response to the state of the control device (pressure switch) without the assistance of an operator.
- When the contacts of the pressure switch close, power is supplied to the starter coil, causing it to energize.
- As a result, the motor is connected to the line through the power contacts.
- The starter coil is de-energized when the contacts of the pressure switch open, switching the motor off.

Figure 2-51 shows an equivalent IEC two-wire control circuit. Although the symbols may be different, the operation of the circuit is basically the same. You need to recognize and understand these symbols in order to follow how the circuit operates.

The two-wire control systems provide **low-voltage release** but not **low-voltage protection.** They use a maintained rather than a momentary-contact type of control device. If the motor is stopped by a power interruption, the starter de-energizes (low-voltage release), but also re-energizes if the control device remains closed when the circuit has power restored. Low-voltage protection is not

Part 5 Manual And Magnetic Motor Starters   43

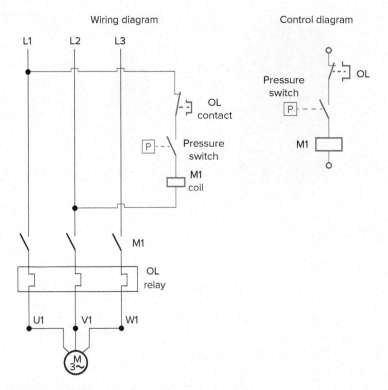

Figure 2-51 Equivalent IEC two-wire control circuit.

provided, as there is no way for the operator to be automatically protected from the circuit once power has been restored. Two-wire control circuits are used to automatically operate machinery where the automatic restarting characteristic is desirable and there is no danger of persons being injured if the equipment should suddenly **restart** after a power failure. Sump pumps and refrigerator compressor controls are two common applications for two-wire control systems.

Three-wire control provides low-voltage *protection*. The starter will drop out when there is a voltage failure, but it will not pick up automatically when voltage returns. Three-wire control uses a momentary-contact control device and a holding circuit to provide the power failure protection. Figure 2-52 shows a typical three-wire control circuit. The operation of the circuit can be summarized as follows:

- **Three-wires** are run from the start/stop pushbutton station to the starter.
- The circuit uses a normally closed (NC) stop push button wired in series with the parallel combination consisting of normally open (NO) start push button and normally open holding contact (M).
- When the momentary-contact start button is closed, line voltage is applied to the starter coil to energize it.
- The three main M contacts close to apply voltage to the motor.

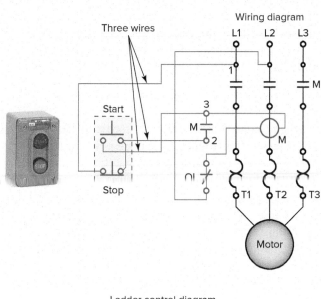

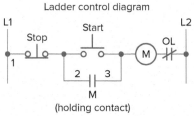

Figure 2-52 Three-wire control circuit.
This material and associated copyrights are proprietary to, and used with the permission of, Schneider Electric.

44   Chapter 2   Understanding Electrical Drawings

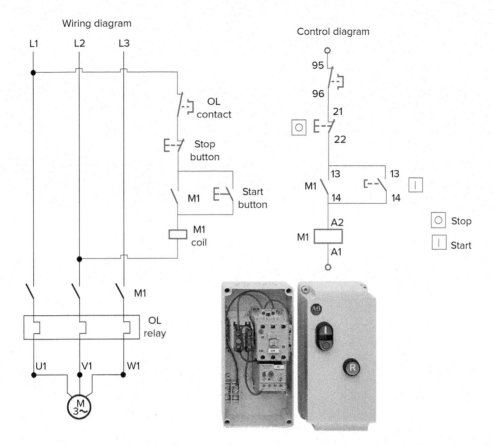

Figure 2-53 Equivalent IEC three-wire control circuit.
Photos courtesy of Rockwell Automation, www.rockwellautomation.com

- The auxiliary M contact closes to establish a circuit around the start button.
- When the start button is released, the starter coil remains energized by the closed M auxiliary contact (also known as the holding, seal-in, or memory contact) and the motor will continue to operate.
- When the momentary-contact stop button is opened, all voltage to the starter coil is lost. The main contacts are opened along with the holding contact and the motor stops.
- The starter drops out at low or no voltage and cannot be re-energized unless line voltage returns and the start button is closed.

Figure 2-53 shows an equivalent IEC three-wire control circuit. Unlike NEMA, the IEC control diagram is drawn vertically instead of horizontally.

Basically, three-wire control uses a maintaining circuit consisting of a **holding contact** wired in parallel with a start button. When the starter drops out, the holding contact opens and breaks the circuit to the coil until the start button is pressed to restart the motor. In the event of a power failure, the maintaining circuit is designed to protect against automatic restarting when the power returns. This type of protection must be used where accidents or damage might result from unexpected starts. All devices that start the circuit are connected in parallel while those that stop the circuit are connected in series.

Part 5 Review Questions

1. How are the contacts of a manual motor starter closed and opened?
2. One advantage of the magnetic motor starter over manual types is that it allows a motor to be controlled from any location. What makes this possible?
3. Power is lost and then returned to a two-wire motor control circuit. What will happen? Why?
4. Trace the current path of the holding circuit found in a three-wire motor control circuit.

Troubleshooting Scenarios

1. Heat is the greatest enemy of a motor. Discuss in what way nonadherence to each of the following motor nameplate parameters could cause a motor to overheat: (*a*) voltage rating; (*b*) current rating; (*c*) ambient temperature; (*d*) duty cycle.

2. Two identical control relay coils are incorrectly connected in series instead of parallel across a 230 V source. Discuss how this might affect the operation of the circuit.

3. A two-wire magnetic motor control circuit controlling a furnace fan uses a thermostat to automatically operate the motor on and off. A single-pole switch is to be installed next to the remote thermostat and wired so that, when closed, it will override the automatic control and allow the fan to operate at all times regardless of the thermostat setting. Draw a ladder control diagram of a circuit that will accomplish this.

4. A three-wire magnetic motor control circuit uses a remote start/stop pushbutton station to operate the motor on and off. Assume the start button is pressed but the starter coil does not energize. List the possible causes of the problem.

5. How is the control voltage obtained in most motor control circuits?

6. Assume you have to purchase a motor to replace the one with the specifications shown below. Visit the website of a motor manufacturer and report on the specifications and price of a replacement motor.

Horsepower	10
Voltage	200
Hertz	60
Phase	3
Full-load amperes	33
RPM	1725
Frame size	215T
Service factor	1.15
Rating	40C AMB-CONT
Locked rotor code	J
NEMA design code	B
Insulation class	B
Full-load efficiency	85.5
Power factor	76
Enclosure	OPEN

Discussion Topics and Critical Thinking Questions

1. Why are contacts from control devices not placed in parallel with loads?

2. Record all the nameplate data for a given motor and write a short description of what each item specifies.

3. Search the Internet for electric motor connection diagrams. Record all information given for the connection of the following types of motors:
   a. DC compound motor
   b. AC single-phase dual-voltage induction motor
   c. AC three-phase two-speed induction motor

4. The AC squirrel-cage induction motor is the dominant motor technology in use today. Why?

5. In general, how do NEMA motor standards compare to IEC standards?

CHAPTER THREE

Motor Transformers and Distribution Systems

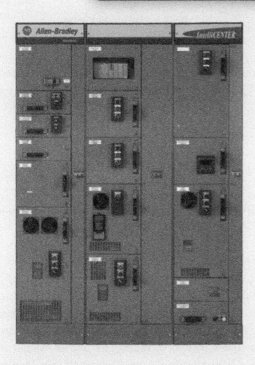

Courtesy of Rockwell Automation

CHAPTER OBJECTIVES

This chapter will help you:

- Understand the principles that are used to efficiently transmit power from the power generating plant to the customer.
- Identify the different sections and functions of a unit substation.
- Discriminate among the service entrance, feeders, and branch circuits of the electrical distribution system within a building.
- Describe the function and types of raceways used in electrical distribution systems.
- Explain the function of switchboards, panelboards, and motor control centers.
- Explain the difference between system and equipment grounding
- Understand the theory of operation of a transformer.
- Properly connect single-phase and three-phase transformers as part of a motor power and control circuit.
- List common transformer testing techniques.

Transformers transfer electric energy from one electric circuit to another by means of electromagnetic mutual induction. In its broadest sense, a distribution system refers to the manner in which electrical energy is transmitted from the generators to its many points of use. In this chapter, we will study the role that transformers play in motor power distribution and control systems.

PART 1 POWER DISTRIBUTION SYSTEMS

Transmission Systems

The **central-station system** of power generation and distribution enables power to be produced at one location for immediate use at another location many miles away. Transmitting large amounts of electric energy over fairly long distances is accomplished most efficiently by using **high voltages.** Figure 3-1 illustrates the typical transformation stages through

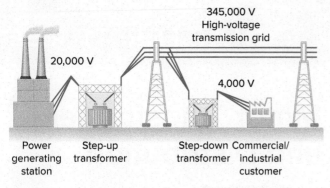

Figure 3-1 Transformation stages of a power distribution system.

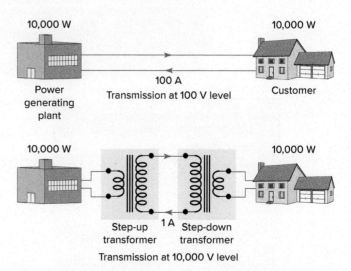

Figure 3-2 High voltage reduces the required amount of transmission current required.

which the distribution system must go in delivering power to a commercial or industrial user.

Without transformers the widespread distribution of electric power would be impractical. **Transformers** are electrical devices that transfer energy from one electrical circuit to another by magnetic coupling. Their purpose in a power distribution system is to convert AC power at one voltage level to AC power of the same frequency at another voltage level.

- High voltages are used in transmission lines to reduce the amount of current flow.
- The power transmitted in a system is equivalent to the voltage multiplied by the current. If the voltage is raised, the current can be reduced to a smaller value, while still transmitting the same amount of power.
- Because of the reduction of current flow, the size and cost of wiring are greatly reduced.
- Reducing the current also minimizes voltage drop (IR) and amount of power lost (I^2R) in the lines.

The circuits of Figure 3-2 illustrate how the use of high voltage reduces the required amount of transmission current required for a given load. Their operation is summarized as follows:

- 10,000 W of power is to be transmitted.
- When transmitted at the 100 V level, the required transmission current would be 100 A:

$$P = V \times I = 100 \text{ V} \times 100 \text{ A} = 10,000 \text{ W}$$

- When the transmission voltage is stepped up to 10,000 V, a current flow of only 1 A is needed to transmit the same 10,000 W of power:

$$P = V \times I = 10,000 \text{ V} \times 1 \text{ A} = 10,000 \text{ W}$$

There are certain limitations to the use of high voltage in power transmission and distribution systems. The higher the voltage, the more difficult and expensive it becomes to safely insulate between line wires, as well as from line wires to ground. The use of transformers in power systems allows generation of electricity at the most suitable voltage level for generation and at the same time allows this voltage to be changed to a higher and more economical voltage for transmission. At the distribution points, transformers allow the voltage to be lowered to a safer and more suitable voltage for a particular load.

Power grid transformers, used to step up or step down voltage, make possible the conversion between high and low voltages and accordingly between low and high currents (Figure 3-3). By use of transformers, each stage of the system can be operated at an appropriate voltage level. **Single-phase** three-wire power is normally supplied to residential customers, while **three-phase** power is supplied to commercial and industrial customers.

Unit Substations

Electric power comes off the transmission lines and is stepped down to the distribution lines. This may happen in several phases. The place where the conversion from **transmission** to **distribution** occurs is in a power **substation**. A substation has transformers that step transmission voltage levels down to distribution voltage levels. Basically a power substation consists of equipment

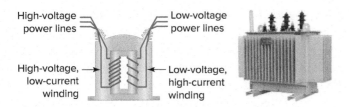

Figure 3-3 Power grid transformer.
Photo courtesy of ABB, www.abb.com

installed for switching, changing, or regulating line voltages. Substations provide a safe point in the electricity system grid for disconnecting the power in the event of trouble, as well as a convenient place to take measurements and check the operation of the system.

The power needs of some users are so great that they are fed through individual substations dedicated to them. These secondary **unit substations** form the heart of an industrial plant's or commercial building's electrical distribution. They receive the electric power from the electric utility and step it down to the utilization voltage level of 600 V nominal or less for distribution throughout the building. Unit substations offer an integrated switchgear and transformer package.

A typical unit substation is shown in Figure 3-4. Substations are factory assembled and tested and therefore require a minimum amount of labor for installation at the site. The unit substation is completely enclosed on all sides with sheet metal (except for the required ventilating openings and viewing windows) so that no live parts are exposed. Access within the enclosure is provided only through interlocked doors or bolted-on removable panels.

Figure 3-4 Factory assembled unit substation.
This material and associated copyrights are proprietary to, and used with the permission of, Schneider Electric.

The single-line diagram for a typical unit substation is illustrated in Figure 3-5. It consists of the following sections:

High-voltage primary switchgear—This section incorporates the terminations for the primary feeder cables and primary switchgear, all housed in one metal-clad enclosure.

Transformer section—This section houses the transformer for stepping down the primary voltage to the low-voltage utilization level. Dry-type, air-cooled transformers are universally used because they do not require any special fireproof vault construction.

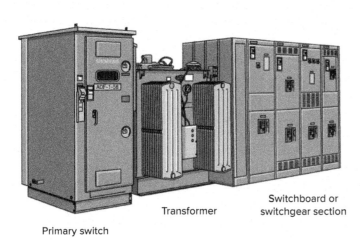
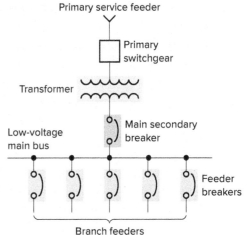

Figure 3-5 Single-line diagram for a typical unit substation.
Photo courtesy of Siemens, www.siemens.com

Part 1 Power Distribution Systems 49

Low-voltage distribution section—This switchboard section provides the protection and control for the low-voltage feeder circuits. It may contain fusible switches or molded-case circuit breakers in addition to metering for the measurement of voltage, current, power, power factor, and energy. The secondary switchgear is intended to be tripped out in the event of overload or faults in the secondary circuit fed from the transformer; the primary gear should trip if a short circuit or ground fault occurs in the transformer itself.

Before attempting to do any work on a unit substation, first the loads should be disconnected from the transformer and locked open. Then the transformer primary should be disconnected, locked out, and grounded temporarily if over 600 V.

Distribution Systems

Distribution systems used to distribute power throughout large commercial and industrial facilities can be complex. Power must be distributed through various switchboards, transformers, and panelboards (Figure 3-6) without any component overheating or unacceptable voltage drops. This power is used for such applications as lighting, heating, cooling, and motor-driven machinery.

The single-line diagram for a typical electrical distribution system within a large premise is shown in Figure 3-7. Typically the distribution system is divided into the following sections:

Service entrance—This section includes conductors for delivering energy from the electricity supply system to the premises being served.

Feeders—A feeder is a set of conductors that originates at a main distribution center and supplies one or more secondary or branch circuit distribution centers. This section includes conductors for delivering the energy from the service equipment location to the final branch circuit overcurrent device; this protects each piece of utilization equipment. Main feeders originate at the service equipment location, and subfeeders

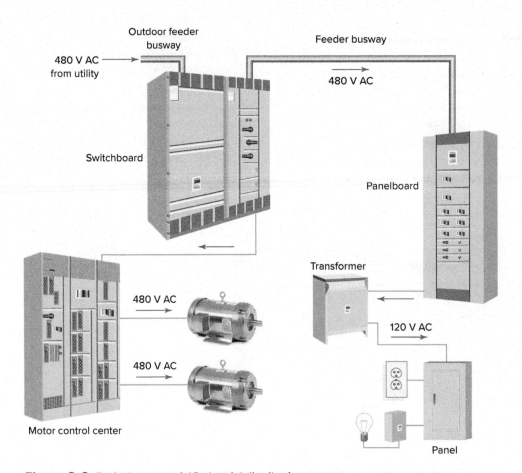

Figure 3-6 Typical commercial/industrial distribution system.
Photo courtesy of Siemens, www.siemens.com

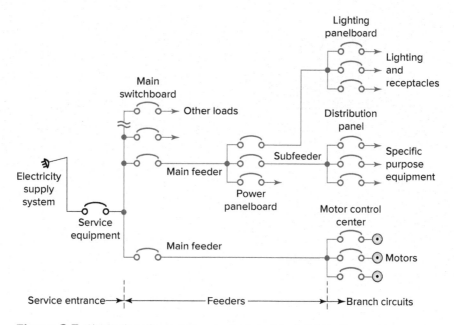

Figure 3-7 Single-line diagram for a typical electrical distribution system.

originate at panelboards or distribution centers at locations other than the service equipment location.

Branch circuits—This section includes conductors for delivering the energy from the point of the final overcurrent device to the utilization equipment. Each feeder, subfeeder, and branch circuit conductor in turn needs its own properly coordinated overcurrent protection in the form of a circuit breaker or fused switch.

Correct selection of conductors for feeders and branch circuits must take into account ampacity, short-circuit, and voltage-drop requirements. Conductor **ampacity** refers to the maximum amount of current the conductor can safely carry without becoming overheated. The ampacity rating of conductors in a raceway depends on the conductor material, gauge size, and temperature rating of the insulation; the number of current-carrying conductors in the raceway; and the ambient temperature.

The National Electrical Code (NEC) contains tables that list the ampacity for approved types of conductor size, insulation, and operating conditions. NEC rules regarding specific motor installations will be covered throughout the text. Installers should always follow the NEC, applicable state and local codes, manufacturers' instructions, and project specifications when installing motors and motor controllers.

All conductors installed in a building must be properly protected, usually by installing them in raceways. *Raceways* provide space, support, and mechanical protection for conductors, and they minimize hazards such as electric shocks and electric fires. Commonly used types of raceways found in motor installations are illustrated in Figure 3-8 and include:

Conduits—Conduits are available in rigid and flexible, metallic and nonmetallic types. They must be properly supported and have sufficient access points to facilitate the installation of the conductors. Conduits must be large enough to accommodate the number of conductors, based generally, on a 40 percent fill ratio.

Cable trays—Cable trays are used to support feeder and branch circuit cables where a number of them are to be run from a common location. They support conductors run in troughs or trays.

Low-impedance busways (bus duct)—The busways are used in buildings for high-current feeders. They consist of heavy bus bars enclosed in ventilated ducts.

Plug-in busways—These busways are used for overhead distribution systems. They provide convenient power tap-offs to the utilization equipment.

Power Losses

The unit of electric energy generated by the power station does not match with that distributed to customers. Some percentage of the units is lost in the power grid. This difference in the generated and distributed units is known as **transmission and distribution loss.**

There are two types of transmission and distribution losses: technical losses and nontechnical losses. **Technical losses** are losses incurred by physical properties of

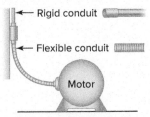

Rigid and flexible conduit

Busway sections bolted together

Cable trays

Plug-in type busway

Figure 3-8 Common types of raceways.
(Upper right and bottom right) Photo courtesy of Siemens, www.siemens.com; (Bottom left) Photo courtesy of Hyperline Systems (www.hyperline.com). The copyrights are owned by the Hyperline Systems or the original creator of the material.

components in the power grid. Common examples of technical losses include:

- Power that is lost in the form of **heat** caused by current passing through the resistance of transmission lines and cables. These losses are equal to the **square** of the electric current (I) flowing through any line times the value of the resistance of the line (R): Losses $= I^2 \times R$
- Power that is lost in **transformers**, which can be categorized into two components: no-load losses and load losses. **No-load losses** are caused by the magnetizing current needed to energize the core of the transformer and do not vary according to the loading on the transformer.
- Power that is lost due to **poor power factor**. If the power factor of a load is low, the current drawn will be high and the losses proportional to the square of the current will be greater. Line losses due to poor power factor can be reduced by the application of **capacitors** into the system.

Nontechnical losses consist of administrative losses, losses due to incorrect metering, nonpaying customers, and theft/fraud. Common examples of such losses include:

- Losses due to metering inaccuracies are defined as the difference between the amount of energy actually delivered through the meters and the amount registered by the meters.
- Theft of power is energy delivered to customers that is not measured by the energy meter for the customer. This can happen as a result of meter tampering or by bypassing the meter.

Switchboards and Panelboards

The Code defines a **switchboard** as a single panel or group of assembled panels with buses, overcurrent devices, and instruments. Figure 3-9 shows a typical combination service entrance and switchboard installed in a commercial building.

- The **service entrance** is the point where electricity enters the building.
- The **switchboard** has the space and mounting provisions required by the local utility for its metering equipment and incoming power.
- The switchboard also controls the power and protects the distribution system through the use of switches, fuses, circuit breakers, and protective relays.
- Switchboards that have more than *six* switches or circuit breakers must include a **main switch** to protect or disconnect all circuits.

Figure 3-9 Combination service entrance switchboard.
Photo courtesy of Siemens, www.siemens.com

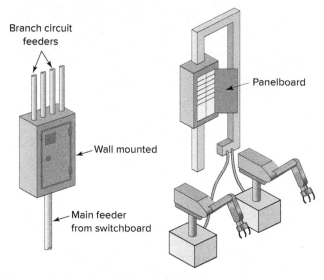

Figure 3-10 Typical panelboard installations.
This material and associated copyrights are proprietary to, and used with the permission of, Schneider Electric.

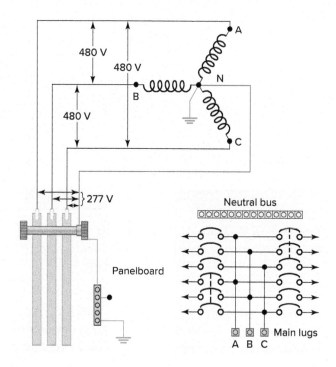

Figure 3-11 Wiring for a 277/480 V, three-phase, four-wire panelboard.

A **panelboard** contains a group of circuit breaker or fuse protective devices for lighting, convenience receptacles, and power distribution branch circuits (Figure 3-10). Panelboards (sometimes referred to as load centers) are placed in a cabinet or cutout box, which is accessible only from the front, and have dead fronts. A **dead front** is defined in the Code as having no exposed live parts on the operating side of the equipment. The panelboard is usually supplied from the switchboard and further divides the power distribution system into smaller parts. Panelboards make up the part of the distribution system that provides the last centrally located protection for the final power run to the load and its control circuitry. Panelboards suitable as service equipment are so marked by the manufacturer.

Figure 3-11 shows the typical internal wiring for a 277/480 V, three-phase, four-wire panelboard equipped with circuit breakers. This popular system used in industrial and commercial installations is capable of supplying both three-phase and single-phase loads. From neutral (N) to any hot line, 277 V single-phase for fluorescent lighting can be obtained. Across the three hot lines (A-B-C), 480 V three-phase is present for supplying motors.

The proper grounding and bonding of electrical distribution systems in general and panelboards in particular are very important. **Grounding** is the connection to earth, while **bonding** is the connection of metal parts to provide a low-impedance path for fault current to aid in clearing the overcurrent protection device and to remove dangerous current from metal that is likely to become energized. The main bonding jumper gives you *system grounding*. If a transformer is immediately upstream of the panelboard, you must bond the neutral bus or neutral conductor to the panel enclosure and to a grounding-electrode conductor, as illustrated in Figure 3-12.

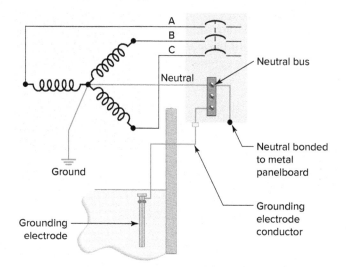

Figure 3-12 Panelboard grounding and bonding.

Part 1 Power Distribution Systems

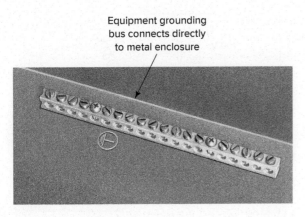

Figure 3-13 Equipment grounding bus.
This material and associated copyrights are proprietary to, and used with the permission of, Schneider Electric.

The Code requires the panelboard cabinets, frames, and the like to be connected to an **equipment grounding conductor,** not merely grounded. A separate equipment grounding terminal bar must be installed and bonded to the panelboard for the termination of feeder and branch circuit equipment grounding conductors (Figure 3-13). The equipment grounding bus is noninsulated and is mounted inside the panelboard and connects directly to the metal enclosure.

A **busbar** can be defined as a common connection for two or more circuits. The Code requires that busbars be located so as to be protected from physical damage and held firmly in place. Three-phase busbars are required to have phases in sequence so that an installer can have the same fixed phase arrangement in each termination point in any panel or switchboard. As established by NEMA, the phase arrangement on three-phase buses shall be **A, B, C** front to back, top to bottom, or left to right as viewed from the front of the switchboard or panelboard (Figure 3-14).

Panelboards are classified as main breaker or main lug types. **Main breaker**–type panelboards have the incoming supply cables connected to the line side of a circuit breaker, which in turn feeds power to the panelboard. The main breaker disconnects power from the panelboard and protects the system from short circuits and overloads.

A **main lug** panelboard does not have a main circuit breaker. The incoming supply cables are connected directly to the busbars. Primary overload protection is not provided as an integral part of the panelboard. It must be externally provided. Normally panelboard circuit terminals are required to be labeled or to have a wiring diagram. One scheme (sometimes called NEMA numbering) uses odd numbers on one side and even on the other, as illustrated in Figure 3-15.

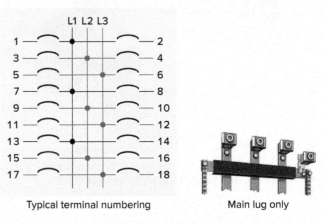

Figure 3-15 Panelboard configurations.
Photo courtesy of Siemens, www.siemens.com

Motor Control Centers (MCCs)

At times a commercial or industrial installation will require that many motors be controlled from a central location. When this is the case, the incoming power, control circuitry, required overload, and overcurrent protection are

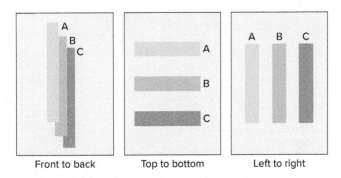

Figure 3-14 Phase arrangement on three-phase buses.

54 Chapter 3 Motor Transformers and Distribution Systems

Figure 3-16 Typical motor control center.
Photo courtesy of Rockwell Automation, www.rockwellautomation.com

combined into one convenient center. This center is called the **motor control center.**

A motor control center is a modular structure designed specifically for plug-in type motor control units. Figure 3-16 illustrates a typical motor control center made up of a compact floor-mounted assembly, composed principally of combination motor starters that contain a safety switch and magnetic starter placed in a common enclosure. The control center is typically constructed with one or more vertical sections, with each section having a number of spaces for motor starters. The sizes of the spaces are determined by the horsepower ratings of the individual starters. Thus, a starter that will control a 10 hp motor will take up less room than a starter that will control a 100 hp motor.

A motor control center is an assembly primarily of motor controllers having a common bus. The structure supports and houses control units, a common bus for distributing power to the control units, and a network of wire troughs for accommodating incoming and outgoing load and control wires. Each unit is mounted in an individual, isolated compartment or **bucket** having its own door. Motor control centers are not limited to housing just motor starters but can typically accommodate many unit types as illustrated in Figure 3-17. These may include:

- Contactors
- Full-voltage nonreversing NEMA and IEC starters
- Full-voltage reversing NEMA and IEC starters
- Soft starters
- AC variable-frequency drives

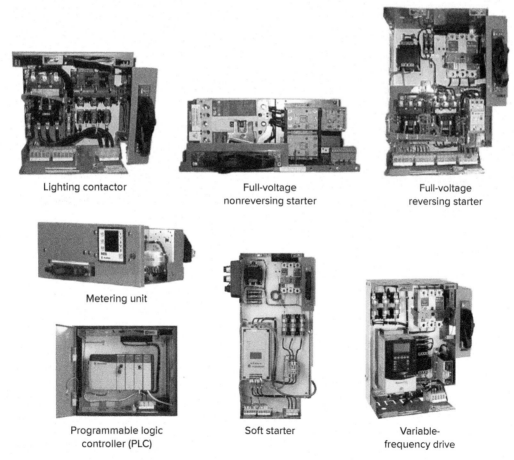

Figure 3-17 Typical motor control center unit types.
Photo courtesy of Rockwell Automation, www.rockwellautomation.com

- Programmable logic controllers (PLCs)
- Solid-state motor controllers
- Transformers
- Analog or digital metering
- Feeder circuit breakers
- Feeder fusible disconnects

The most common type of motor control center bucket is a three-phase full-voltage starter (Figure 3-18) consisting of a contactor and an overload relay. The contactor portion provides the means to remotely start and stop a motor while the overload relay protects the motor from overload conditions. This type of arrangement electrically isolates the starter for servicing without impacting the operation of the other equipment installed in the control center.

Electrical Grounding

Electrical grounding can be classified as either system grounding or equipment grounding. **System grounding** involves grounding of the power supply neutral in order that the circuit protective devices will remove a faulty circuit from the system quickly and effectively. System grounding conductors are solidly connected to earth so that any fault current will safely flow through grounding wires, building steel, conduit, and water pipes.

Equipment grounding involves grounding of the non-current-carrying conductive part of electrical equipment and of enclosures that contain electrical equipment for personnel safety. In order to protect personnel from electric

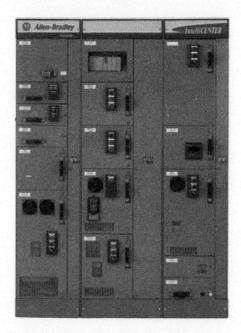

Figure 3-18 Motor control center 3-phase full-voltage starter.
Courtesy of Rockwell Automation

shock, enclosures that house electrical devices that might become energized because of unintentional contact with energized electrical conductors should be effectively grounded. If the enclosure is grounded adequately, stray voltage will be reduced to safe levels. If the enclosures are not grounded properly, unsafe voltages could exist, which could be fatal to the operating personnel.

Part 1 Review Questions

1. a. Why are high voltages used for transmitting electric power over long distances?
   b. What limitation is there to the use of high-voltage transmission systems?
2. a. If 1 MW of electric power is to be transmitted at a voltage of 100 V, calculate the amount of current the conductors would be required to carry.
   b. Calculate the amount of conductor current flow for the same amount of power and a transmission voltage of 100,000 V.
3. Compare the type of AC power normally supplied to residential customers with that supplied to commercial and industrial customers.
4. a. Outline the basic function of a unit substation.
   b. What three separate sections are contained within a typical unit substation?
5. List three factors taken into account in selecting conductors for feeders and branch circuits.
6. When motors and motor controllers are installed, what regulations must be followed?
7. a. What types of conduit raceways are commonly used in motor installations?
   b. List several installation requirements for conduit runs.
8. Compare the function of a switchboard, panelboard, and motor control center as part of the electrical distribution system.
9. Compare the purpose of the contactor and overload relay portions of a motor control center three-phase motor starter bucket.
10. What is the difference between system and equipment grounding.

PART 2 TRANSFORMER PRINCIPLES

Transformer Operation

A transformer is used to transfer AC energy from one circuit to another. The two circuits are coupled by a magnetic field that is linked to both instead of a conductive electrical path. This transfer of energy may involve an increase or decrease in voltage, but the frequency will be the same in both circuits. In addition, a transformer doesn't change **power** levels between circuits. If you put 100 VA into a transformer, 100 VA (minus a small amount of losses) comes out. The average efficiency of a transformer is well over 90 percent, in part because a transformer has no moving parts. A transformer can be operated only with AC voltage because no voltage is induced if there is no change in the magnetic field. Trying to operate a transformer from a constant DC voltage source will cause a large amount of DC current to flow, which can destroy the transformer.

Figure 3-19 illustrates a simplified version of a single-phase (1ϕ) transformer. The transformer consists of two electrical conductors, called the primary winding and the secondary winding. The primary winding is fed from a varying alternating current, which creates a varying magnetic field around it. According to the principle of **mutual inductance,** the secondary winding, which is in this varying magnetic field, will have a voltage induced into it. In its most basic form, a transformer is made up of the:

- **Core,** which provides a path for the magnetic lines of force.
- **Primary winding,** which receives energy from the source.
- **Secondary winding,** which receives energy from the primary winding and delivers it to the load.
- **Enclosure,** which protects the components from dirt, moisture, and mechanical damage.

The essentials that govern the operation of a transformer are summarized as follows:

- If the primary has more turns than the secondary, you have a step-down transformer that reduces the voltage.
- If the primary has fewer turns than the secondary, you have a step-up transformer that increases the voltage.

Single-phase motor control transformers

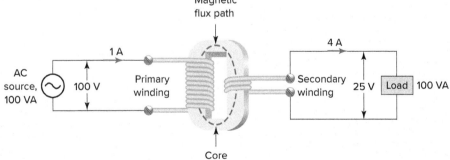

Figure 3-19 Simplified version of a single-phase (1ø) transformer.
Photo courtesy of Acme Electric, www.acmetransformer.com

- If the primary has the same number of turns as the secondary, the outgoing secondary voltage will be the same as the incoming primary voltage. This is the case for an *isolation* transformer.
- In certain cases, one large coil of wire can serve as both the primary and secondary. This is the case with *autotransformers.*
- The primary volt-amperes (VA) or kilovolt-amperes (kVA) of a transformer will be equal to that of the secondary less a small amount of losses.

Transformer Voltage, Current, and Turns Ratio

The ratio of turns in a transformer's primary winding to those in its secondary winding is known as the **turns ratio** and is the same as the transformer's **voltage ratio.** For example, if a transformer has a 10:1 turns ratio, then for every 10 turns on the primary winding there will be one turn on the secondary winding. Inputting 10 V to the primary winding steps down the voltage and will produce a 1 V output at the secondary winding. The exact opposite is true for a transformer with a 1:10 turns ratio. A transformer with a 1:10 turns ratio would have one turn on the primary winding for every 10 turns on the secondary winding. In this case, inputting 10 V to the primary winding steps up the voltage and will produce 100 volts at the secondary winding. The actual number of turns is not important, just the turns ratio. A transformer turns ratio test set, such as that shown in Figure 3-20, can directly measure the turns ratio of single-phase transformers as well as three-phase transformers. Any deviations from rated values will indicate problems in transformer windings and in the magnetic core circuits.

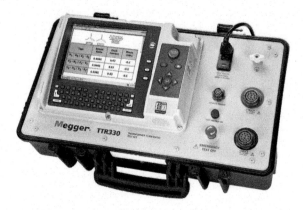

Figure 3-20 Transformer turns ratio test set.
Photo courtesy of Megger, www.megger.com/us

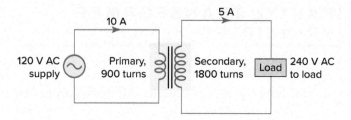

Figure 3-21 Step-up transformer.

The voltage ratio of an ideal transformer (one with no losses) is directly related to the turns ratio, while the current ratio is inversely related to the turns ratio:

$$\frac{\text{Turns primary}}{\text{Turns secondary}} = \frac{\text{Voltage primary}}{\text{Voltage secondary}}$$

$$\frac{\text{Turns primary}}{\text{Turns secondary}} = \frac{\text{Current secondary}}{\text{Current primary}}$$

The following table shows examples of some common single-phase transformer turns ratios based on primary and secondary voltage ratings.

Primary Voltage	Secondary Voltage	Turns Ratio
480 V	240 V	2:1
480 V	120 V	4:1
480 V	24 V	20:1
600 V	120 V	5:1
600 V	208 V	2.88:1
208 V	120 V	1.73:1

Figure 3-21 shows the schematic diagram of a **step-up** transformer wound with 900 turns on the primary winding and 1800 turns on the secondary winding. As a step-up unit, this transformer converts low-voltage, high-current power into high-voltage, low-current power. The transformer equations that apply to this circuit are as follows:

$$\text{Turns ratio} = \frac{\text{Number of turns on the primary}}{\text{Number of turns on the secondary}}$$

$$= \frac{900}{1800} = \frac{1}{2} = 1:2 \text{ Turns ratio}$$

If the voltage of one winding and the turns ratio are known, the voltage of the other winding can be determined.

Primary voltage = Secondary voltage × Turns ratio

$$= 240 \text{ V} \times \frac{1}{2} = 120 \text{ V}$$

Secondary voltage = $\dfrac{\text{Primary voltage}}{\text{Turns ratio}}$

$$= \frac{120}{\frac{1}{2}} = 120 \times 2 = 240 \text{ V}$$

If the current of one winding and the turns ratio are known, the current of the other winding can be determined.

Primary current = $\dfrac{\text{Secondary current}}{\text{Turns ratio}}$

$$= \frac{5 \text{ A}}{\frac{1}{2}} = 5 \times 2 = 10 \text{ A}$$

Secondary current = Primary current × Turns ratio

$$= 10 \text{ A} \times \frac{1}{2} = 5 \text{ A}$$

Figure 3-22 shows the schematic diagram of a **step-down** transformer wound with 1,000 turns on the primary winding and 50 turns on the secondary winding. As a step-down unit, this transformer converts high-voltage, low-current power into low-voltage, high-current power. A larger-diameter wire is used in the secondary winding to handle the increase in current. The primary winding, which doesn't have to conduct as much current, may be made of a smaller-diameter wire. The transformer equations that apply to this circuit are the same as those for a step-up transformer:

Turns ratio = $\dfrac{\text{Number of turns on the primary}}{\text{Number of turns on the secondary}}$

$$= \frac{1000}{50} = \frac{20}{1} = 20{:}1 \text{ turns ratio}$$

If the voltage of one winding and the turns ratio are known, the voltage of the other winding can be determined.

Primary voltage = Secondary voltage × Turns ratio

$$= 12 \text{ V} \times \frac{20}{1} = 240 \text{ V}$$

Secondary voltage = $\dfrac{\text{Primary voltage}}{\text{Turns ratio}}$

$$= \frac{240}{\frac{20}{1}} = 240 \times \frac{1}{20} = 12 \text{ V}$$

If the current of one winding and the turns ratio are known, the current of the other winding can be determined.

Primary current = $\dfrac{\text{Secondary current}}{\text{Turns ratio}}$

$$= \frac{60 \text{ A}}{\frac{20}{1}} = 60 \times \frac{1}{20} = 3 \text{ A}$$

Secondary current = Primary current × Turns ratio

$$= 3 \text{ A} \times \frac{20}{1} = 60 \text{ A}$$

A transformer automatically adjusts its input current to meet the requirements of its output or load current. If no load is connected to the secondary winding, only a small amount of current, known as the **magnetizing current** (also known as exciting current), flows through the primary winding. Typically, the transformer is designed in such a way that the power consumed by the magnetizing current is only enough to overcome the losses in the iron core and in the resistance of the wire with which the primary is wound. If the secondary circuit of the transformer becomes overloaded or shorted, primary current increases dramatically. It is for this reason that a fuse is placed in series with the primary winding to protect both the primary and secondary circuits from excessive current. The most critical parameter of a transformer is its insulation qualities. Failure of a transformer, in most instances, can be traced to a breakdown of the insulation of one or more of the windings.

For a purely resistive load, according to Ohm's law, the amount of secondary winding current equals the secondary

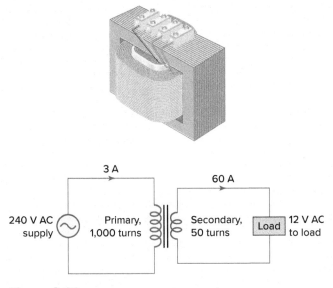

Figure 3-22 Step-down transformer.

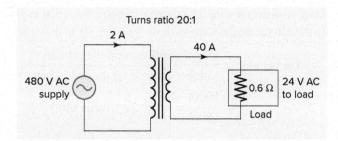

Figure 3-23 Step-down transformer connected to a resistive load.

voltage divided by the value of the load resistance connected to the secondary circuit (a negligible coil winding resistance is assumed). Figure 3-23 shows the schematic diagram of a step-down transformer with a 20:1 turns ratio connected to a 0.6-Ω resistive load. The transformer equations that apply to this circuit are as follows:

$$\text{Secondary winding current} = \frac{\text{Secondary voltage}}{\text{Load resistance}}$$

$$= \frac{24\ \text{V}}{0.6\ \Omega} = 40\ \text{A}$$

$$\text{Primary winding current} = \frac{\text{Secondary winding current}}{\text{Turns ratio}}$$

$$= \frac{40\ \text{A}}{\frac{20}{1}} = 40 \times \frac{1}{20} = 2\ \text{A}$$

Transformer Power Rating

Just as horsepower ratings designate the power capacity of an electric motor, a transformer's **kVA rating** indicates its maximum power output capacity. Transformers' kVA ratings are calculated as follows:

Single-phase loads: $\text{kVA} = \dfrac{I \times E}{1{,}000}$

Three-phase loads: $\text{kVA} = \dfrac{I \times E \times 1.732}{1{,}000}$

The maximum power rating of a transformer can be found on the transformer's nameplate. Transformers are rated in volt-amperes (VA) or kilovolt-amperes (kVA). You may recall that volt-amperes is the total power supplied to the circuit from the source and includes real (watts) and reactive (VAR) power. The primary and secondary full-load currents usually are not given. If the volt-ampere rating is given along with the primary voltage, then the primary full-load current can be determined using the following equations:

Single-phase: $\text{Full-load current} = \dfrac{\text{VA}}{\text{Voltage}}$ or $\dfrac{\text{kVA} \times 1{,}000}{\text{Voltage}}$

Three-phase: $\text{Full-load current} = \dfrac{\text{kVA} \times 1{,}000}{1.73 \times \text{Voltage}}$

Figure 3-24 shows the diagram for a single-phase 25 kVA transformer, rated 480 V primary and 240 V secondary. The rated full-load primary and secondary currents are calculated as follows:

$$\text{Primary full-load current} = \frac{\text{kVA} \times 1{,}000}{\text{Voltage}}$$

$$= \frac{25\ \text{kVA} \times 1{,}000}{480\ \text{V}} = 52\ \text{A}$$

$$\text{Secondary full-load current} = \frac{\text{kVA} \times 1{,}000}{\text{Voltage}}$$

$$= \frac{25\ \text{kVA} \times 1{,}000}{240\ \text{V}}$$

$$= 104\ \text{A}$$

Figure 3-25 shows the diagram for a three-phase 37.5 kVA transformer, rated 480 volts primary and 208 V secondary. The rated full-load primary and secondary currents are calculated as follows:

$$\text{Primary full-load current} = \frac{\text{kVA} \times 1{,}000}{1.73 \times \text{Voltage}}$$

$$= \frac{37.5\ \text{kVA} \times 1{,}000}{1.73 \times 480\ \text{V}}$$

$$= 45\ \text{A}$$

$$\text{Secondary full-load current} = \frac{\text{kVA} \times 1{,}000}{1.73 \times \text{Voltage}}$$

$$= \frac{37.5\ \text{kVA} \times 1{,}000}{1.73 \times 208\ \text{V}}$$

$$= 104\ \text{A}$$

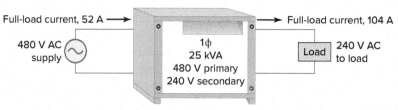

Figure 3-24 Single-phase 25 kVA transformer, rated 480/240 V full-load current.

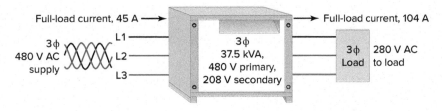

Figure 3-25 Three-phase 37.5 kVA transformer, rated 480/240 V full-load current.

Transformer Performance

The **efficiency** of a transformer is a measure of the proportion of the applied energy that is transferred to the load. A transformer does not require any moving parts to transfer energy, and for this reason, power losses associated with them are generally relatively small. Transformer losses in the form of waste heat are a combination of core losses and coil losses.

- **Core losses** (also known as iron or hysteresis losses) consist of those generated by energizing the laminated steel core. Whenever a transformer is energized, the core losses remain essentially **constant** regardless of how much electric power is flowing through it.
- **Coil losses** (also known as copper or I^2R losses) result from resistance of the windings and **vary** with the square of the electric current flowing through the windings.
- At low loads, core losses account for the majority of losses. As the load increases, winding losses represent the greatest loss.
- To reduce core losses, high-efficiency transformers are designed with a better grade of core steel with thinner core laminations than standard-efficiency types.
- Coil losses can be reduced through improved conductor design and increases in the size of the copper conductor used.
- An ideal transformer would have no energy losses, and would be 100 percent efficient. Real transformers, when operated at full load, have an efficiency in the 94 to 98 percent range.

Transformer **temperature rise** is defined as the average temperature rise of the windings above the ambient (surrounding) temperature when the transformer is loaded at its nameplate rating. Transformers rated for a lower temperature rise:

- Generate less heat.
- Operate more efficiently.
- Last longer than transformers with a higher temperature rise rating.

The output secondary voltage of a transformer varies some from no-load to the full-load condition, even with a constant input voltage applied to the primary. This is due in part to:

- Primary and secondary winding resistance and inductance.
- The degree of mutual inductance (magnetic coupling) between the primary and secondary windings.

The measure of how well a power transformer maintains constant secondary voltage over a range of load currents is called the transformer's **voltage regulation** and is calculated as follows:

$$\text{Voltage regulation percentage} = \frac{V_{\text{no-load}} - V_{\text{full-load}}}{V_{\text{full-load}}} \times 100$$

where

No-load means no load connected to the secondary.

Full-load means the point at which the transformer is operating at maximum rated secondary current.

The steady-state magnetizing current for a transformer is very low, but the momentary, primary **inrush current** when it is first energized can be quite high. This may result during the first few cycles of the AC voltage. The magnitude of the inrush current depends on the point on the AC wave where the transformer is switched on. Inrush currents can be typically many times the normal full-load current and result in nuisance tripping of the input over current protection device. A general approach to the problem of dealing with transformer inrush currents is to choose an input over current protection device with the right time delay characteristic to avoid nuisance tripping.

Part 2 Review Questions

1. Define the terms *primary* and *secondary* as they apply to a transformer winding.
2. On what basis is a transformer classified as being a *step-up* or *step-down* type?
3. Explain how the transfer of energy takes place in a transformer.
4. In an ideal transformer, what is the relationship between:
   a. The turns ratio and the voltage ratio?
   b. The voltage ratio and current ratio?
   c. The primary power and secondary power?
5. A step-down transformer with a turns ratio of 10:1 has 120 V AC applied to its primary coil winding. A 3-Ω load resistor is connected across the secondary coil. Assuming ideal transformer conditions, calculate the following:
   a. Secondary coil winding voltage
   b. Secondary coil winding current
   c. Primary winding coil current
6. A step-up transformer has a primary current of 32 A and an applied voltage of 240 V. The secondary coil has a current of 2 A. Assuming ideal transformer conditions, calculate the following:
   a. Power input of the primary winding coil
   b. Power output of the secondary winding coil
   c. Secondary coil winding voltage
   d. Turns ratio
7. What is meant by the term *transformer magnetizing*, or *exciting*, *current*?
8. Why will a fuse placed in series with the primary coil winding protect both the primary and secondary coil winding circuits from excessive current?
9. Is transformer power rated in watts or volt-amperes? Why?
10. A transformer primary winding has 900 turns and the secondary winding has 90 turns. Which winding of the transformer has the larger-diameter conductor? Why?
11. The primary of a transformer is rated for 480 V and the secondary for 240 V. Which winding of the transformer has the larger-diameter conductor? Why?
12. A single-phase transformer is rated for 0.5 kVA, a primary voltage of 480 V, and a secondary voltage of 120 V. What is the maximum full load current that can be supplied by the secondary?

PART 3 TRANSFORMER CONNECTIONS AND SYSTEMS

Transformer Polarity

Transformer **polarity** refers to the relative direction or polarity of the induced voltage between the high-voltage and low-voltage terminals of a transformer. An understanding of transformer polarity markings is essential in making three-phase and single-phase transformer connections. Knowledge of polarity is also required to connect potential and current transformers to power metering and protective relays.

On power transformers, the high-voltage winding leads are marked **H1 and H2** and the low-voltage winding leads are marked **X1 and X2** (Figure 3-26). By convention, H1 and X1 have the same polarity, which means that when H1 is instantaneously positive, X1 also is instantaneously positive. These markings are used in establishing the proper terminal connections when single-phase

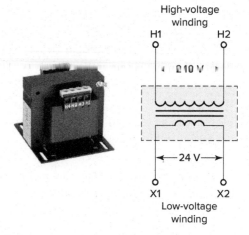

Figure 3-26 Transformer polarity markings.
Photo courtesy of Rockwell Automation, www.rockwellautomation.com

transformers are connected in parallel, series, and three-phase configurations.

In practice, the four terminals on a single-phase transformer are mounted in a standard way so the transformer

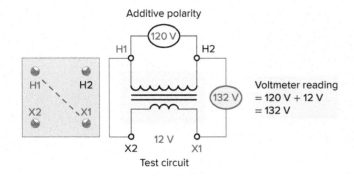

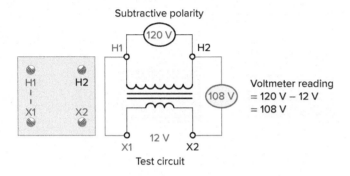

Figure 3-27 Additive and subtractive transformer terminal markings.

Photo courtesy of Tesco, www.tesco-advent.com

has either **additive** or **subtractive** polarity. Whether the polarity is additive or subtractive depends on the location of the H and X terminals. A transformer is said to have additive polarity when terminal H1 is diagonally opposite terminal X1. Similarly, a transformer has subtractive polarity when terminal H1 is adjacent to terminal X1. Figure 3-27 illustrates additive and subtractive transformer terminal markings along with a test circuit that can be used to verify markings. Also shown is a battery-operated transformer polarity checker that can perform the same test.

Single-Phase Transformers

Motor control transformers are designed to reduce supply voltages to motor control circuits. Most AC commercial and industrial motors are operated from three-phase AC supply systems in the 208 to 600 V range. However, the control systems for these motors generally operate at 120 V. The major disadvantage to higher-voltage control schemes is that these higher voltages can be much more lethal than 120 V. Additionally, on a higher-voltage control system tied directly to the supply lines, when a short circuit occurs in the control circuit, a line fuse will blow or a circuit breaker will trip; however, it may not do so right away. In some cases, light-duty contacts, such as those in stop buttons or relay contacts, can weld together before the protective device trips or blows.

Step-down control transformers are installed when the control circuit components are not rated for the line voltage. Figure 3-28 shows the typical connection for a step-down motor control transformer. The primary side (H1 and H2) of the control transformer will be the line voltage (240 V), while the secondary voltage (X1 and X2) will be the voltage required for the control components (120 V).

Single-, dual-, and multitap primary control transformers are available. The versatile dual- and multitap primary transformers allow reduced control voltage from a variety of voltage sources to meet a wide array of applications. Figure 3-29 shows the connections for a typical dual

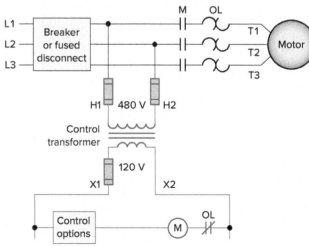

Figure 3-28 Motor control transformer wiring.

Photo by Steve Armstrong

Part 3 Transformer Connections and Systems 63

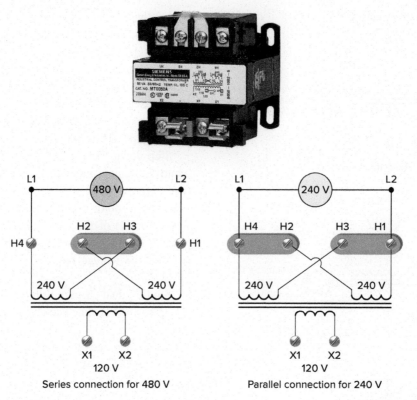

Figure 3-29 Typical dual-voltage 480 V and 240 V transformer connections.
Photo courtesy of Siemens, www.siemens.com

primary transformer used to step 240 or 480 V down to 120 V.

- The primary connections on the transformer are identified as H1, H2, H3, and H4.
- The transformer coil between H1 and H2 and the one between H3 and H4 are rated for 240 V each.
- The low-voltage secondary connections on the transformer, X1 and X2, can have 120 V from either a 480 or 240 V line.
- If the transformer is to be used to step *480 V* down to 120 V, the primary windings are connected in **series** by a jumper wire or metal link.
- When the transformer is to be used to step *240 V* down to 120 V, the two primary windings must be connected in **parallel** with each other.

The control transformer secondary can be grounded or ungrounded. Where grounding is provided, the X2 side of the circuit common to the coils must be grounded at the control transformer. This will ensure that an accidental ground in the control circuit will not start the motor or make the stop button or control inoperative. An additional requirement for all control transformers is that they be protected by fuses or circuit breakers. Depending on the installation, this protection can be placed on the primary, secondary, or both sides of the transformer. Figure 3-30 shows fuse protection for both the primary and secondary of the transformer and the correct ground connection for a grounded control system. The fuses must be properly sized

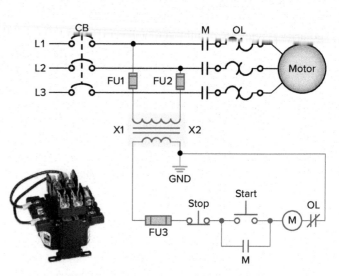

Figure 3-30 Fuse protection for both the primary and secondary of the transformer and the correct ground connection for a grounded control system.
Photo courtesy of SolaHD, www.solahd.com

for the control circuit. **Section 430.72** of the Code lists requirements for the protection of transformers used in motor control circuits.

Three-Phase Transformers

Large amounts of power are generated and transmitted using high-voltage three-phase systems. Transmission voltages may be stepped down several times before they reach the motor load. This transformation is accomplished using three-phase **wye-** or **delta**-connected transformers or a combination of the two. Figure 3-31 illustrates some of the

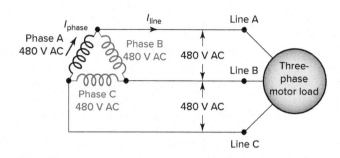

Figure 3-32 Three-phase, three-wire delta transformer connection supplying power to a three-phase motor load.

common three-phase wye and delta transformer connections. The connections are named after the way the windings are connected inside the transformer. Polarity markings are fixed on any transformer, and the connections are made in accordance with them.

The transformers supplying motor loads can be connected on the load (secondary) side either in delta or in wye configuration. Two types of secondary distribution systems commonly used are the three-phase three-wire system and three-phase four-wire system. In both, the secondary voltages are the same for all three phases. The three-phase three-wire delta system is used for balanced loads and consists of three transformer windings connected end to end. Figure 3-32 shows a typical three-phase, three-wire delta transformer connection supplying power to a three-phase motor load. For a delta-connected transformer:

- The phase voltage (E_{phase}) of the transformer secondary is always the same as the line voltage (E_{line}) of the load.
- The line current (I_{line}) of the load is equal to the phase current (I_{phase}) of the transformer secondary multiplied by 1.73.

$$\textbf{kVA (transformer)} = \frac{I_{line} \times E_{line} \times 1.732}{1,000}$$

- The constant 1.73 is the square root of 3 and is used because the transformer phase windings are 120 electrical degrees apart.

The other commonly used three-phase distribution is the three-phase, four-wire system. Figure 3-33 shows a typical wye-connected three-phase, four-wire distribution system. The three phases connect at a common point, which is called the neutral. Because of this, none of the windings are affected by the other windings. Therefore, the wye three-phase, four-wire system is used for unbalanced loads. The phases are 120 electrical degrees

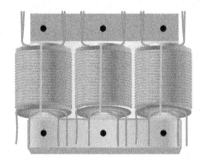

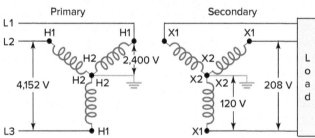

Wye-wye three-phase transformer connection

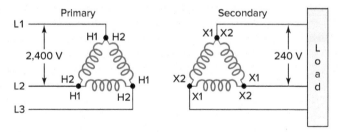

Delta-delta three-phase transformer connection

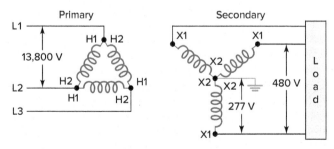

Delta-wye three-phase transformer connection

Figure 3-31 Common wye and delta transformer connections.

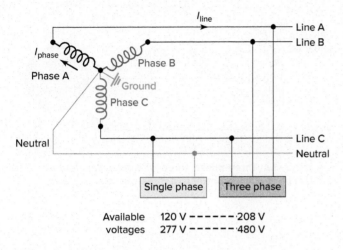

Figure 3-33 Wye-connected, three-phase, four-wire distribution system.

apart; however, they have a common point. For a wye-connected transformer:

- The phase-to-phase voltage is equal to the phase-to-neutral voltage multiplied by 1.73.
- The line current is equal to the phase current.

$$\text{kVA (transformer)} = \frac{I_{\text{line}} \times E_{\text{line}} \times 1.732}{1,000}$$

- Common arrangements are 480Y/277 V and 208Y/120 V.

The delta-to-wye configuration is the most commonly used three-phase transformer connection. A typical delta-to-wye voltage transformation is illustrated in Figure 3-34. The secondary provides a **neutral point** for supplying line-to-neutral power to single-phase loads. The neutral point is also grounded for safety reasons. Three-phase loads are

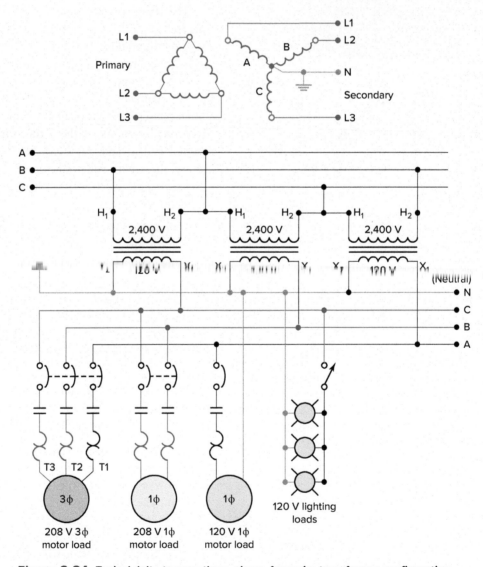

Figure 3-34 Typical delta-to-wye, three-phase, four-wire transformer configuration.

Variable autotransformer

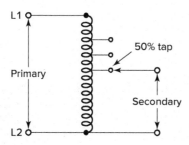

Figure 3-35 Autotransformer.
Photo courtesy of Superior Electric, www.superiorelectric.com.

supplied at 208 V, while the voltage for single-phase loads is 208 V or 120 V. When the transformer secondary supplies large amounts of unbalanced loads, the delta primary winding provides a better current balance for the primary source.

The *autotransformer*, shown in Figure 3-35, is a transformer consisting of a single winding with electrical connection points called *taps*.

- Each tap corresponds to a **different voltage** so that, in effect, a portion of the same inductor acts as part of both the primary and secondary windings.
- There is **no electrical isolation** between the input and output circuits, unlike the traditional two-winding transformer.
- The ratio of secondary to primary voltages is equal to the ratio of the number of turns of the tap they connect to. For example, connecting at the 50 percent tap (middle) and bottom of the autotransformer output will halve the input voltage.
- Because it requires both fewer windings and a smaller core, an autotransformer for some power applications is typically **lighter and less costly** than a two-winding transformer.
- A **variable autotransformer** is one in which the output connection is made through a sliding brush and is widely used where adjustable AC voltages are required.

An autotransformer motor starter, such as shown in Figure 3-36, reduces inrush motor current by using a three-coil autotransformer in the line just ahead of the motor to step down the voltage applied to the motor terminals. By reducing the voltage, the current drawn from the line is reduced during start-up. During the starting period, the motor is connected to the reduced-voltage taps on the autotransformer. Once the motor has accelerated, it is automatically connected to full-line voltage.

Instrument Transformers

Instrument transformers are small transformers used in conjunction with instruments such as ammeters, voltmeters,

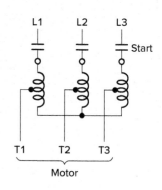

Figure 3-36 Autotransformer motor starter.
Photo courtesy of Rockwell Automation, www.rockwellautomation.com

power meters, and relays used for protective purposes. These transformers step down the voltage or current of a circuit to a low value that can be effectively and safely used for the operation of instruments. Instrument transformers also provide isolation between the instrument and the high voltage of the power circuit.

A typical current and potential transformer connection diagram is shown in Figure 3-37.

- **Polarity marks** are shown to ensure correct wiring.
- Low-resistance **current elements** of the instruments are connected in **series,** while **voltage inputs** are connected in **parallel.**
- Polarity is **not** a consideration on single-element devices such as an ammeter or voltmeter but is essential for proper operation of **three-phase** metering and other protective and control devices that require proper polarity connections.

A **potential** (voltage) transformer operates on the same principle as a standard power transformer. The main difference is that the capacity of a potential transformer is relatively small compared to power transformers. Potential transformers have typical power ratings of from 100 VA to 500 VA. The secondary low-voltage side is usually wound for 120 V, which makes it possible to use standard instruments with potential coil ratings of 120 V. The primary side is designed to be connected in parallel with the circuit to be monitored.

A **current transformer (CT)** is a type of instrument transformer that is designed to provide a current in its **secondary** that is accurately **proportional** to the current flowing in its **primary.** Current transformers are used when the magnitude of AC currents or voltage exceeds the safe value of current of measuring instruments. Figure 3-38 illustrates how a current transformer operates.

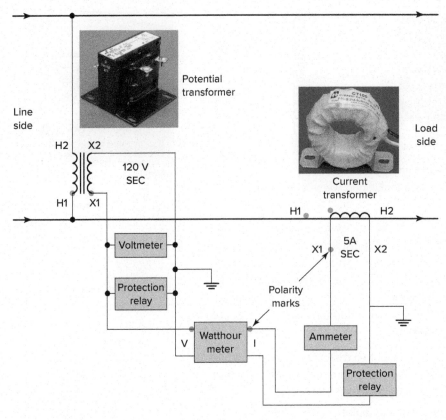

Figure 3-37 Instrument transformers.
Photos courtesy of Hammond Mfg. Co., www.hammondmfg.com.

- The **primary** is a line conductor connected in series that passes through the center of the transformer.
- The **secondary** winding consisting of many turns is designed to produce a **standard 5 A** output when its rated current is flowing in the primary.
- Current transformers are specified by a current ratio such as 5:200, where 200 A of primary current will result in 5 A of secondary output current.
- The secondary load (ammeter) of the current transformer is called the **burden** to distinguish it from the load of the circuit whose current is being measured.

All current transformers have **subtractive polarity.** Recall that on subtractive polarity transformers, the **H1** primary lead and the **X1** secondary lead will be on the **same side** of the transformer, as illustrated in Figure 3-39. The secondary circuit of a current transformer should never be **opened** when there is current in the primary winding. If the secondary is not connected to a burden load, the transformer acts to **step up the secondary voltage to a dangerous level**, because of the high turns ratio. For this reason, a current transformer should always have its secondary (X1–X2) **shorted** when not connected to an external load.

In applications that depend on the interaction of two instruments, such as a watthour meter or protective relay, it is essential that the polarity of both current and potential transformers be known and that definite relationships be maintained.

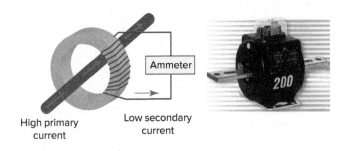

Figure 3-38 Current transformer.
Photo courtesy of ABB, www.abb.com

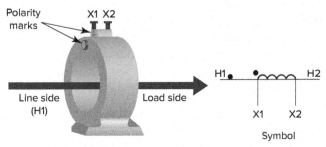

Figure 3-39 Current transformer polarity.

Transformer Testing

Failures in transformers include:

- **Winding failures** due to open windings and/or short circuits between coil turns, coil individual phases and coil to ground.
- **Core faults** due to core insulation failure and shorted laminations.
- **Terminal failures** due to open leads, loose connections, and short circuits.

A **resistance check** of the transformer coil windings can be made using an ohmmeter. An infinite resistance reading across a coil indicates that the winding is open. The resistance value of windings may differ substantially due to the difference in the number of turns and the cross-sectional area of the conductor. In this case, the manufactures specifications would have to be checked to determine if the resistance values are correct.

An **insulation check** of the transformer windings is made using a **megger.** The megger insulation tester is a small, portable instrument that gives you a direct reading of insulation resistance in ohms or megohms. The circuit diagrams of Figure 3-40 illustrate how insulation tests of transformer windings are made using a megger. Both the resistance insulation to ground as well as between coil windings should be tested. When making the insulation to ground test, all windings should be grounded except the winding being tested. When making the insulation between coils test, be sure to test all possible coil winding combinations.

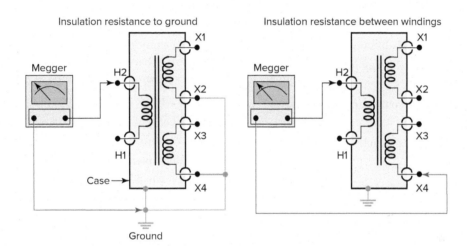

Figure 3-40 Insulation testing of transformer windings.

Part 3 Review Questions

1. Explain the way in which the high-voltage and low-voltage leads of a single-phase power transformer are identified.
2. A polarity test is being made on the transformer shown in Figure 3-41.
   a. What type of polarity is indicated?
   b. What is the value of the voltage across the secondary winding?
   c. Redraw the diagram with the unmarked leads of the transformer correctly labeled.
3. The control circuit for a three-phase 480 V motor is normally operated at what voltage? Why?
4. A 240/480 V dual-primary control transformer is to be operated from a 480 V three-phase system. How would the two primary windings be connected relative to each other? Why?
5. For the motor control circuit of Figure 3-42, assume the circuit is *incorrectly* grounded at X1 instead of correctly grounded as shown at X2. With this incorrect connection, explain how the control circuit would operate if point 2 of the stop or start push button were to become accidentally grounded.
6. What are the two basic types of three-phase transformer configurations?

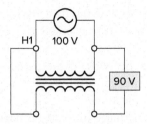

Figure 3-41 Circuit for review question 2.

7. The phase-to-neutral voltage of a wye-connected, three-phase, four-wire distribution system is rated for 277 V. What would its phase-to-phase rating be?
8. Why is it necessary to apply the constant 1.73 ($\sqrt{3}$) in three-phase circuit calculations?
9. Explain the basic difference between the primary and secondary circuits of a standard voltage transformer and an autotransformer.
10. How are autotransformers used to reduce the starting current for large three-phase motors?
11. Give two examples of the way in which instrument transformers are used.
12. Compare the primary connection of a potential transformer with that of a current transformer.
13. What important safety precaution should be followed when operating current transformers in live circuits?

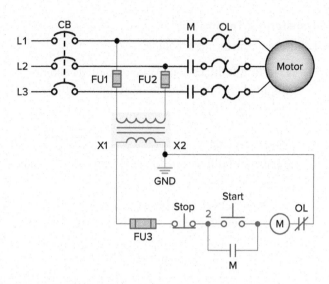

Figure 3-42 Circuit for review question 5.

14. The current rating of the primary winding of a current transformer is 100 A and its secondary rating is 5 A. An ammeter connected across the secondary indicates 4 A. What is the value of the current flow in the primary?
15. List three common types of transformer failures.
16. What type of fault does an infinite resistance reading across a transformer winding indicate?
17. What instrument would be used to take an insulation check of a transformer winding?

Troubleshooting Scenarios

1. The control transformer for an across-the-line three-phase motor starter is tested and found to have an open in the secondary winding. Discuss what would occur if an attempt where made to temporarily operate the control system directly from two of the three-phase supply lines.
2. The two primary windings of a dual-primary control transformer (240 V or 480 V) are to be connected in parallel to step the line voltage of 240 V down to a control voltage of 120 V. Assuming the two primary windings are incorrectly connected in series instead of parallel, what effect would this have on the control circuit?
3. A local fire department responded to an automatic trouble call at a large office building. Unable to determine which breaker controlled which branch circuit, the fire chief issued an order to open all of the breakers. This put two floors of the building in the dark, rather than just the one room in question. How could this unnecessary interruption in operation been prevented?
4. A dry-type general-purpose power transformer is observed to be operating above its rated temperature. What are some possible causes?
5. A current transformer is to be tested in circuit using a digital multimeter and clamp-on ammeter. Outline the procedure to be followed.
6. A DC resistance check is made on the secondary winding of a control transformer and indicates a very low resistance reading. Why can you not automatically conclude that the winding has a short circuit fault?

Discussion Topics and Critical Thinking Questions

1. Discuss how electric power might be distributed within a small commercial or industrial site.

2. Research the specifications for a typical four-wire electrical power panelboard capable of feeding single-phase and three-phase loads. Include in your findings:
   - All electrical specifications
   - Internal bus layout
   - Connections for single-phase and three-phase loads

3. A block of several transformers are fed from individual circuit breakers. What steps should be taken to limit the transformer inrush current to the system?

4. How would you proceed with a DC resistance check of the windings of a three-phase transformer?

5. The high-voltage and low-voltage sides of a single-phase transformer are to be verified using an ohmmeter. How would you proceed with this test?

CHAPTER FOUR

Motor Control Devices

Source: www.valin.com

CHAPTER OBJECTIVES

This chapter will help you:

- Identify manually operated switches commonly found in motor control circuits and explain their operation.
- Interpret the operation of an IEC break-make pushbutton control circuit.
- Identify mechanically operated switches commonly found in motor control circuits and explain their operation.
- Compare the operation of Bourdon tube-, bellows-, and diaphragm-type pressure switches.
- Identify different types of sensors and explain how they detect and measure the presence of something.
- Compare the characteristics of inductive and capacitive proximity sensors.
- Compare the characteristics of through beam, retroreflective, and diffuse scanning.
- Describe the operating characteristics of a relay, solenoid, solenoid valve, stepper motor, and brushless DC motor.

Control devices are components that govern the power delivered to an electrical load. Motor control systems make use of a wide variety of control devices. The motor control devices introduced in this chapter range from simple pushbutton switches to more complex solid-state sensors. The terms and practical applications presented here illustrate how selection of a control device depends on the specific application.

PART 1 MANUALLY OPERATED SWITCHES

Primary and Pilot Control Devices

A control device is a component that governs the power delivered to an electrical load. All components used in motor control circuits may be classed as either primary control devices or pilot control devices. A **primary control device,** such as a motor contactor, starter, or controller, connects the load to the line. A **pilot control device,** such as a relay or switch contact, is used to activate the primary control device. Pilot-duty devices

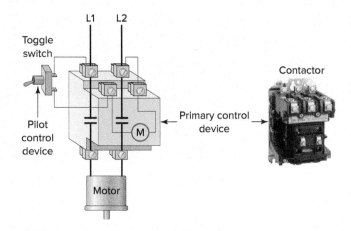

Figure 4-1 Primary and pilot control devices.
Photo courtesy of Rockwell Automation,
www.rockwellautomation.com

should not be used to switch horsepower loads unless they are specifically rated to do so. Contacts selected for both primary and pilot control devices must be capable of handling the voltage and current to be switched. Figure 4-1 shows a typical motor control circuit that includes both primary and pilot control devices. In the application shown, the closing of the toggle switch contact completes the circuit to energize contactor coil M. This in turn closes the contacts of the contactor to complete the main power circuit to the motor.

Toggle Switches

A manually operated switch is one that is controlled by hand. The **toggle switches** illustrated in Figure 4-2 are examples of manually operated switches. A toggle switch uses a mechanical lever mechanism to implement a positive snap action for switching of electrical contacts. This type of switching or contact arrangement is specified by the appropriate abbreviation as follows:

SPST—Single pole, single throw
SPDT—Single pole, double throw
DPST—Double pole, single throw
DPDT—Double pole, double throw

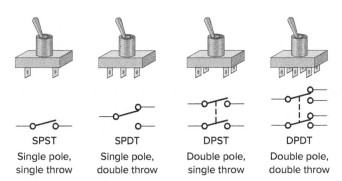

Figure 4-2 Toggle switches.

Electrical ratings for switches are expressed in terms of the maximum interrupting voltage and current they can safely handle. AC and DC contact current ratings are not the same for a given switch. The AC current rating will be higher than its DC rating for an equivalent amount of voltage. The reason for this is that AC current is at zero level twice during each cycle, which reduces the likelihood of an electric arc forming across the contacts. Also, higher decaying voltages are generated in DC circuits that contain inductive-type load devices. Switch voltage and current ratings represent maximum values and may be used in circuits with voltages and currents below these levels but never above.

Pushbutton Switches

Pushbutton switches are commonly used in motor control applications to start and stop motors, as well as to control and override process functions. A push button operates by opening or closing contacts when pressed. Figure 4-3 shows commonly used types of pushbutton symbols and switching action. Abbreviations NO (normally open) and NC (normally closed) represent the state of the switch contacts when the switch is not activated. The NO push button makes a circuit when it is pressed and returns to its open position when the button is released. The NC push button opens the circuit when it is pressed and returns to the closed position when the button is released.

With a **break-make** push button, the top section contacts are NC and the bottom section contacts are NO When the button is pressed, the bottom contacts are closed after the top contacts open. Figure 4-4 shows a break-make push button used in a jog-start-stop motor control

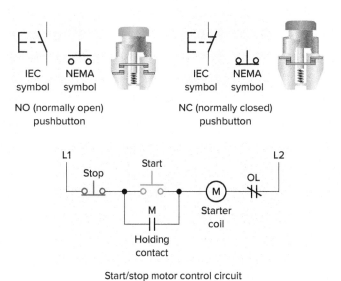

Figure 4-3 Pushbutton symbols and switching action.

Part 1 Manually Operated Switches

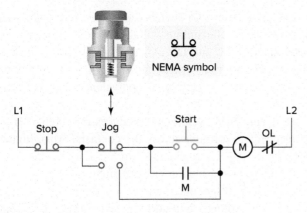

Figure 4-4 Break-make push button and typical motor control circuit.

circuit. The operation of the circuit can be summarized as follows:

- Pressing the start push button completes a circuit for the M coil, causing the motor to start, and the M contact holds in the M coil circuit.
- With the M coil de-energized and the jog push button then pressed, a circuit is completed for the M coil around the M contact. The M contact closes, but the holding circuit is incomplete, as the NC contact of the jog button is open.

Figure 4-5 illustrates the IEC version of the break-make push button and typical motor control circuit. Although the symbols are different, the operation of the

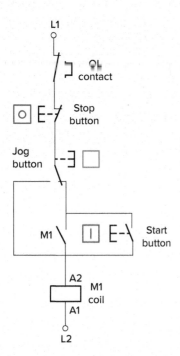

Figure 4-5 IEC version break-make push button and typical motor control circuit.

Figure 4-6 Pushbutton station—NEMA Type1.
This material and associated copyrights are proprietary to, and used with the permission of Schneider Electric.

circuit is exactly the same. Note that IEC-type circuit schematics are meant to be read vertically from top to bottom in comparison to those of NEMA, which are read horizontally from left to right.

When you have one or more push buttons in a common enclosure, it is referred to as a pushbutton station (Figure 4-6). Electrical **enclosures** are designed to protect their contents from troublesome operating environmental conditions such as dust, dirt, oil, water, corrosive materials, and extreme variations in temperature. The types of enclosures are standardized by the National Electrical Manufacturers Association (NEMA). NEMA enclosure types are selected according to the environment in which the equipment is installed. A partial listing of specific enclosure types is given on the next page.

Pushbutton assemblies are manufactured in both 30-mm NEMA and smaller 22-mm IEC types, as illustrated in Figure 4-7. The size is related to the diameter of the circular hole the push button is mounted in—either 30 millimeters or 22 millimeters in diameter. Pushbutton assemblies basically consist of an operator, legend plate, and contact block.

The **operator** is the part of the pushbutton assembly that is pressed, pulled, or rotated to activate the push button's contacts. Operators come in many different colors, shapes, and sizes designed for specific control applications.

- **Flush** push buttons (Figure 4-8) have the actuator flush with the mounting ring and are often used for start buttons that need to be protected from accidental initiation.
- **Extended** push buttons (Figure 4-9) have the actuator protruding about ¼ inch beyond the mounting ring and allow easier activation because the machine

74 Chapter 4 Motor Control Devices

NEMA Enclosure Types

Type	Use	Service Conditions
1	Indoor	No unusual
3	Indoor/Outdoor	Windblown dust, rain, sleet, and ice on enclosure
3R	Indoor/Outdoor	Falling rain and ice on enclosure
4	Indoor/outdoor	Windblown dust and rain, splashing water, hose-directed water, and ice on enclosure
4X	Indoor/outdoor	Corrosion, windblown dust and rain, splashing water, hose-directed water, and ice on enclosure
6	Indoor/outdoor	Occasional temporary submersion at a limited depth
6P	Indoor/outdoor	Prolonged submersion at a limited depth
7	Indoor locations classified as Class I, Groups A, B, C, or D, as defined in the NEC	Withstand and contain an internal explosion sufficiently so an explosive gas-air mixture in the atmosphere is not ignited
9	Indoor locations classified as Class II, Groups E or G, as defined in the NEC	Dust
12	Indoor	Dust, falling dirt, and dripping noncorrosive liquids
13	Indoor	Dust, spraying water, oil, and noncorrosive coolant

Figure 4-7 Typical pushbutton assembly.
Photo courtesy of Rockwell Automation, www.rockwellautomation.com

Figure 4-8 Flush pushbutton operator.
Photo courtesy of Rockwell Automation, www.rockwell-automation.com

Figure 4-9 Extended pushbutton operator.
Photo courtesy of Rockwell Automation, www.rockwell-automation.com

operator does not need to place a finger squarely on the actuator to operate the push button.

- **Mushroom-head** push buttons (Figure 4-10) have an actuator that extends over the edges of the mounting ring and have a diameter larger than a standard push button. Because of their size and shape, mushroom-head push buttons are more easily seen and actuated and, for these reasons, are used as emergency stop buttons.

- **Half-shrouded** pushbutton operators (Figure 4-11) contain a guard ring, which extends over the top half of the button. This helps prevent accidental operation while allowing easy access, particularly with the thumb. Machine operators wearing gloves find half-shrouded push buttons easier to activate than flush push buttons.

- **Illuminated** pushbutton operator units (Figure 4-12) are used as a visual aid for operators and users in industrial or commercial control applications. Often, a light-emitting diodes (LED) is integrated into the pushbutton operator to provide the desired illumination.

Figure 4-10 Mushroom-head pushbutton operator.
This material and associated copyrights are proprietary to, and used with the permission of Schneider Electric.

Figure 4-11 Half-shrouded pushbutton operator.
Photo courtesy of Eaton, www.eaton.com

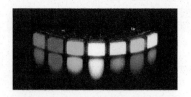

Figure 4-12 Illuminated pushbutton operators.
Photo courtesy of Omron Automation and Safety, www.Omron247.com

Figure 4-13 RUN legend plate.

Legend plates are the labels that are installed around a push button to identify its purpose. They come in many sizes, colors, and languages. Examples of label text include START, STOP, FWD, REV, JOG, UP, DOWN, ON, OFF, RESET, and RUN (Figure 4-13).

The **contact block** is the part of the pushbutton assembly (Figure 4-14) that is activated when the button is pressed. The contact block may house many sets of contacts that open and close when you operate the push button. The normal contact configuration allows for one normally open and one normally closed set of contacts within a contact block. A push button may contain stacked contacts that change state with a single push of a single button.

- The contacts of the contact block itself are spring loaded and return to their normal open or closed state when the operator is released.
- When contact blocks are **attached to a pushbutton operator,** their switching action is determined in part by that of the operator.

- Pushbutton operators are available for momentary or maintained operation.
- **Momentary**-type pushbutton operators return to their normal open or closed state as soon as the operator is released.
- **Maintained**-type pushbutton operators require you to press and release the operator to switch the contacts to their closed state and to press and release the operator a second time to return the contacts to their open state.

Standard three-wire motor control circuits use a holding circuit in conjunction with momentary start/stop pushbutton operators for starting and stopping a motor. **Emergency stop** switches are devices that users manipulate to initiate the complete shutdown of a machine, system, or process. Emergency stop push buttons installed in motor control circuits are normally maintained-contact types with mushroom-type heads. Using maintained contacts in emergency stop push buttons prevents a motor control process from restarting until the maintained push button is physically reset. The Occupational Safety and Health Administration (OSHA) regulations require that once the emergency stop switch has activated, the control process cannot be started again until the actuating stop switch has been reset to the on position. Figure 4-15 shows a typical motor control circuit that includes an emergency stop push button.

- The normally closed maintained contacts of the emergency stop push button will open when the push button is pressed and remain open until it is **manually reset.**

Figure 4-14 Pushbutton assembly.
Photo courtesy of Eaton, www.eaton.com.

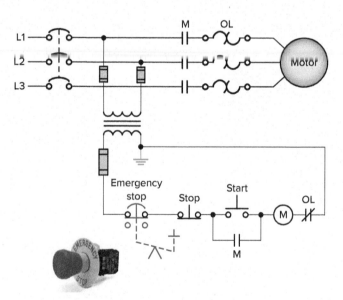

Figure 4-15 Emergency stop push button.
Photo courtesy of Rockwell Automation, www.rockwellautomation.com

- Because the emergency stop contacts are held open by the pushbutton operator mechanism, the motor will not operate if the start button is pressed.
- In order to restart the motor after the emergency stop push button has been activated, you must first **reset the emergency stop** push button and then press the start button.

Figure 4-16 shows an example of a motor control that combines the operation of two motors with pilot lights and an emergency stop button. The operation of the circuit can be summarized as follows:

- The normal operation of motors M1 and M2 are individually controlled by their associated start and stop buttons PBS#1 and PBS#2.
- When illuminated, red plot light M1 indicates that motor M1 is operating.
- When illuminated, green plot light M2 indicates that motor M2 is operating.
- If the emergency stop button is activated, both motors will be de-energized regardless of their current state.
- Once the emergency stop button is activated, it must be manually reset in order to commence normal operation.

Pilot Lights

Pilot lights provide visual indication of the status for many motor-controlled processes permitting personnel at remote locations to observe the current state of the operation. They are commonly used to indicate whether or not a motor is operating. Figure 4-17 shows the circuit for a start/stop pushbutton station with a pilot light connected to indicate when the starter is energized. For this application, the red pilot light is energized to show when the motor is operating, as the motor and drive are located at some remote location not within view of the pushbutton station.

Pilot lights are available in both full-voltage and low-voltage types. A **transformer pilot light,** such as shown

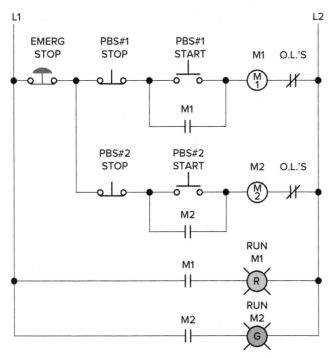

Figure 4-16 Motor control circuit that combines the operation of two motors with pilot lights and an emergency stop button.

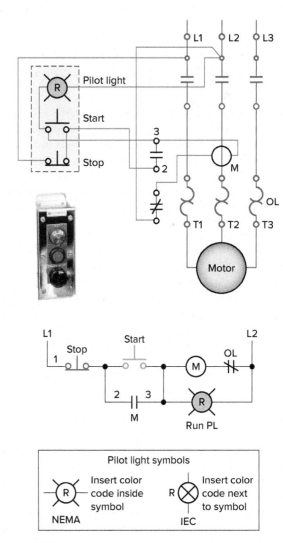

Figure 4-17 Remote start/stop station with run pilot light.
Photo courtesy of Rockwell Automation,
www.rockwellautomation.com

Part 1 Manually Operated Switches

Figure 4-18 Transformer pilot light.
Photo courtesy of Rockwell Automation, www.rockwellautomation.com.

in Figure 4-18, uses a step-down transformer to reduce the operating voltage supplied to the lamp. The primary voltage of the transformer is matched to the input voltage of L1 and L2, while the secondary voltage is matched to that of the lamp. The lower lamp voltage can provide a margin of safety if the lamp requires replacement while the control circuit is energized. Also available are illuminated units utilizing integrated LEDs, which operate at 6 to 24 V, AC or DC.

Dual-input **"push-to-test"** pilot lights are designed to reduce the time required to troubleshoot a suspected faulty lamp. Push-to-test pilot lights can be energized from two separate input signals of the same voltage. This is done by wiring the "test" terminal to the second input signal as illustrated in the push-to-test circuits of Figure 4-19. Pressing the push-to-test pilot light opens the normal signal input to the light while at the same time completing a path directly to L1 and illuminating the lamp if the unit is not at fault.

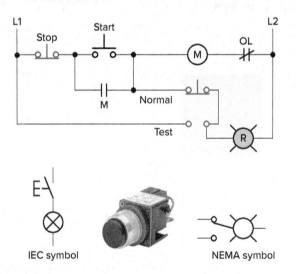

Figure 4-19 Push-to-test pilot light.
Photo courtesy of Rockwell Automation,
www.rockwellautomation.com

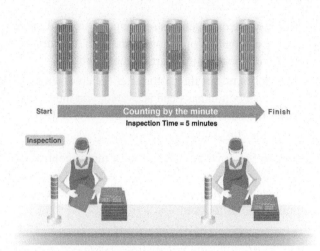

Figure 4-20 Signal tower designed to prevent defective product outflow during inspection.
Source: www.valin.com

Tower Light Indicators

Light towers are a vital component for any industrial manufacturing and/or process control application. They provide a highly visual LED indication of machine statuses. Figure 4-20 illustrates an example of a signal tower designed to prevent defective product outflow during inspection. Sensors detect inspectors as they enter the process line, triggers the light tower to begin the count and inspectors carry out inspection until the tower light turns all blue.

Selector Switch

The difference between a push button and selector switch is the operator mechanism. A selector switch operator is **rotated** (instead of pushed) to open and close contacts of the attached contact block. Switch positions are established by turning the operator knob right or left. These switches may have two or more selector positions, with either maintained contact position or spring return to give momentary contact operation.

The circuit of Figure 4-21 is an example of a three-position selector switch used to select three different operating modes for control of a pump motor. The operation of the circuit can be summarized as follows:

- In the HAND position, the pump can be started by closing the manual control switch. It can be stopped by opening the manual control switch or selecting the OFF position of the selector switch. The liquid-level switch has no effect in either the HAND or OFF position.

- When AUTO is selected, the liquid-level switch controls the pump. At a predetermined level, the liquid-level switch will close, starting the pump. At another

Selector switch

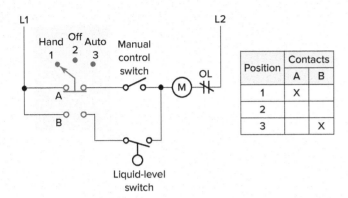

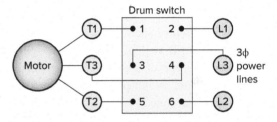

Figure 4-21 Three-position selector switch.
Photo courtesy of Rockwell Automation,
www.rockwellautomation.com

predetermined level, the liquid-level switch will open, stopping the pump.
- The selector switch contact position and resultant state are identified by means of the table shown. Contacts are marked as A and B, while positions are marked as 1, 2, and 3. An X in the table indicates the contact is closed at that particular position.

Drum Switch

A **drum switch** consists of a set of moving contacts and a set of stationary contacts that open and close as the shaft is rotated. Reversing drum switches are designed to start and reverse motors by connecting them directly across the line. The drum switch may be used with squirrel-cage motors; single-phase motors designed for reversing service; and series, shunt, and compound DC motors. Figure 4-22 shows how a drum switch is wired to reverse the direction of rotation of a three-phase motor.

Figure 4-22 Drum switch used for reversing the direction of rotation of a three-phase motor.
This material and associated copyrights are proprietary to, and used with the permission of Schneider Electric.

- Reversal of the direction of rotation is accomplished by interchanging two of the three main power lines to the motor.
- The internal switching arrangements and resulting motor connections, for forward and reverse, are shown in the tables.
- The drum switch is used only as a means for controlling the direction of rotation of the motor and does not provide overcurrent or overload protection.
- A good rule for most motors is that they should be allowed to come to a complete stop before you reverse their direction.

Part 1 Review Questions

1. List three examples of primary motor control devices.
2. List three examples of pilot motor control devices.
3. What do the terms *normally open* and *normally closed* refer to when used in defining the switching action of a pushbutton switch?
4. The types of enclosures used to house motor control devices have been standardized by NEMA. What criteria are used to classify NEMA enclosure types?
5. Name the three basic parts of a standard pushbutton assembly.

6. Compare the operation of *momentary* and *maintained* pushbutton operators.
7. What is the OSHA requirement for resetting emergency stop switches?
8. A pilot light is to be connected to indicate when a magnetic starter is energized. Across what component of the circuit should the pilot light be connected?
9. Explain how a push-to-test pilot light operates.
10. Compare the way in which pushbutton and selector switch operators actuate contacts.
11. When a drum switch is used for starting and reversing a three-phase squirrel-cage motor, how is the reversing action accomplished?
12. A motor control operation consists of multiple motors with individual controls as well as an emergency stop. During which mode of the operation must the emergency stop function be available?

PART 2 MECHANICALLY OPERATED SWITCHES

Limit Switches

A mechanically operated switch is one that is controlled automatically by factors such as pressure, position, and temperature. The **limit switch,** illustrated in Figure 4-23, is a very common type of mechanically operated motor control device. Limit switches are designed to operate only when a predetermined limit is reached, and they are usually actuated by contact with an object such as a cam. These devices take the place of human operators. They are often used in the control circuits of machine processes to govern the starting, stopping, or reversal of motors.

Limit switches are constructed of two main parts: the body and the operator head (also called the actuator). The body houses the contacts that are opened or closed in response to the movement of the actuator. Contacts may consist of normally open, normally closed, momentary (spring-returned), or maintained-contact types. The terms **normally open** and **normally closed** refer to the state of the contacts when the switch is in its normally deactivated state. Figure 4-24 shows the standard symbols used to represent limit switch contacts as follows:

- The NO held closed symbol indicates that the contact is wired as an NO contact, but when the circuit is in its normal off state, some part of the machine holds the contact closed.

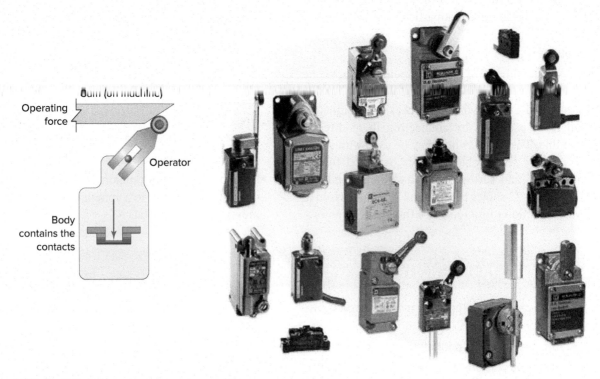

Figure 4-23 Limit switch.
This material and associated copyrights are proprietary to, and used with the permission of, Schneider Electric.

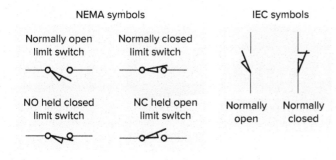

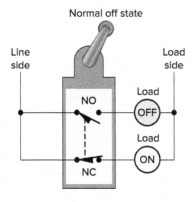

Figure 4-24 Limit switch symbols and configuration.

- The NC held open symbol indicates that the contact is wired as an NC contact, and some part of the machine in its normal off state holds the contact open.
- A contact block with one NO and one NC set of contacts is the most common configuration.
- When you have two or more sets of contacts in a limit switch that are electrically isolated, you must wire the loads that these contacts are controlling on the same side of the line.

Limit switches come with a wide variety of operators (Figure 4-25) designed for a broad range of applications. These include:

- **Lever type,** which consists of a single arm with a roller attached at the end to help prevent wear. The length of the lever may be fixed or adjustable.

Adjustable types are used in applications that require adjustment of the actuator length or travel.

- **Fork lever,** which is designed for applications where the actuating object travels in two directions. A typical application is a machine bed that automatically alternates back and forth.
- **Wobble stick,** which is used in applications that require detection of a moving object from any direction rather than in one or two directions along a single plane. They may be constructed of steel, plastic, Teflon, or nylon and are connected to the limit switch by a flexible spring attachment.
- **Push roller type,** which operates by a direct forward movement into the limit switch. This type has the least amount of travel compared to other types and is commonly used to prevent overtravel of a machine part or object. The limit switch contacts are connected so as to stop the forward movement of the object when it makes contact with it.

A common application for limit switches is to limit the travel of electrically operated doors, conveyors, hoists, machine tool worktables, and similar devices. Figure 4-26 shows the control circuit for starting and stopping a motor in the forward and reverse directions with two limit switches providing overtravel protection. The operation of the circuit can be summarized as follows:

- Pressing the momentary forward push button completes the circuit for the F coil, closing the normally open maintaining contact and sealing in the circuit for the forward starter coil.
- At the same instant, the normally closed interlock contact F opens to prevent the reverse direction of the motor.
- To reverse the motor direction, the operator must first press the stop button to de-energize the F coil and then press the reverse push button.

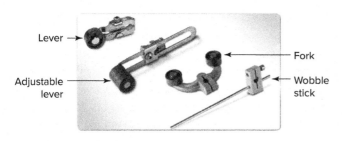

Figure 4-25 Limit switch operators.
Photo courtesy of Eaton, www.eaton.com.

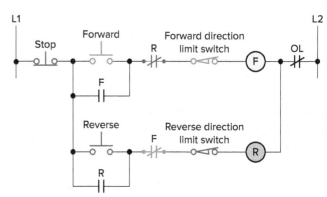

Figure 4-26 Limit switches providing overtravel protection.

Part 2 Mechanically Operated Switches

Figure 4-27 Snap-action micro limit switch.
Photo courtesy of Cherry, www.cherrycorp.com.

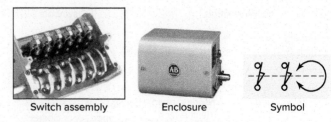

Switch assembly   Enclosure   Symbol

Figure 4-28 Rotating cam limit switch.
Photo courtesy of Rockwell Automation,
www.rockwellautomation.com

- If overtravel position should be reached in either the forward or reverse direction, the respective NC limit switch will open to prevent any further travel in that direction.
- The forward direction is also interlocked with a normally closed R contact.

The **micro limit switch,** shown in Figure 4-27, is a snap-action switch housed in a small enclosure. Snap action switches are mechanical switches that produce a very rapid transfer of contacts from one position to another. They are useful in situations that require a fast opening or closing of a circuit. In a snap-action switch, the actual switching of the circuit takes place at a fixed speed no matter how quickly or slowly the activating mechanism moves.

One difference between traditional limit switches and micro limit switches is the electrical configuration of the switch contacts. Micro switches use a single-pole, double-throw, contact arrangement that has one terminal connected as a common between the normally open and normally closed contacts instead of two electrically isolated contacts. The micro switch body is normally constructed of molded plastic, which offers a limited amount of electrical isolation and physical protection for contacts. Because of this, these switches are normally mounted within enclosures where there is a lower risk of physical damage. When used in conjunction with equipment doors, micro limit switches function as safety devices that are interlocked with control circuitry to prevent the process from operating if the door is not in place.

The **rotating cam limit switch,** shown in Figure 4-28, is a control device that senses angular shaft rotation within 360 degrees and then activates contacts. They are typically used with machinery having a repetitive cycle of operation, where motion is correlated to shaft rotation. The switch assembly consists of one or more snap-action switches that are operated by cams assembled on a shaft. Cams are independently adjustable for operating at different locations within a complete 360-degree rotation.

The relative characteristics of limit switch applications are summarized as follows:

- Can be used in almost any industrial environment because of their rugged design.
- Capable of safely switching high inductive loads up to 10 amperes.
- Very precise in terms of accuracy and repeatability.
- Moving mechanical parts wear out.
- Requires physical contact with target to sense.
- Slower response time compared to equivalent electronic devices.

Temperature Control Devices

Temperature control devices (also called **thermostats,** depending on the application) monitor the temperature or changes in temperature for a particular process. Although there are many types available, they are all actuated by some specific environmental temperature change. Temperature switches open or close when a designated temperature is reached. Temperature control devices are used in heating or cooling applications where temperature must be maintained within preset limits. Symbols used to represent temperature switches are shown in Figure 4-29.

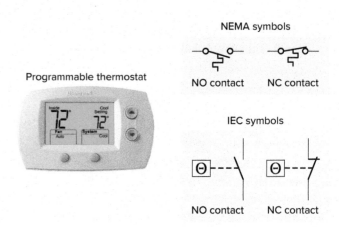

Figure 4-29 Temperature switch symbols.
Photo courtesy of Honeywell, www.honeywell.com

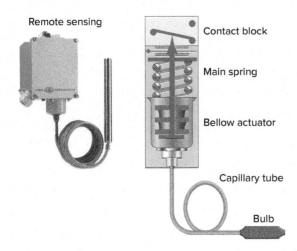

Figure 4-30 Capillary tube temperature switch.
Photo courtesy of Georgin, www.georgin.com

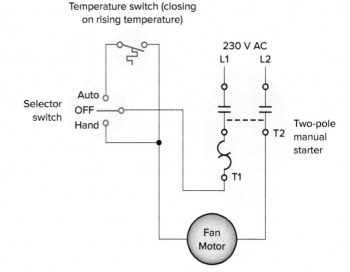

Figure 4-31 Temperature switch utilized as part of a motor control circuit.

Temperature switches are designed to work with a number of different operating principles. These devices typically comprise sensing elements and switching contacts housed in a single mechanical assembly. Switches may open or close on temperature rise, depending on their internal construction. The **capillary tube** temperature switch, illustrated in Figure 4-30, operates on the principle that a temperature-sensitive liquid will expand and contract with a change in temperature.

- Pressure in the system changes in proportion to temperature and is transmitted to the bellows through a bulb and capillary tube.
- As the temperature rises, the pressure in the tube increases.
- Similarly, as the temperature decreases, the pressure in the tube decreases.
- Bellows motion, in turn, is transmitted through a mechanical linkage to actuate a precision switch at a predetermined setting.
- Capillary tube–type temperature switches can be connected to a remote fluid-containing bulb, allowing the switch element to be remote from the sensing bulb and environment or process under control.

Figure 4-31 shows the wiring diagram for the automatic control of a 230 V fractional-horsepower fan motor that utilizes a temperature switch. Temperature switches rated to carry motor current may be used with fractional-horsepower manual starters. Note that a double-pole manual starter is used. This type of starter is required when both line leads to the motor must be switched, such as for a 230 V, single-phase source. When the three-position selector switch is in the AUTO position, the temperature switch reacts to a preset rising temperature to automatically switch the motor on. When the temperature drops below the preset value, the contacts open to turn the motor off.

Pressure Switches

Pressure switches are used to monitor and control the pressure of liquids and gases. They are commonly used to monitor a system and, in the event that pressure reaches a dangerous level, open relief valves or shut the system down. The three categories of pressure switches used to activate electrical contacts are positive pressure, vacuum (negative pressure), and differential pressure. Symbols used to represent pressure switches are shown in Figure 4-32.

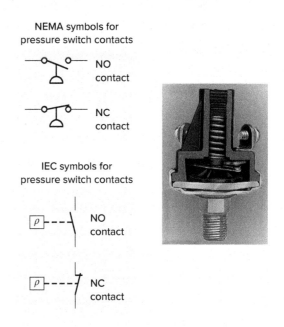

Figure 4-32 Pressure switch symbols.
Photo courtesy of Honeywell, www.honeywell.com

Part 2 Mechanically Operated Switches

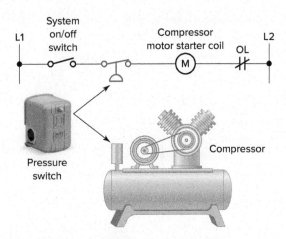

Figure 4-33 Pressure switch used as part of an air compressor control system.
This material and associated copyrights are proprietary to, and used with the permission of, Schneider Electric.

Pressure switches are used in many different types of industries and applications. They can be used to control pneumatic systems, maintaining preset pressures between two values. The compressor shown in Figure 4-33 consists of a compressor drive motor, a compressor unit, and tank. The operation of the circuit can be summarized as follows:

- The pressure switch is used to stop the motor when tank pressure reaches a preset limit. The differential between the high and low cutout settings is known as the span or dead band.
- When the preset system pressure is reached, the NC contacts of the pressure switch open to de-energize the motor starter coil and automatically switch the compressor motor off.
- To prevent frequent stopping and starting of the motor, this type of pressure switch has a built-in pressure differential between the set stopping and starting pressures. This differential is referred to as the **dead band.**

Mechanically operated pressure switches contain electrical contacts that are operated by a **pressure-measuring sensor.** Common types of pressure sensors (Figure 4-34) include:

Bourdon tube-type pressure switches use a coiled tube to sense pressure. One end of the tube is fixed and provides an inlet for fluid pressure. The other end is sealed and is free to move. As the fluid pressure inside the tube is increased, the tube bends and this movement is used to activate a set of electrical contacts.

Bellows-type pressure switches use a sealed chamber that has multiple ridges like the pleats of an accordion that are compressed slightly when the sensor is manufactured. When pressure is applied to the chamber, it expands and activates a set of electrical contacts. Bellow types are generally more sensitive than the bourdon tube types and can be pressure-adjusted by varying the tension of a heavy coil spring.

Diaphragm-type pressure switches consist of a circular membrane fitted into a housing. As pressure is applied to one side of the diaphragm, the central portion is pushed away from the pressure. An arm attached to the diaphragm activates a set of electrical contacts. By admitting pressure to both sides of the diaphragm, it can operate as a differential-pressure sensor.

Float and Flow Switches

A **float switch** (also called a level switch) is used to sense the height of a liquid. Float switches provide automatic control for motors that pump liquid from a sump or into a tank. The switch must be installed above the tank or sump, and the float must be in the liquid for the float switch to operate. For tank operation, a float operator assembly is attached to the float switch by a rod, chain, or cable. The float switch is actuated according to the location of the float in the liquid.

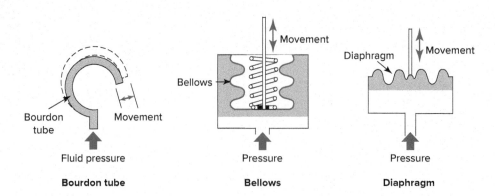

Figure 4-34 Types of mechanically operated pressure sensors.

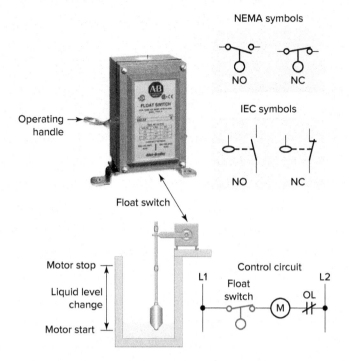

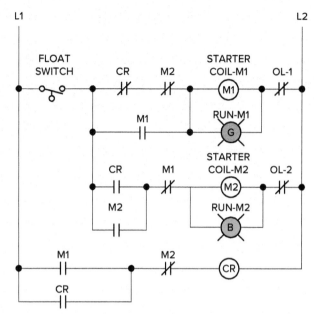

Figure 4-35 Float switch symbols and circuit.
Photo courtesy of Rockwell Automation,
www.rockwellautomation.com

Figure 4-36 Alternating pumping operation control circuit.

There are several styles of float switches. One type uses a rod that has a float mounted on one end, as illustrated in Figure 4-35. In this application, a float switch is used to control the pump motor in an automatic tank-filling operation. The operation of the circuit can be summarized as follows:

- The float switch contacts are open when the float forces the operating lever to the UP position.
- As the liquid level falls, the float and rod move downward.
- When the float reaches a preset low level, the float switch contacts close, activating the circuit and starting the pump motor to refill the tank.
- Adjustable stops on the rod determine the amount of movement that must take place before the switch contacts open or close.

When a pump operates frequently, too much stress may be subjected to a single motor pump. In such cases a second pump motor designed to operate alternately between two pumps may be used. The circuit of Figure 4-36 shows a float switch control of alternating pumps for removing fluid from a tank. A **control relay (CR)** consisting of an electromagnetic coil with normally open and closed contacts is used to switch low-level currents. The steps in the operation of the circuit can be summarized as follows:

- The float switch is a normally open type.
- At initiation, with the tank empty, starter coils M1 and M2 as well the control relay coil CR are de-energized.
- When the fluid in the tank reaches the full level, the float switch contacts close to energize starter coil M1 and start pump motor 1.
- M1 auxiliary contacts then change state to energize control coil CR and complete the seal-in path to starter coil M1.
- When the fluid reaches the empty level, the float switch contacts return to their normally open state.
- This de-energizes starter coil M1 to halt the pumping operation.
- At the same time, control relay CR remains energized through its seal-in circuit.
- When the fluid in the tank reaches the full level for the second time, the float switch contacts close to energize starter coil M2 through the CR contact and start pump motor 2. Normally closed M2 opens to de-energize CR.
- When the fluid reaches the empty level, the float switch contacts return to their normally open state to de-energizes starter coil M2 to halt the pumping operation.

Part 2 Mechanically Operated Switches

- The sequence is reset by each activation and deactivation of the float switch.
- Pilot lights provide visual indication of the status of the process.

A **flow switch** is used to detect the movement of air or liquid through a duct or pipe. In certain applications, it is essential to be able to determine whether fluid is flowing in a pipeline, duct, or other conduit and to respond accordingly to such a determination. One of the simplest types of flow switch is the paddle type illustrated in Figure 4-37. The paddle extends into the pipe and moves to close the electrical contacts of the flow switch when the fluid flow is sufficient to overcome the spring tension on the paddle. When the flow stops, the contacts open. On most paddle-type flow switches the spring tension is adjustable, allowing for different flow rate adjustments. It is important to install the flow switch properly. A flow-direction arrow indicates the correct positioning of the flow switch into the fluid line.

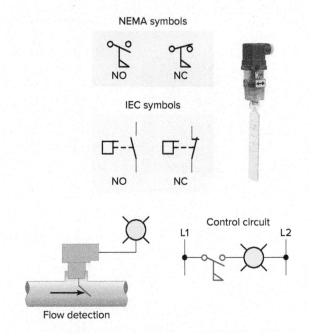

Figure 4-37 Flow switch symbols and circuit.
Photo courtesy of Kobold Instruments, www.kobold.com

Part 2 Review Questions

1. Define the term *mechanically operated switch*.
2. In what way are limit switches normally actuated?
3. A control application calls for an NC held open limit switch. What type of connection does this imply?
4. List four common types of limit switch operator heads.
5. What is an important operating feature of a snap-action micro limit switch?
6. In what way is the contact configuration of a traditional limit switch different from that of a micro limit switch?
7. For what types of machine control applications are rotating cam limit switches best suited?
8. How does a fluid capillary tube temperature switch actuate its electrical contact block?
9. In what types of applications are pressure switches used?
10. Compare the function of a float switch with that of a flow switch.
11. List two limitations of limit switches compared to equivalent electronic-type devices.
12. What advantages do bellow-type pressure switches have over bourdon tube types?

PART 3 SENSORS

Sensors are devices that are used to detect, and often to measure, the magnitude of something. They basically operate by converting mechanical, magnetic, thermal, optical, and chemical variations into electric voltages and currents. Sensors are usually categorized by what they measure, and play an important role in modern manufacturing process control. Typical applications for sensors are illustrated in Figure 4-38 and include light sensor, pressure sensor, and bar code sensor.

Proximity Sensors

Proximity sensors detect the presence of an object (usually called the target) without physical contact. Detection of the presence of solids such as metal, glass, and plastics, as well as most liquids, is achieved by a means of sensing changes in magnetic or electrostatic fields. These electronic sensors are completely encapsulated to protect against excessive vibration, liquids, chemicals, and corrosive agents found in the industrial environment.

Proximity sensors are available in various sizes and configurations to meet different application requirements. One of

Light sensor

Pressure sensor

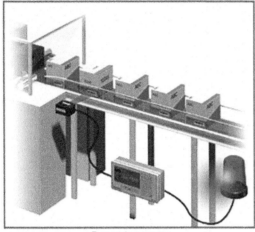
Bar code sensor

Figure 4-38 Typical sensor applications.
Photo courtesy of Keyence Canada Inc., www.keyence.com

Figure 4-39 Proximity sensor and symbols.
Photo courtesy of Turck, Inc., www.turck.com.

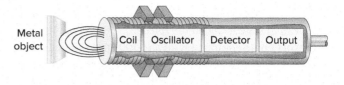

Figure 4-40 Inductive proximity sensor.

Inductive proximity sensors operate under the electrical principle of inductance, where a fluctuating current induces an electromotive force (emf) in a target object. The block diagram for an inductive proximity sensor is shown in Figure 4-40 and its operation can be summarized as follows:

- The oscillator circuit generates a high-frequency electromagnetic field that radiates from the end of the sensor.
- When a metal object enters the field, eddy currents are induced in the surface of the object.
- The eddy currents on the object absorb some of radiated energy from the sensor, resulting in a loss of energy and change of strength of the oscillator.
- The sensor's detection circuit monitors the oscillator's strength and triggers a solid-state output at a specific level.
- Once the metal object leaves the sensing area, the oscillator returns to its initial value.

The type of metal and size of the target are important factors that determine the effective sensing range of the sensor. Ferrous metals may be detected up to 2 inches away, while most nonferrous metals require a shorter distance, usually within an inch of the device. The point at which the proximity sensor recognizes an incoming target is called the **operating point** (Figure 4-41). The point at

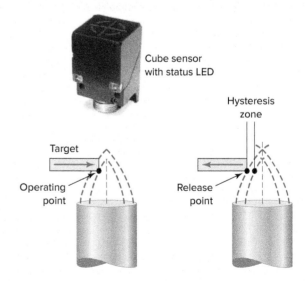

Figure 4-41 Proximity sensor sensing range.
Photo courtesy of Eaton, www.eaton.com.

the most common configurations is the barrel type, which houses the sensor in a metal or polymer barrel with threads on the outside of the housing. Figure 4-39 shows a barrel type proximity switch along with the symbols used to represent it. The threaded housing allows the sensor to be easily adjusted on a mounting frame.

Inductive Proximity Sensors Proximity sensors operate on different principles, depending on the type of matter being detected. When an application calls for noncontact metallic target sensing, an **inductive-type proximity sensor** is used. Inductive proximity sensors are used to detect both ferrous metals (containing iron) and nonferrous metals (such as copper, aluminum, and brass).

which an outgoing target causes the device to switch back to its normal state is called the *release point*. Most proximity sensors come equipped with an LED status indicator to verify the output switching action. The area between operating and release points is known as the *hysteresis zone*. Hysteresis is specified as a percentage of the nominal sensing range and is needed to keep proximity sensors from chattering when subjected to shock and vibration, slow-moving targets, or minor disturbances such as electrical noise and temperature drift.

Most sensor applications operate either at 24V DC or at 120V AC. Figure 4-42 illustrates typical two-wire and three-wire sensor connections. The three-wire DC proximity sensor (Figure 4-42a) has the positive and negative line leads connected directly to it. When the sensor is actuated, the circuit will connect the signal wire to the positive side of the line if operating normally open. If operating normally closed, the circuit will disconnect the signal wire from the positive side of the line.

Figure 4-42b illustrates a typical two-wire proximity sensor connection intended to be connected in series with the load. They are manufactured for either AC or DC supply voltages. In the off state, enough current must flow through the circuit to keep the sensor active. This off state current is called leakage current and typically may range from 1 to 2 mA. When the switch is actuated, it will conduct the normal load circuit current. Keep in mind that sensors are basically pilot devices for loads such as starters, contactors, and solenoids, and should not be used to directly operate a motor.

The relative merits of inductive proximity sensor applications are summarized as follows:

- Immune to adverse environmental conditions.
- High switching rate for rapid response applications.
- Can detect metallic targets through nonmetallic barriers.

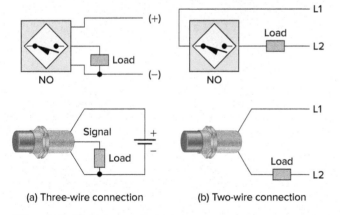

Figure 4-42 Typical two-wire and three-wire sensor connections.

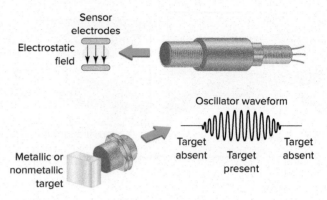

Figure 4-43 Capacitive proximity sensor.

- Long operational life with virtually unlimited operating cycles.
- Provides a bounce-free input signal to PLCs and other solid-state logic control devices
- The sensing range is limited.
- Can detect only metal objects.
- Its operation may be affected by metal particles accumulating on sensor face.

Capacitive Proximity Sensors **Capacitive proximity sensors** are similar to inductive proximity sensors. The main differences between the two types are that capacitive proximity sensors produce an electrostatic field instead of an electromagnetic field and are actuated by both conductive and nonconductive materials. Capacitive sensors contain a high-frequency oscillator along with a sensing surface formed by two metal electrodes (Figure 4-43).

- When the target nears the sensing surface, it enters the electrostatic field of the electrodes and changes the capacitance of the oscillator.
- As a result, the oscillator circuit begins oscillating and changes the output state of the sensor when it reaches a certain amplitude.
- As the target moves away from the sensor, the oscillator's amplitude decreases, switching the sensor back to its original state.

Capacitive proximity sensors will sense metal objects as well as nonmetallic materials such as paper, glass, liquids, and cloth. They typically have a short sensing range of about 1 inch, regardless of the type of material being sensed. The larger the dielectric constant of a target, the easier it is for the capacitive sensor to detect. This makes possible the detection of materials inside nonmetallic containers as illustrated in Figure 4-44. In this example, the liquid has a much higher dielectric constant than the cardboard container, which gives the sensor the ability to see

Figure 4-44 Capacitive proximity sensor liquid detection.
Photo courtesy of Omron Automation and Safety, www.Omron247.com

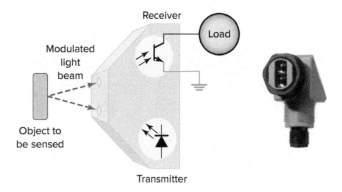

Figure 4-45 Photoelectric sensor.
Photo courtesy of SICK, Inc., www.sickusa.com

through the container and detect the liquid. In the process shown, detected empty containers are automatically diverted via the push rod.

The relative merits of capacitive proximity sensor applications are summarized as follows:

- Able to detect both metallic and nonmetallic objects.
- High switching rate for rapid response applications such as counting.
- Able to detect liquid targets through nonmetallic barriers such as glass and plastic.
- Provides a bounce-free input signal for solid-state logic control devices
- Long operational life with virtually unlimited operating cycles.
- Affected by varying temperature, humidity, and moisture conditions.
- Not as accurate as inductive proximity sensors.

Photoelectric Sensors

A **photoelectric sensor** is an optical control device that operates by detecting a visible or invisible beam of light, and responding to a change in the received light intensity. Photoelectric sensors are composed of two basic components: a transmitter (light source) and a receiver (sensor), as shown in Figure 4-45. These two components may or may not be housed in separate units. The basic operation of a photoelectric sensor can be summarized as follows:

- The transmitter contains a light source, usually an LED along with an oscillator.
- The oscillator modulates or turns the LED on and off at a high rate of speed.
- The transmitter sends this modulated light beam to the receiver.

- The receiver decodes the light beam and switches the output device, which interfaces with the load.
- The receiver is tuned to its emitter's modulation frequency and will only amplify the light signal that pulses at the specific frequency.
- Most sensors allow adjustment of how much light will cause the output of the sensor to change state.
- Response time is related to the frequency of the light pulses. Response times may become important when an application calls for the detection of very small objects, objects moving at a high rate of speed, or both.

The scan technique refers to the method used by photoelectric sensors to detect an object. Common scan techniques include through-beam, retroreflective, and diffuse scan. Understanding the differences among the available photoelectric sensing techniques is important in determining which sensor will work best in a specific application.

Through-Beam Scanning The **through-beam** scan technique (also called direct scan) places the transmitter and receiver in direct line with each other, as illustrated in Figure 4-46. The operation of the system can be summarized as follows:

- The receiver is aligned with the transmitter beam to capture the maximum amount of light emitted from the transmitter.
- The object to be detected, placed in the path of the light beam, blocks the light to the receiver and causes the receiver's output to change state.
- Because the light beam travels in only one direction, through-beam scanning provides long-range sensing. The maximum sensing range is about 300 feet.
- This scan technique is a more reliable method in areas of heavy dust, mist, and other types of airborne contaminants that may disperse the beam and for monitoring large areas.

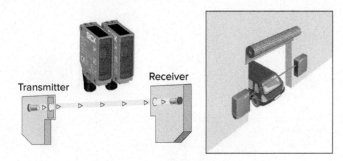

Figure 4-46 Through-beam scan.
Photo courtesy of SICK, Inc., www.sickusa.com

- Quite often, a garage door opener has a through-beam photoelectric sensor mounted near the floor, across the width of the door. For this application, the sensor senses that nothing is in the path of the door when it is closing.

The relative characteristics of photoelectric through-beam scanning sensor applications are summarized as follows:

- Capable long sensing distances.
- Highly reliable operation.
- Can detect opaque objects.
- Requires two components to mount and wire.
- Alignment could be difficult with a longer distance detection zone.

Retroreflective Scanning In a **retroreflective scan,** the transmitter and receiver are housed in the same enclosure. This arrangement requires the use of a separate reflector or reflective tape mounted across from the sensor to return light back to the receiver. This sensor is designed to respond to objects that interrupt the beam normally maintained between the transmitter and receiver, as illustrated in Figure 4-47. In contrast to a through-beam

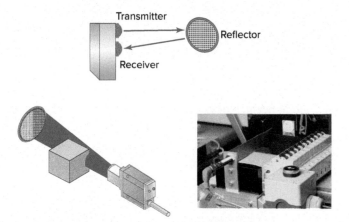

Figure 4-47 Retroreflective scan sensor.
Photo courtesy of ifm efector, www.ifm.com/us

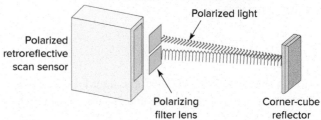

Figure 4-48 Polarized retroreflective scan sensor.
Photo courtesy of Banner Engineering Corp., www.bannerengineering.com.

application, retroreflective sensors are used for medium-range applications.

Retroreflective scan sensors may not be able to detect shiny targets because they tend to reflect light back to the sensor. In this case the sensor is unable to differentiate between light reflected from the target and that from the reflector. A variation of retroreflective scan, the **polarized retroreflective scan** sensor is designed to overcome this problem. Polarizing filters are placed in front of the emitter and receiver lenses as illustrated in Figure 4-48. The polarizing filter projects the emitter's beam in one plane only. As a result, this light is considered to be polarized. A corner-cube reflector must be used to rotate the light reflected back to the receiver. The polarizing filter on the receiver allows rotated light to pass through to the receiver.

The relative characteristics of photoelectric retrorefective scanning sensor applications are summarized as follows:

- Medium range sensing distance.
- Ease of installation.
- Alignment does not need to be exact.
- Polarizing filter ignores unwanted light.
- Reflector must be mounted.
- Problems detecting clear objects.
- Dirt on retroreflector can hamper operation.
- Not suitable for detecting small objects

Diffuse Scanning In a **diffuse scan sensor** (also called proximity scan), the transmitter and receiver are housed in

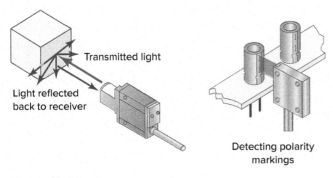

Figure 4-49 Diffuse scan sensor.

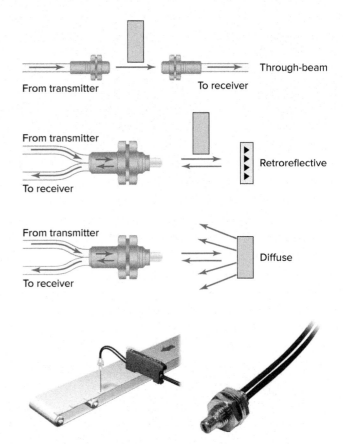

Figure 4-50 Fiber optic sensors.
Photo courtesy of Omron Automation and Safety, www.Omron247.com

the same enclosure, but unlike similar retroreflective devices, they do not rely on any type of reflector to return the light signal to the receiver. Instead, light from the transmitter strikes the target and the receiver picks up some of the diffused (scattered) light. When the receiver receives enough reflected light, the output will switch states. Because only a small amount of light will reach the receiver, its operating range is limited to a maximum of about 40 inches. The sensitivity of the sensor may be set to simply detect an object or to detect a certain point on an object that may be more reflective. Often, this is accomplished using various colors with different reflective properties. In the application shown in Figure 4-49, a diffuse scan sensor is used to inspect for the presence of the polarity mark on a capacitor.

The relative characteristics of photoelectric diffused scanning sensor applications are summarized as follows:

- Application flexibility.
- Easy installation and alignment.
- Relatively short sensing distance.
- Sensing distance depends on target size, surface, and shape.

Fiber Optics Fiber optics is not a scan technique, but another method for transmitting light. **Fiber optic sensors** use a flexible cable containing tiny fibers that channel light from emitter to receiver. Fiber optics can be used with through-beam, retroreflective scan, or diffuse scan sensors, as illustrated in Figure 4-50. In through-beam scan, light is emitted and received with individual cables. In retroreflective and diffuse scan, light is emitted and received with the same cable.

Fiber optic sensor systems are completely immune to all forms of electrical interference. The fact that an optical fiber does not contain any moving parts and carries only light means that there is no possibility of a spark. This means that it can be safely used even in the most hazardous sensing environments such as a refinery for producing gases, grain bins, mining, pharmaceutical manufacturing, and chemical processing. Another advantage of using optical fibers is the luxury it affords users to route them through extremely tight areas to the sensing location. Certain fiber optic materials, particularly the glass fibers, have very high operating temperatures (450°F and higher).

Hall Effect Sensors

Hall effect sensors are used to detect the proximity and strength of a magnetic field. When a current-carrying conductor is placed into a magnetic field, a voltage will be generated perpendicular to both the current and the field. This principle is known as the Hall effect. A Hall effect sensor switch is constructed from a small integrated circuit (IC) chip like that shown in Figure 4-51.

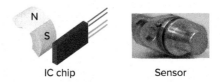

Figure 4-51 Hall effect sensor.
Photo courtesy of Motion Sensors, Inc., www.motionsensors.com.

Part 3 Sensors

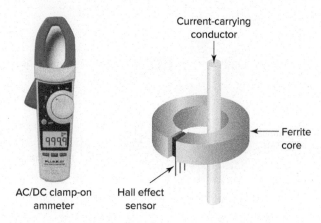

Figure 4-52 Hall effect sensor for current measurement.
Photo courtesy of Fluke, www.fluke.com. Reproduced with Permission.

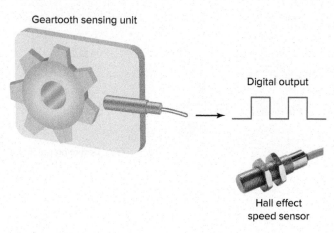

Figure 4-53 Monitoring speed using a Hall effect sensor.
Photo courtesy of Hamlin, www.hamlin.com

A permanent magnet or electromagnet is used to trigger the sensor on and off. The sensor is off with no magnetic field and triggered on in the presence of a magnetic field. Hall effect sensors are designed in a variety of body styles. Selection of a sensor based on body style will vary by application.

Analog-type Hall effect sensors put out a continuous signal proportional to the sensed magnetic field. An analog linear Hall effect sensor may be used in conjunction with a split ferrite core for current measurement, as illustrated in Figure 4-52. The magnetic field across the gap in the ferrite core is proportional to the current through the wire, and therefore the voltage reported by the Hall effect sensor will be proportional to the current. Clamp-on ammeters that can measure both AC and DC current use a Hall effect sensor to detect the DC magnetic field induced into the clamp. The signal from the Hall effect device is then amplified and displayed.

Digital-type Hall effect devices are used in magnetically operated proximity sensors. In industrial applications they may serve to determine shaft or gear speed or direction by detecting fluctuations in the magnetic field. One such application, which involves the monitoring of the speed of a motor, is illustrated in Figure 4-53. The operation of the device can be summarized as follows:

- When the sensor is aligned with the rotating ferrous gear tooth, the magnetic field will be at its maximum strength.
- When the sensor is aligned with the gap between the teeth, the strength of the magnetic field is weakened.
- Each time the tooth of the target passes the sensor, the digital Hall switch activates, and a digital pulse is generated.
- By measuring the frequency of the pulses, the shaft speed can be determined.

- The Hall effect sensor is sensitive to the magnitude of flux, not its rate of change, and as a result the digital output pulse produced is of constant amplitude regardless of speed variations.
- This feature of Hall effect technology allows you to make speed sensors that can detect targets moving at arbitrarily slow speeds, or even the presence or absence of nonmoving targets.

Ultrasonic Sensors

An **ultrasonic sensor** operates by sending high-frequency sound waves toward the target and measuring the time it takes for the pulses to bounce back. The time taken for this echo to return to the sensor is directly proportional to the distance or height of the object.

Figure 4-54 illustrates a practical application in which the returning echo signal is electronically converted to a **4 to 20 mA output,** which supplies a monitored flow rate to external control devices. The 4 to 20 mA range represents the sensor's measurement span. The 4 mA set point is typically placed near the bottom of the empty tank, or the greatest measurement distance from the sensor. The 20 mA set point is typically placed near the top of the full tank, or the shortest measurement distance from the sensor. The sensor will proportionately generate a 4 mA signal when the tank is empty and a 20 mA signal when the tank is full. Ultrasonic sensors can detect solids, fluids, granular objects, and textiles. In addition, they enable the detection of different objects irrespective of color and transparency and therefore are ideal for monitoring transparent objects.

Ultrasonic wind sensors are used on wind turbine systems to determine wind speed and direction. They measure the time taken for an ultrasonic pulse of sound to

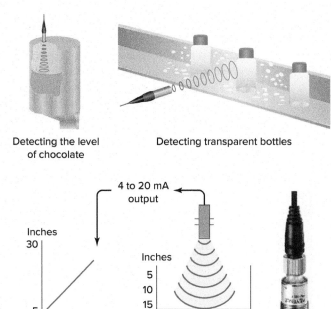

Figure 4-54 Ultrasonic sensor.
Photo courtesy of Keyence Canada Inc., www.keyence.com

Temperature Sensors

There are many types of temperature sensors that will use various technologies and have different configurations. The four basic types of temperature sensors commonly used today are thermocouple, resistance temperature detector, thermistor, and IC sensor.

Thermocouple A **thermocouple** (TC) is a sensor that measures temperature. The thermocouple is by far the most widely used temperature sensor for industrial control. Thermocouples operate on the principle that when two dissimilar metals are joined, a predictable DC voltage will be generated that relates to the difference in temperature between the hot junction and the cold junction (Figure 4-56). The **hot junction** (measuring junction) is the joined end of a thermocouple that is exposed to the process where the temperature measurement is desired. The **cold junction** (reference junction) is the end of a thermocouple that is kept at a constant temperature to provide a reference point. For example, a K-type thermocouple, when heated to a temperature of 300°C at the hot junction, will produce 12.2 mV at the cold junction.

The signal produced by a thermocouple is a function of the **difference** in temperature between the probe tip (hot junction) and the other end of the thermocouple wire (cold junction). For this reason, it is important that the cold (or reference) junction be maintained at a constant known temperature to produce accurate temperature measurements. In most applications, the cold junction is maintained at a known (reference) temperature, while the other end is attached to a probe.

A thermocouple probe consists of thermocouple wire housed inside a metallic tube. The wall of the tube is

Figure 4-55 Ultrasonic wind sensor.
Gill Instruments Limited, 2014.

travel between sensors. Figure 4-55 shows an ultrasonic wind-sensing device made up of four sensors, one at each major compass point (N-S, E-W), which operate in pairs.

- Sensors fire ultrasonic sound pulses to opposing sensors.
- In still air, all pulses' times of flight are equal.
- When the wind blows, it increases the time of flight for pulses traveling against it.
- From the changes in time of flight, the system calculates the wind speed and direction.

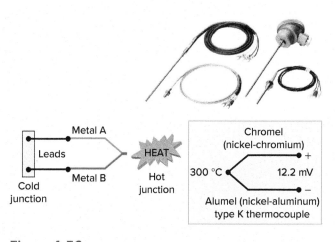

Figure 4-56 Thermocouple heat sensor.
Photo courtesy of Omron Automation and Safety, www.Omron247.com

Part 3 Sensors   93

referred to as the sheath of the probe. The tip of the thermocouple probe is available in three different styles: grounded, ungrounded, and exposed, as illustrated in Figure 4-57.

Thermowells are tubular fittings used to protect temperature sensors installed in industrial processes. The thermowell installation shown in Figure 4-58 consists of a tube closed at one end and mounted in the process fluid stream. The thermocouple sensor is inserted in the open end of the tube, which is located outside the process being monitored. The process fluid transfers heat to the thermowell wall, which in turn transfers heat to the sensor. If the sensor fails, it can be easily replaced without draining the vessel or piping.

The types of metals used in a thermocouple are based on intended operating conditions, such as temperature range and working atmosphere. Different thermocouple types have very different voltage output curves. When a replacement thermocouple is required, it is important that the thermocouple type used in the replacement matches the original. It is also required that thermocouple or thermocouple extension wire of the proper type be used all the way from the sensing element to the measuring element. Large errors can develop if this practice is not followed.

Resistance Temperature Detectors **Resistance temperature detectors** (RTDs) are wire-wound temperature-sensing devices that operate on the principle of the **positive** temperature coefficient (PTC) of metals. That means the electrical resistance of metals is directly proportional to temperature. The hotter they become, the larger or higher the value of their electrical resistance. This proportional variation is precise and repeatable, and therefore allows the consistent measurement of temperature through electrical resistance detection. Platinum is the material most often used in RTDs because of its superiority regarding temperature limit, linearity, and stability. RTDs are among the most precise temperature sensors available and are normally found encapsulated in probes for external temperature sensing and measurement or enclosed inside devices where they measure temperature as a part of the device's function. Figure 4-59 illustrates how an RTD is used as part of a temperature control system. A controller uses the signal from the RTD sensor to monitor the temperature of the liquid in the tank and thereby control heating and cooling lines.

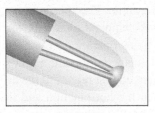
Grounded

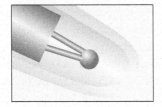
Ungrounded

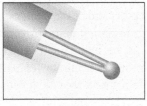
Exposed

Figure 4-57 Thermocouple tip styles.

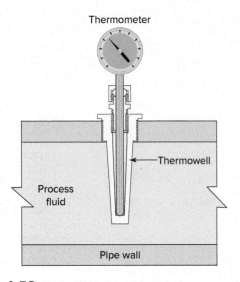
Figure 4-58 Typical thermowell installation.

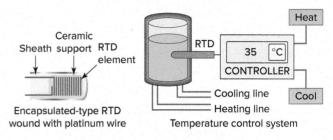
Figure 4-59 Resistance temperature detector.

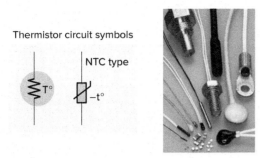

Figure 4-60 Thermistors.
Photo courtesy of TE Connectivity, www.te.com

Thermistors **Thermistors** are generally described as thermally sensitive resistors that exhibit changes in resistance with changes in temperature. This change of resistance with temperature can result in a **negative temperature coefficient** (NTC) of resistance, where the resistance decreases with an increase in temperature (NTC thermistor). When the resistance increases with an increase in temperature, the result is a **positive temperature coefficient** (PTC) thermistor. Thermistors tend to be more accurate than RTDs and thermocouples, but they have a much more limited temperature range. Their sensing area is small, and their low mass allows a fairly fast response time of measurement. Figure 4-60 shows the circuit symbols used to represent thermistors, along with common configurations. A thermistor placed inside a motor housing is used to supplement the standard overload protection by monitoring the motor winding temperature.

Integrated Circuit Integrated circuit (IC) temperature sensors (Figure 4-61) use a silicon chip for the sensing element. Most are quite small, and their principle of operation is based on the fact that semiconductor diodes have temperature-sensitive voltage versus current characteristics. Although limited in temperature range (below 200°C), IC temperature sensors produce a very linear output over their operating range. There are two main types of IC temperature sensors: analog and digital. **Analog** sensors can produce a voltage or current proportional to temperature. **Digital** temperature sensors are similar to analog temperature sensors, but instead of outputting the data in current or voltage, they convert the data into a digital format of 1s and 0s. Digital-output temperature sensors are therefore particularly useful when interfacing to a microcontroller.

Velocity and Position Sensors

Tachometer **Tachometer generators** provide a convenient means of converting rotational speed into an analog voltage signal that can be used for motor speed indication and control applications. A tachometer generator is a small AC or DC generator that develops an output voltage (proportional to its rpm) whose phase or polarity depends on the rotor's direction of rotation. The DC tachometer generator usually has permanent magnetic field excitation. The AC tachometer generator field is excited by a constant AC supply. In either case, the rotor of the tachometer is mechanically connected, directly or indirectly, to the load. Figure 4-62 illustrates motor speed control applications in which a tachometer generator is used to provide a feedback voltage to the motor controller that is proportional to motor speed. The control motor and tachometer generator may be contained in the same or separate housings.

Magnetic Pickup A **magnetic pickup** is essentially a coil wound around a permanently magnetized probe. When a ferromagnetic object, such as gear teeth, is passed through the probe's magnetic field, the flux density is modulated. This induces AC voltages in the coil. One

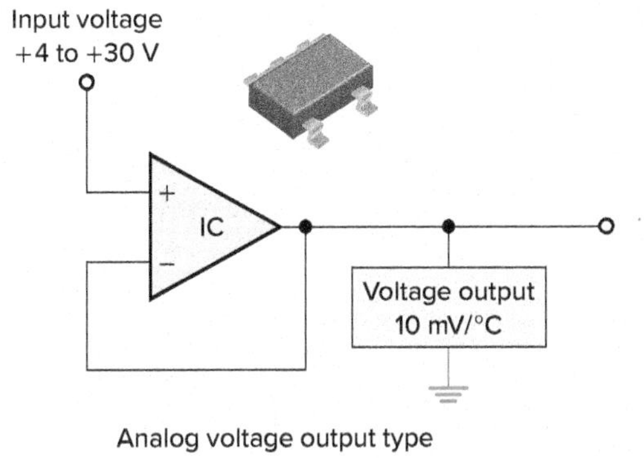

Figure 4-61 Integrated circuit temperature sensor.

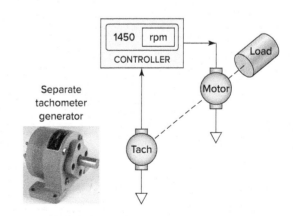

Figure 4-62 Tachometer generator.
Photo courtesy of ATC Digitec, www.atcdigitec.com

Part 3 Sensors

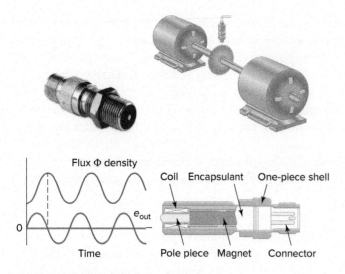

Figure 4-63 Magnetic pickup sensor.
Photo courtesy of Daytronic, www.daytronic.com.

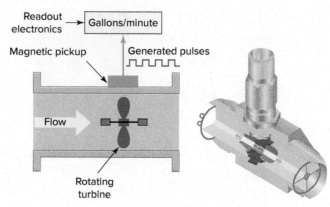

Figure 4-65 Turbine flowmeter.

complete cycle of voltage is generated for each object passed. By measuring the frequency of this signal voltage, the shaft speed can be determined. Figure 4-63 shows a magnetic pickup used in conjunction with a 60-tooth gear to measure the rpm of a rotating shaft.

Encoder An **encoder** is used to convert linear or rotary motion into a binary digital signal. Encoders are used in applications such as robotic control where positions have to be precisely determined. The optical encoder illustrated in Figure 4-64 uses a light source shining on an optical disk with lines or slots that interrupt the beam of light to an optical sensor. An electronic circuit counts the interruptions of the beam and generates the encoder's digital output pulses.

Flow Measurement

Many processes depend on accurate measurement of fluid flow. Although there are a variety of ways to measure fluid flow, the usual approach is to convert the kinetic energy that the fluid has into some other measurable form. This can be as simple as connecting a paddle to a potentiometer or as complex as connecting rotating vanes to a pulse-sensing system or tachometer.

Turbine Flowmeters **Turbine**-type flowmeters are a popular means of measurement and control of liquid products in industrial, chemical, and petroleum operations. Turbine flowmeters, like windmills, utilize their angular velocity (rotation speed) to indicate the flow velocity. The operation of a turbine flowmeter is illustrated in Figure 4-65.

- Its basic construction consists of a bladed turbine rotor installed in a flow tube.
- The bladed rotor rotates on its axis in proportion to the rate of the liquid flow through the tube.
- A magnetic pickup sensor is positioned as close to the rotor as practical.
- Fluid passing through the flow tube causes the rotor to rotate, which generates pulses in the pickup coil.
- The frequency of the pulses is then transmitted to readout electronics and displayed to indicate the flow velocity.

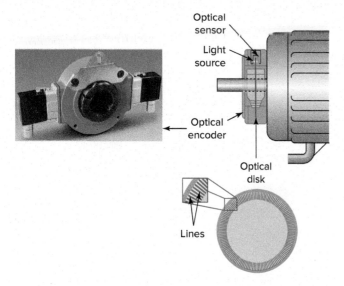

Figure 4-64 Optical encoder.
Photo courtesy of Nidec Avtron Automation Corporation

Target Flowmeters **Target**-type flowmeters insert a target, usually a flat disk with an extension rod, oriented perpendicularly to the direction of the flow. They then measure the drag force on the inserted target and convert it to the flow velocity. One advantage of the target flowmeter over other types is its ability to measure fluids that

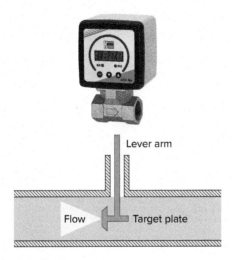

Figure 4-66 Target-type flowmeter.
Photo courtesy of Kobold Instruments, www.kobold.com

Figure 4-67 Magnetic flowmeter.

Magnetic Flowmeters

Magnetic flowmeters, also known as electromagnetic flowmeters or induction flowmeters, obtain the flow velocity by measuring the changes of induced voltage of the conductive fluid passing across a controlled magnetic field. Figure 4-67 shows a magnetic flowmeter that can be used with electrically conducting fluids and offers no restriction to flow. A coil in the unit sets up a magnetic field. If a conductive liquid flows through this magnetic field, a voltage is induced that is proportional to the average flow velocity—the faster the flow rate, the higher the voltage. This voltage is picked up by sensing electrodes and used to calculate the flow rate.

are corrosive or extremely dirty. Figure 4-66 shows a typical target flowmeter. Fluid flow causes the target plate and lever arm to deflect against a spring. A permanent magnet attached to the lever arm and a Hall effect sensor mounted inside the display unit translate the angular motion of the target to an electrical signal that operates a flow rate display.

Part 3 Review Questions

1. In general, how do sensor pilot devices operate?
2. What is the main feature of a proximity sensor?
3. List the main components of an inductive proximity sensor.
4. Explain the term *hysteresis* as it applies to a proximity sensor.
5. How is a two-wire sensor connected relative to the load it controls?
6. In what way is the sensing field of a capacitive proximity sensor different from that of the inductive proximity sensor?
7. For what type of target would a capacitive proximity sensor be selected over an inductive type?
8. Outline the principle of operation of a photoelectric sensor.
9. Name the three most common scan techniques for photoelectric sensors.
10. What are the advantages of fiber optic sensing systems?
11. Outline the principle of operation of a Hall effect sensor.
12. Outline the principle of operation of an ultrasonic sensor.
13. List the four basic types of temperature sensors and describe the principle of operation of each.
14. Compare the way in which a tachometer and magnetic pickup are used in speed measurement.
15. Outline the principle of operation of an optical encoder.
16. What approach is usually taken to measurement of fluid flow?
17. List three common types of flowmeters.
18. What advantage do proximity sensors have as input signals to solid-state logic control devices?
19. Which type of photoelectric scanning is best suited for long-distance scanning?
20. Describe three operating scenarios that may cause a problem with a retroreflective-type scanning sensor.

PART 4 ACTUATORS

Relays

An **actuator,** in the electrical sense, is any device that converts an electrical signal into mechanical movement. An electromechanical **relay** is a type of actuator that mechanically switches electric circuits. Relays play an important role in many motor control systems. In addition to providing control logic by switching multiple control circuits, they are also used for controlling low-current pilot loads such as contactor and starter coils, pilot lights, and audible alarms.

Figure 4-68 shows a typical electromechanical control relay. This relay consists of a coil, wound on an iron core, to form an electromagnet. When the coil is energized by a control signal, the core becomes magnetized and sets up a magnetic field that attracts the iron arm of the armature to it. As a result, the contacts on the armature close. When the current to the coil is switched off, the armature is spring-returned to its normal de-energized position and the contacts on the armature open.

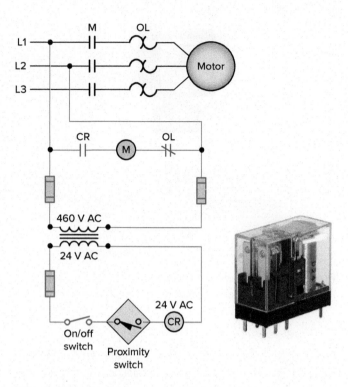

Figure 4-69 Relay motor control circuit.
Photo courtesy of IDEC Corporation, www.IDEC.com/usa, RJ Relay

Figure 4-69 illustrates one simple application of a control relay used in a motor control circuit. The relay enables the energy in the motor high-power line circuit to be switched by a low-power two-wire proximity switch. In this example, the proximity switch operates a relay coil whose contacts operate motor starter coil M. The operation of the control circuit can be summarized as follows:

- With the switch on, anytime the proximity switch contacts close, relay coil CR will become energized.
- This in turn will cause the normally open CR contacts to close and complete the path of current to the M motor starter. The M motor starter energizes and closes the M contacts in the power circuit, starting the motor.
- When the proximity switch opens, the CR coil deenergizes, opening CR contacts that in turn de-energizes coil M and opens the M contacts in the power circuit to stop the motor.

Double-break relay contacts use two pair of contacts that open the circuit in two places, creating two air gaps, as illustrated in Figure 4-70. It is analogous to having two contacts in series. Double-break contacts dissipate heat more readily, providing longer contact life and enabling the contact to handle higher voltages. Other benefits include greater DC load–breaking capability and better isolation of contacts.

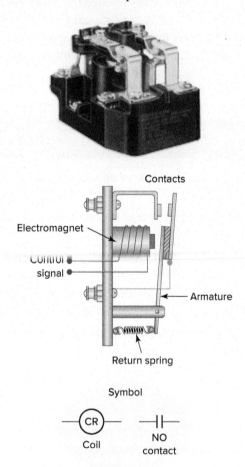

Figure 4-68 Electromechanical relay.
Photo courtesy of Tyco Electronics Connectivity, www.tycoelectronics.com

98   Chapter 4   Motor Control Devices

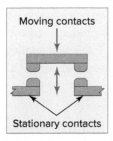

Figure 4-70 Double-break contacts.

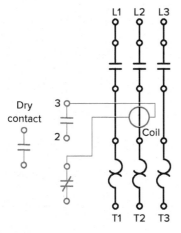

Figure 4-71 Dry contact.

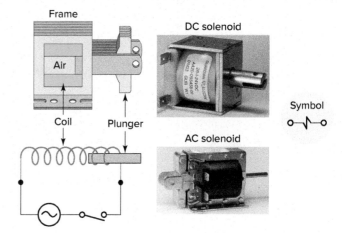

Figure 4-72 Solenoid construction and operation. Photo courtesy of Guardian Electric, www.guardian-electric.com.

The choice of using DC- or AC-operated solenoid coils is usually predetermined by the type of supply voltage available. Most solenoid applications use DC. Differences between DC- and AC-operated solenoids include:

- AC solenoids tend to be more powerful in the fully open position than DC. This is due to the inrush current, which at maximum stroke can be more than 10 times the closed current.
- The coil current for DC solenoids is limited by coil resistance only. The resistance of an AC solenoid coil is very low, so current flow is primarily limited by the inductive reactance of the coil.
- AC solenoids must close completely so that the inrush current falls to its normal value. If an AC solenoid plunger sticks in the open position, a burnout of the coil is likely. DC solenoids take the same current throughout their stroke and cannot overheat through incomplete closing.
- AC-operated solenoids are usually faster than DC, but with a few milliseconds variation in response time, depending on the point of the cycle when the solenoid is energized. DC solenoids are slower, but they repeat their closing times accurately against a given load.
- A good AC solenoid, correctly used, should be quiet when closed but only because its fundamental tendency to hum has been overcome by correct design and accurate assembly. Dirt on the mating faces or mechanical overload may make it noisy. A DC solenoid is naturally quiet.

A **dry contact** refers to one that has both terminals available and in which neither contact is initially connected to a voltage source. Figure 4-71 shows a magnetic motor starter with an extra dry contact that is not prewired to the control circuit that starts and stops the motor. When the starter coil is energized, the dry contact changes state and can be wired to devices or circuits that are separate from the motor starter circuit.

Solenoids

An electromechanical **solenoid** is a device that uses electrical energy to magnetically cause mechanical control action. A solenoid consists of a coil, frame, and plunger (or armature, as it is sometimes called). Figure 4-72 shows the basic construction and operation of a solenoid. The coil and frame form the fixed part. When the coil is **energized,** it produces a magnetic field that attracts the plunger, pulling it into the frame and thus creating mechanical motion. When the coil is **de-energized,** the plunger returns to its normal position through gravity or assistance from spring assemblies within the solenoid. The frame and plunger of an AC-operated solenoid are constructed with laminated pieces instead of a solid piece of iron to limit eddy currents induced by the magnetic field.

There are two main categories of solenoids: linear and rotary. The direction of movement, either rotary or linear, is based on the mechanical assembly within which the electromagnetic circuit is encased. **Rotary** solenoids

Part 4 Actuators 99

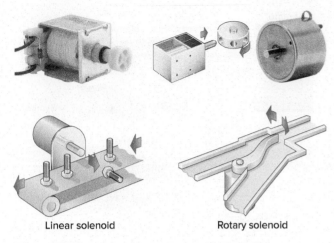

Figure 4-73 Linear and rotary solenoid applications.
Photos and art courtesy of Ledex, www.ledex.com.

incorporate a mechanical design that converts linear motion to rotary motion. **Linear** solenoids are usually classified as pull (the electromagnetic path pulls a plunger into the solenoid body) or push (the plunger shaft is pushed out of the frame case).

Figure 4-73 illustrates common applications for linear and rotary solenoids. The linear solenoid application shown is used in part rejection processes in which electronic interfacing with a sensor produces an actuation signal to the solenoid. In the rotary solenoid application, the solenoid is used in a sorting conveyor to control a diverter gate.

Solenoid Valves

Solenoid valves are electromechanical devices that work by passing an electrical current through a solenoid, thereby changing the state of the valve. Normally, there is a mechanical element, which is often a spring, that holds the valve in its default position. A solenoid valve is a combination of a solenoid **coil** operator and **valve,** which controls the flow of liquids, gases, steam, and other media. When electrically energized, they either open, shut off, or direct the flow of media.

Figure 4-74 illustrates the construction and principle of operation of a typical fluid solenoid valve. The valve body contains an orifice in which a disk or plug is positioned to restrict or allow flow. Flow through the orifice is either restricted or allowed depending on whether the solenoid coil is energized or de-energized. When the coil is energized, the core is drawn into the solenoid coil to close the valve. The spring returns the valve to its original open position when the current coil is de-energized.

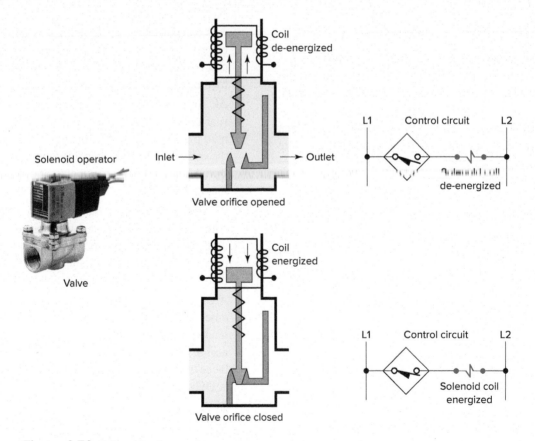

Figure 4-74 Solenoid valve.
Photo courtesy of ASCO Valve Inc., www.ascovalve.com

100   **Chapter 4**   Motor Control Devices

A valve must be installed with direction of flow in accordance with the arrow cast on the side of the valve body.

Solenoid valves are commonly used as part of tank filling and emptying processes. Figure 4-75 shows the circuit for a tank filling and emptying operation. The operation of the control circuit can be summarized as follows:

- Assuming the liquid level of the tank is at or below the full level mark, momentarily pressing the fill push button will energize control relay 1CR.
- Contacts $1CR_1$ and $1CR_2$ will both close to seal in the 1CR coil and energize normally closed solenoid valve A to start filling the tank.
- As the tank fills, the normally open empty-level sensor switch closes.
- When the liquid reaches the full level, the normally closed full-level sensor switch opens to open the circuit to the 1CR relay coil and switch solenoid valve A to its de-energized closed state.
- Any time the liquid level of the tank is above the empty level mark, momentarily pressing the empty push button will energize control relay 2CR.
- Contacts $2CR_1$ and $2CR_2$ will both close to seal in the 2CR coil and energize normally closed solenoid valve B to start emptying the tank.
- When the liquid reaches the empty level, the normally open empty level sensor switch opens to open the circuit to the 2CR relay coil and switch solenoid valve B to its de-energized closed state.
- The stop button may be pressed at any time to halt the process.

Stepper Motors

Stepper motors operate differently than standard types, which rotate continuously when voltage is applied to their terminals. The shaft of a stepper motor rotates in discrete **increments** when electrical command pulses are applied to it in the proper sequence. Every revolution is divided into a number of steps, and the motor must be sent a voltage pulse for each step. The amount of rotation is directly proportional to the number of pulses, and the speed of rotation is relative to the frequency of those pulses. A 1-degree-per-step motor will require 360 pulses to move through one revolution; the degrees per step is known as the *resolution*. When stopped, a stepper motor inherently holds its position. Stepper systems are used most often in "open-loop" control systems, where the controller tells the motor only how many steps to move and how fast to move, but does not have any way of knowing what position the motor is at.

The movement created by each pulse is precise and repeatable, which is why stepper motors are so effective for load-positioning applications. Conversion of rotary to linear motion inside a linear actuator is accomplished through a threaded nut and lead screw. Generally, stepper motors produce less than 1 hp and are therefore frequently used in low-power position control applications. Figure 4-76

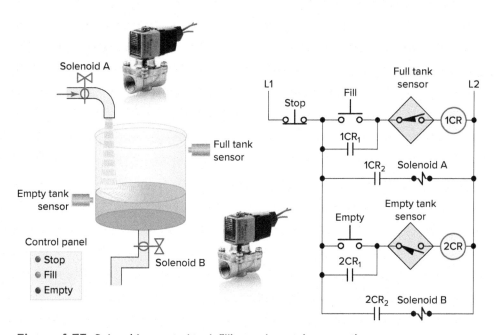

Figure 4-75 Solenoid-operated tank filling and emptying operation.
Photos courtesy of ASCO Valve Inc., www.ascovalve.com

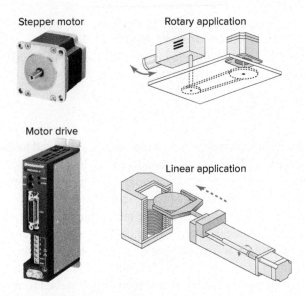

Figure 4-76 Stepper motor/drive unit.
Photos courtesy of Oriental Motor, www.orientalmotor.com

shows a stepper motor/drive unit along with typical rotary and linear applications.

As all motors, the stepper motor consists of a stator and a rotor. The rotor carries a set of permanent magnets, and the stator has the coils. A simplified operation of a stepper motor is illustrated in Figure 4-77 and summarized as follows:

- There are four coils with a 90-degree angle between each other fixed on the stator.
- The motor has a 90-degree rotation step.
- The coils are activated in a cyclic order, one by one.
- As each coil is energized by a pulse of current, the shaft rotates 90 degrees.
- The rotation direction of the shaft is determined by the order in which the coils are activated.

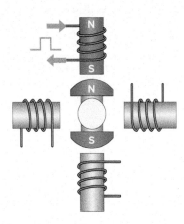

Figure 4-77 Stepper motor operation.

Servo Motors

All servo motors operate in closed-loop mode, whereas most stepper motors operate in open-loop mode. Closed- and open-loop control schemes are illustrated in Figure 4-78. **Open loop** uses control without feedback, for example, when the controller tells the stepper motor how many steps to move and how fast to move, but does not verify where the motor is. **Closed-loop** control compares speed or position feedback with the commanded speed or position and generates a modified command to make the error smaller. The error is the difference between the required speed or position and the actual speed or position.

Figure 4-79 shows a typical closed-loop servo motor system. The motor controller directs operation of the servo motor by sending speed or position command signals to the amplifier, which drives the servo motor. A feedback device such as an encoder for position and a tachometer for speed are either incorporated within the servo motor or are remotely mounted, often on the load itself. These provide the servo motor's position and speed feedback information that the controller compares to its programmed motion profile and uses to alter its position or speed.

While stepper motors are DC operated, a servo motor can be either DC or AC operated. Three basic types of servo motors are used in modern servo systems: AC servo motors, based on induction motor designs; DC servo motors, based on DC motor designs; and DC or AC brushless servo motors.

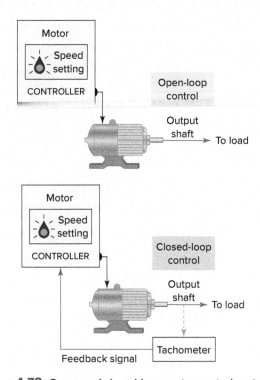

Figure 4-78 Open- and closed-loop motor control systems.

102   Chapter 4   Motor Control Devices

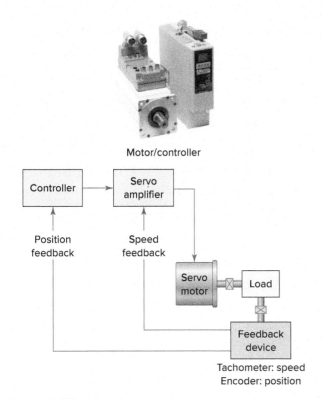

Figure 4-79 Closed-loop servo system.
Photo courtesy of Rockwell Automation,
www.rockwellautomation.com

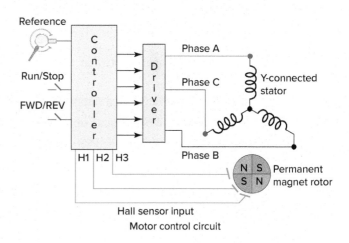

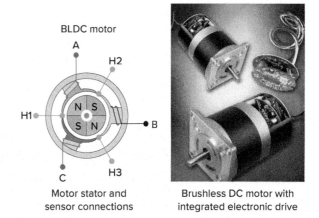

Figure 4-80 Brushless DC motor with integrated drive.
Photo courtesy of ElectroCraft, www.electrocraft.com

A **DC brushless** servo motor is shown in Figure 4-80. As the name implies, brushless DC motors (BLDCs) have no brush or commutation mechanism; instead, they are electronically commutated.

- The stator is normally a three-phase stator (A-B-C) like that of an induction motor, and the rotor has surface-mounted permanent magnets.
- Three Hall sensors (H1-H2-H3) are used to detect the rotor position, and commutation is performed electronically, based on signals from the Hall sensor inputs.

- The signals from the Hall effect sensors are decoded by the controller and used to control the driver circuit, which energizes the stator coils in the proper rotational sequence.
- Therefore, the motor requires an electronic drive in order to operate.

Part 4 Review Questions

1. Define the term *actuator* as it applies to an electric circuit.
2. In what ways are electromagnetic relays employed in motor control systems?
3. What are the two main parts of an electromagnetic relay?
4. Describe how an electric solenoid operates.
5. Which type of solenoid (AC or DC) is constructed with laminated instead of solid pieces of steel? Why?
6. Why are AC solenoid coils likely to overheat if the plunger sticks in the open position when energized?

7. In what way is the design of a rotary solenoid different from that of a linear solenoid?
8. A solenoid valve is a combination of what two elements?
9. Explain how rotation is achieved in a stepper motor.
10. What is the basic difference between an open-loop and closed-loop positioning or motor speed control system?
11. What do all servo motors have in common?
12. What replaces brushes in a brushless DC motor?
13. In what way does a double-break contact differ from that of a single-break type?
14. What type of contact would be classified as a dry type?

Troubleshooting Scenarios

1. A defective switch rated for 10 A DC at a given voltage is replaced with one rated for 10 A AC at the same voltage. What is most likely to happen? Why?
2. The resistance of a suspected AC solenoid coil, rated for 2 A at 120 V, is measured with an ohmmeter and found to have a resistance of 1 Ω. Does this mean the coil is shorted? Why?
3. The NO and NC contacts of a relay with a coil operating voltage of 12 V DC are to be bench-checked for faults by using an ohmmeter. Develop a complete outline, including circuit schematics, of the procedure you would follow.
4. A 12 V pilot light is incorrectly replaced with one rated for 120 V. What would be the result?
5. What voltage values are typically produced by thermocouples?
6. A through-beam photoelectric sensor appears to be missing detection of small bottles on a high-speed conveyor line. What could be creating this problem?

Discussion Topics and Critical Thinking Questions

1. List typical electrical and mechanical problems that may cause failed operation of a mechanically operated limit switch.
2. How might a flow switch be used in a building fire protection system?
3. Should a resistance check of a good thermocouple yield a "low resistance" or "infinite" reading? Why?
4. What does the range adjustment on a float switch accomplish?
5. A stepper motor cannot be bench-checked directly from a power source. Why?

CHAPTER FIVE

Electric Motors

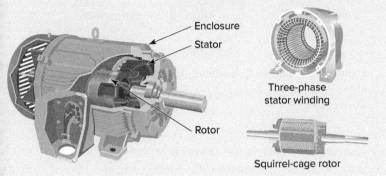

CHAPTER OBJECTIVES

This chapter will help you:

- Understand the basic electric motor operating principle.
- Demonstrate a working knowledge of the construction, connection, and operating characteristics of different types of DC motors.
- Explain the operating principle of DC brushless motors.
- Demonstrate a working knowledge of the construction, connection, and operating characteristics of different types of AC motors.
- Perform testing procedures used to troubleshoot motor problems.

An electric motor converts electric energy into mechanical energy by using interacting magnetic fields. Electric motors are used for a wide variety of residential, commercial, and industrial operations. This chapter deals with the operating principles of different types of DC, universal, and AC electric motors.

PART 1 MOTOR PRINCIPLE

Magnetism

Electric motors are used to convert electric energy into mechanical energy. The motor represents one of the most useful and labor-saving inventions in the electrical industry. Over 50 percent of the electricity produced in the United States is used to power motors.

An electric motor uses magnetism and electric currents to operate. There are two basic categories of motors, AC and DC. Both use the same fundamental parts but with variations to allow them to operate using two different kinds of electrical power supply.

Magnetism is the force that creates rotation for a motor to operate. Therefore, before we discuss basic motor operation, a short review of magnetism is in order. Recall that a permanent magnet will attract and hold magnetic materials such as iron and steel when such objects are near or in contact with the magnet. The permanent magnet is able to do this because of its inherent magnetic force, which is referred to as a

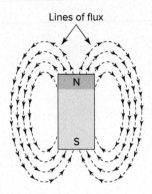

Figure 5-1 Magnetic field of a permanent bar magnet.

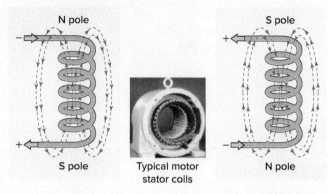

Figure 5-3 Magnetic field produced by a current-carrying coil.
Photo courtesy of Electrical Apparatus Service Association, www.easa.com

magnetic field. In Figure 5-1, the magnetic field of a permanent bar magnet is represented by **lines of flux.**

- Lines of flux help us visualize the magnetic field of any magnet even though they actually represent an invisible phenomenon.
- The number of flux lines varies from one magnetic field to another, and the stronger the magnetic field, the greater the number of lines of flux.
- Lines of flux are assumed to travel from the north pole to the south pole outside the magnet and from the south pole to the north pole inside the magnet.

Electromagnetism

A similar type of magnetic field is produced around a current-carrying conductor. The strength of the magnetic field is directly proportional to the amount of current flowing through the conductor and takes the form of concentric circles around the wire. Figure 5-2 illustrates the magnetic field around a straight current-carrying conductor. A relationship exists between the direction of current flow through a conductor and the direction of the magnetic field created. Known as the **left-hand conductor rule,** it uses electron flow from negative to positive as the basis for the current direction. When you place your left hand so that your thumb points in the direction of the electron flow, your curled fingers point in the direction of the lines of flux that circle the conductor.

When a current-carrying conductor is shaped into a coil, the individual flux lines produced by each of the turns form one stronger magnetic field. The magnetic field produced by a current-carrying coil resembles that of a permanent magnet (Figure 5-3). As with the permanent magnet, these flux lines leave the north of the coil and re-enter the coil at its south pole. The magnetic field of a wire coil is much greater than the magnetic field around the wire before it is formed into a coil and can be further strengthened by placing a core of iron in the center of the coil. The iron core presents less resistance to the lines of flux than the air, thereby causing the field strength to increase. This is exactly how a motor stator coil is constructed: using a coil of wire with an iron core. The polarity of the poles of a coil reverses whenever the current flow through the coil reverses. Without this phenomenon, the operation of electric motors would not be possible.

Generators

An electric generator is a machine that uses magnetism to convert mechanical energy into electric energy. Generators operate on the principle that a voltage is induced in a conductor whenever the conductor is moved through a magnetic field so as to cut lines of force.

An AC generator (alternator) generates a voltage by rotation of a loop of wire (armature) within a magnetic field, as illustrated in Figure 5-4. Its operation can be summarized as follows:

- A device called a prime mover provides the mechanical energy input, which physically turns the armature coil.

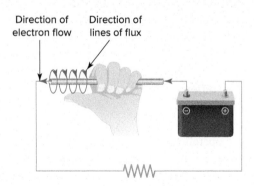

Figure 5-2 Magnetic field around a straight current-carrying conductor.

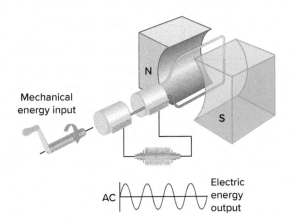

Figure 5-4 Simplified AC generator.

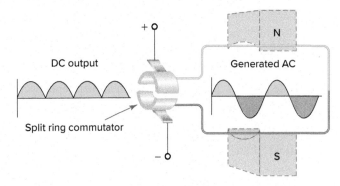

Figure 5-5 Simplified DC generator.

- Relative motion between the wire and the magnetic field causes a voltage to be induced between the ends of the wire.
- The generated voltage changes in magnitude as the loop cuts the magnetic lines of force at different angles.
- The generated voltage changes in polarity as the loop cuts the magnetic lines of force in two different directions.
- Two slip rings are attached to the armature wire and rotate with it.
- Carbon brushes ride against the slip rings to conduct AC current from the armature to the resistor load.
- A complete rotation of 360 degrees results in the generation of one cycle of alternating current.

Generating DC is basically the same as generating AC. The only difference is the manner in which the generated voltage is supplied to the output terminals. Figure 5-5 shows a simplified DC generator. Its operation can be summarized as follows:

- Voltage is induced into the coil when it cuts magnetic flux lines.
- The shape of the voltage generated in the loop is still that of an AC sine wave.
- The two slip rings of the AC generator are replaced by a single split ring called a commutator.
- The commutator acts like a mechanical switch, which converts the generated AC voltage into DC voltage.
- As the armature starts to develop a negative alternation, the commutator switches the polarity of the output terminals via the brushes. This keeps all positive alternations on one terminal and all negative alternations on the other.

- The only essential difference between an AC and a DC generator is the use of slip rings on the one and a commutator on the other.
- Brushes move from one segment of the commutator to the next during the period of zero induced voltage when the coil is moving parallel to the magnetic flux, or in what is called the *neutral plane*.
- Any DC machine can act either as a generator or as a motor.

Motor Rotation

An electric motor rotates as the result of the interaction of two magnetic fields. One of the well-known laws of magnetism is that "like" poles (N-N or S-S) repel while "unlike" poles (N-S) attract. Figure 5-6 illustrates how this attraction and repulsion of magnetic poles can be used to produce a rotating force. The operation can be summarized as follows:

- The electromagnet is the moving armature part and the permanent magnet the fixed stator part.
- Like magnetic poles repel each other, causing the armature to begin to turn.
- After it turns part way around, the force of attraction between the unlike poles becomes strong enough to keep the electromagnet rotating.
- The rotating electromagnet continues to turn until the unlike poles are lined up. At this point the rotor would normally stop because of the attraction between the unlike poles.

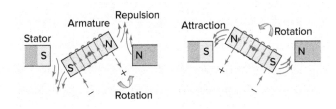

Figure 5-6 Motor principle.

- Commutation is the process of reversing armature current at the moment when unlike poles of the armature and field are facing each other, thereby reversing the polarity of the armature field.
- Like poles of the armature and field then repel each other, causing armature rotation to continue.

When a current-carrying conductor is placed in a magnetic field, there is an interaction between the magnetic field produced by the current and the permanent field, which leads to a force being experienced by the conductor. The magnitude of the force on the conductor will be directly proportional to the current which it carries. A current-carrying conductor, placed in a magnetic field and at right angles to it, tends to move at right angles to the field, as illustrated in Figure 5-7.

A simple method used to determine the direction of movement of a conductor carrying current in a magnetic field is the **right-hand motor rule**. To apply the right hand motor rule:

- The thumb and first two fingers of the right hand are arranged to be at right angles to each other.
- The forefinger is pointed in the direction of the magnetic lines of force of the field from N pole to S pole.
- The middle finger is pointed in the direction of electron current flow (− to +) in the conductor.
- The thumb will then be pointing in the direction of movement of the conductor.
- Applying the right-hand motor rule to Figure 5-8, the conductor will move upward through the magnetic field.
- If the current through the conductor were to be reversed, the conductor would move downward.
- Note that the conductor current is at a right angle to the magnetic field. This is required to bring about motion because no force is felt by a conductor if the current and the field direction are parallel.

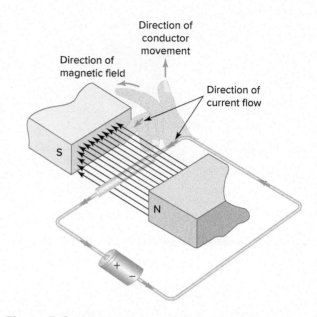

Figure 5-8 Right-hand motor rule.

Figure 5-9a illustrates how motor **torque** (rotational force) is produced by a current-carrying coil or loop of wire placed in a magnetic field. Rotation is the result of the interaction of the magnetic fields generated by the permanent magnets and current flow through the armature coil.

- This interaction of the two magnetic fields causes a bending of the lines of force.
- When the lines tend to straighten out, they cause the loop to undergo a rotating motion.
- The left conductor is forced downward, and the right conductor is forced upward, causing a counterclockwise rotation of the armature.

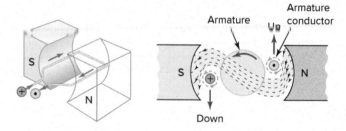

(a) Torque produced by a single-coil armature.

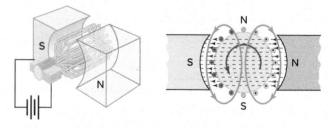

(b) Torque produced by a multicoil armature.

Figure 5-9 Developing motor torque.

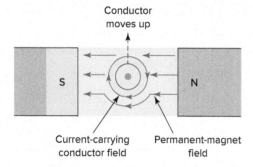

Figure 5-7 A current-carrying conductor, placed in a magnetic field.

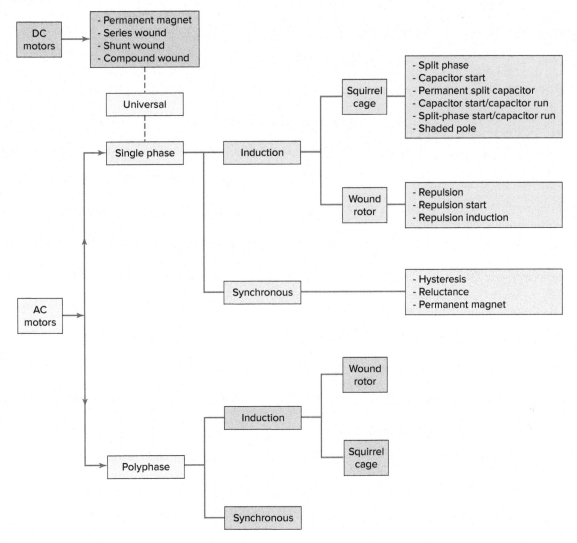

Figure 5-10 **Family tree of common motors.**

- A practical motor armature is made up of many coils of conductors as illustrated in Figure 5-9b.
- The magnetic fields of these conductors combine to form a resultant armature field with north and south poles that interact with those of the main stator field to exert a continuous torque on the armature.

In general, motors are classified according to the type of power used (AC or DC) and the motor's principle of operation. There are several major classifications of the motors in common use; each will specify characteristics that suit it to particular applications. Figure 5-10 shows a family tree of common types of motors.

Part 1 Review Questions

1. State the operating principle of a generator.
2. What name is given to the device that drives a generator?
3. Explain the essential difference between construction of an AC and a DC generator.
4. What is the basic purpose of an electric motor?
5. According to power requirements, in what ways are motors classified?
6. In what direction are the lines of flux of a magnet assumed to travel?
7. How does electricity produce magnetism?
8. Why is the motor stator coil constructed with an iron core?
9. How is the polarity of the poles of a given coil reversed?
10. In general, what causes an electric motor to rotate?

11. In what direction will a current-carrying conductor, placed in and at right angles to a magnetic field, move?
12. Applying the right-hand motor rule to a given current-carrying conductor placed in a magnetic field indicates movement in the downward direction. What could be done to reverse the direction in which the conductor moves?
13. What two main criteria are used to classify motors?

PART 2 DIRECT CURRENT MOTORS

Direct current motors are not used as much as alternating-current types because all electric utility systems deliver alternating current. For special applications, however, it is advantageous to transform the alternating current into direct current in order to use DC motors. Direct current motors are ideally suited to a multitude of industrial and marine applications in which high torque and variable speed are required. These applications include mine hoists, steel rolling mills, ship propulsion, cranes, conveyors, and elevators (Figure 5-11).

The construction of a DC motor (Figure 5-12) is considerably more complicated and expensive than that of an AC motor, primarily because of the commutator, brushes, and armature windings. Maintenance of the brush/commutator assembly found on DC motors is significant compared to that of AC motor designs. An AC induction motor requires no commutator or brushes, and most use cast squirrel-cage rotor bars instead of wound copper wire windings. There are several types of DC motors, classified according to field type. These are permanent magnet, series, shunt, and compound.

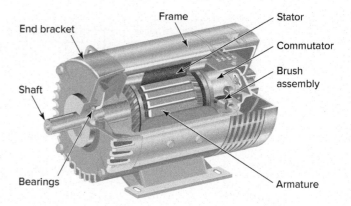

Figure 5-12 Major components of a DC motor.

Motor speed, torque, and horsepower (hp) are important parameters used to predict DC motor performance:

Speed: Refers to the rotational speed of the motor's shaft and is measured in revolutions per minute (rpm).

Torque: Refers to the turning force supplied by the motor's shaft. Torque consists of force acting on a radius. The standard units of torque as used in the motor control industry are pound-inches (lb-in), or pound-feet (lb-ft).

Horsepower: Refers to the rate at which work is done. As an example, 1 horsepower is equivalent to lifting 33,000 pounds to a height of 1 foot in 1 minute. One horsepower is also equivalent to 746 watts of electrical power. Therefore, you can use watts to calculate horsepower and vice versa.

Permanent-Magnet DC Motor

Permanent-magnet DC motors use permanent magnets to supply the main field flux and electromagnets to provide the armature flux. Movement of the magnetic field of the armature is achieved by switching current between coils within the motor. This action is called *commutation*. Figure 5-13 illustrates the operation of a simple permanent-magnet motor. The operation of the circuit can be summarized as follows:

- Current flow through the armature coil from the DC voltage supply causes the armature to act as an electromagnet.

Figure 5-11 Typical DC industrial motor application.
Photo courtesy of TECO-Westinghouse Motor Co., http://www.tecowestinghouse.com/

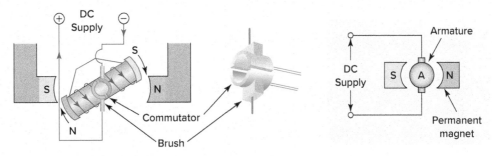

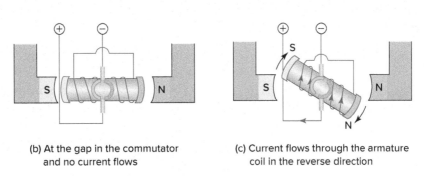

Figure 5-13 Permanent-magnet DC motor operation.

- The armature poles are attracted to field poles of opposite polarity causing the armature to rotate in a clockwise direction (Figure 5-13a).
- When the armature poles are in line with the field poles, the brushes are at the gap in the commutator and no current flows in the armature (Figure 5-13b). At this point the forces of magnetic attraction and repulsion stop and inertia carries the armature past this neutral point.
- Once past the neutral point, current flows through the armature coil in the reverse direction because of the commutator's reversing action (Figure 5-13c). This in turn reverses the polarity of the armature poles, resulting in repulsion of the like poles and further rotation in a clockwise direction.
- The cycle is repeated with the current flow through the armature reversed by the commutator once each cycle to produce a continuous rotation of the armature in a clockwise direction.

Figure 5-14 shows a permanent-magnet (PM) motor. The motor is made up of two main parts: a housing containing the field magnets and an armature consisting of coils of wire wound in slots in an iron core and connected to a commutator. Brushes, in contact with the commutator, carry current to the coils. PM motors produce high torque compared to wound-field motors. However, permanent-magnet motors are limited in load-handling ability and are therefore used mainly for low-horse-power applications.

The force that rotates the motor armature is the result of the interaction between two magnetic fields (the stator field and the armature field). To produce a constant torque from the motor, these two fields must remain constant in magnitude and in relative orientation. This is achieved by constructing the armature as a series of small sections connected to the segments of a commutator, as illustrated in Figure 5-15.

- Electrical connection is made to the commutator by means of two brushes.
- It can be seen that if the armature rotates through one-sixth of a revolution clockwise, the current in coils 3 and 6 will have changed direction.
- As successive commutator segments pass the brushes, the current in the coils connected to those segments changes direction.

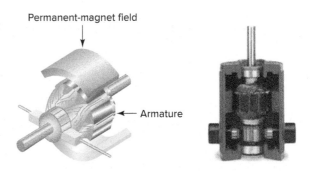

Figure 5-14 Permanent-magnet DC motor.
Photo courtesy of Leeson, www.leeson.com

Part 2 Direct Current Motors

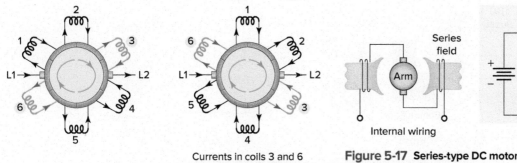

Figure 5-15 Armature commutation or switching effect.

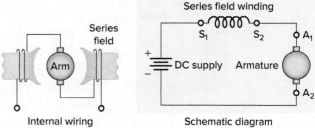

Figure 5-17 Series-type DC motor.

- The commutator can be regarded as a switch that maintains the proper direction of current in the armature coils to produce constant unidirectional torque.

The direction of rotation of a permanent-magnet DC motor is determined by the direction of the current flow through the armature. Reversing the polarity of the voltage applied to the armature will reverse the direction of rotation, as illustrated in Figure 5-16. Variable-speed control of a PM motor is accomplished by varying the value of the voltage applied to the armature. The speed of the motor varies directly with the amount of armature voltage applied. The higher the value of the armature voltage, the faster the motor will run.

Series DC Motor

Wound-field DC motors are usually classified as series-wound, shunt-wound, or compound-wound. The connection for a **series-type** DC motor is illustrated in Figure 5-17. A series-wound DC motor consists of a series field winding (identified by the symbols S1 and S2) connected in series with the armature (identified by the symbols A1 and A2). Since the series field winding is connected in series with the armature, it will carry the same amount of current that passes through the armature. For this reason the windings of the series field are made from heavy-gauge wire that is large enough to carry the full motor load current. Because of the large diameter of the series winding, the winding will have only a few turns of wire and a very low resistance value.

A series-wound DC motor has a low resistance field and low resistance armature circuit. Because of this, when voltage is first applied to it, the current is high ($I = E/R$). The advantage of high current is that the magnetic fields inside the motor are strong, producing high torque (turning force), so it is ideal for starting very heavy mechanical loads.

Figure 5-18 shows the speed–torque characteristic curves for a series DC motor. Note that the speed varies widely between no load and rated load. Therefore, these motors cannot be used where a constant speed is required with variable loads. Also the motor runs fast with a light load (low current) and runs substantially slower as the motor load increases. Because of their ability to start very heavy loads, series motors are often used in cranes, hoists, and elevators, which can draw thousands of amperes on starting. **Caution: The no-load speed of a series motor can increase to the point of damaging the motor. For this reason, it should never be operated without a load of some type coupled to it.**

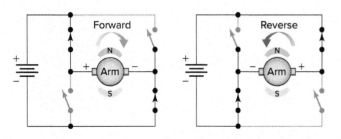

Figure 5-16 Reversing the direction of rotation of a PM motor.

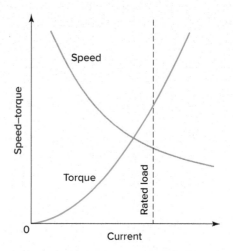

Figure 5-18 Speed–torque characteristic curves for a series DC motor.

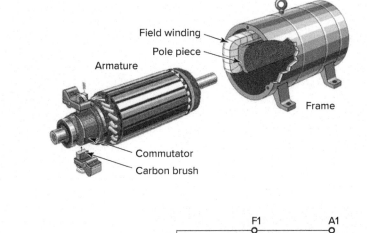

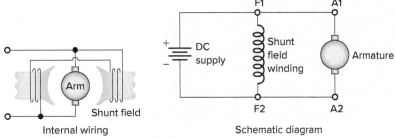

Figure 5-19 Shunt-type DC motor.
Photo courtesy of Siemens, www.siemens.com

Shunt DC Motor

The connection for a **shunt-type** DC motor is illustrated in Figure 5-19. A shunt-wound DC motor consists of a shunt field (identified by the symbols F1 and F2) connected in parallel with the armature. This motor is called a shunt motor because the field is in parallel to, or "shunts," the armature. The shunt field winding is made up of many turns of small-gauge wire and has a much higher resistance and lower current flow compared to a series field winding.

Figure 5-20 shows the speed–torque characteristic curves for a shunt DC motor. Since the field winding is connected directly across the power supply, the current through the field is constant. The field current does not vary with motor speed, as in the series motor and, therefore, the torque of the shunt motor will vary only with the current through the armature.

- When the motor is starting and the speed is very low, the motor has very little torque.
- After the motor reaches full rpm, its torque is at its fullest potential.
- One of the main advantages of a shunt motor is its constant speed. It runs almost as fast fully loaded as it does with no load.
- Unlike series motor, the shunt motor will not accelerate to a high speed when no load is coupled to it.
- Shunt motors are particularly suitable for applications such as conveyors, where constant speed is desired and high starting torque is not needed.

A separately excited DC motor has the armature and field coils fed from **separate** supply sources. This type of motor has a field coil similar to that of a self-excited shunt motor. In the most common configuration, armature voltage control is used in conjunction with a constant or variable voltage field excitation. Figure 5-21 shows a typical DC drive application for a separately excited DC motor. An advantage to separately exciting the field is that the

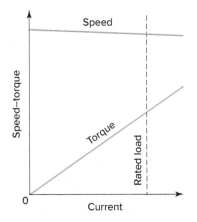

Figure 5-20 Speed–torque characteristic curves for a shunt DC motor.

Part 2 Direct Current Motors   113

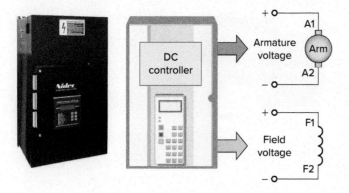

Figure 5-21 Separately excited motor.
Photo courtesy of Nidec Avtron

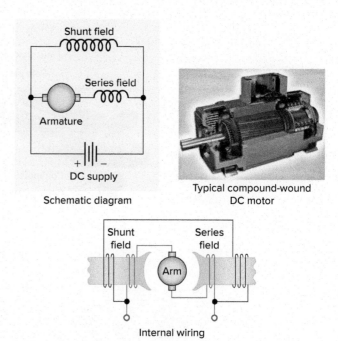

Figure 5-22 Compound-type DC motor.
Photo courtesy of ABB, www.abb.com

variable-speed DC drive can be used to provide independent control of the field and armature.

Compound DC Motor

A compound-wound DC motor is a combination of the shunt-wound and series-wound types. This type of DC motor has two field windings, as shown in Figure 5-22. One is a shunt field connected in parallel with the armature; the other is a series field that is connected in series with the armature. The **shunt field** gives this type of motor the constant-speed advantage of a regular shunt motor. The **series field** gives it the advantage of being able to develop a large torque when the motor is started under a heavy load. This motor is normally connected **cumulative-compound** so that under load the series field flux and shunt field act in the same direction to strengthen the total field flux.

Figure 5-23 shows a comparison of speed–torque characteristic curves for a cumulative-compound DC motor versus series and shunt types.

- The speed of the compound motor varies a little more than that of shunt motors, but not as much as that of series motors.
- Compound-type DC motors have a fairly large starting torque—much more than shunt motors, but less than series motors.

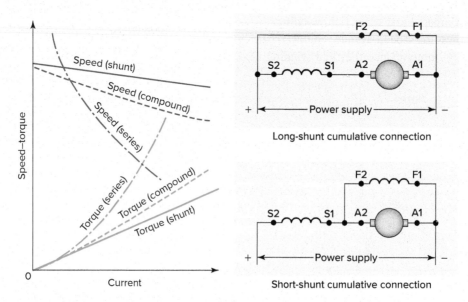

Figure 5-23 DC cumulative-compound motor connections and speed–torque characteristics.

114   Chapter 5   Electric Motors

- The shunt winding can be wired as a cumulative long-shunt or as a short-shunt compound motor.
- For short-shunt, the shunt field is connected in parallel with only the armature, whereas with long-shunt, the shunt field is connected in parallel with both the series field and the armature.
- There is very little difference in the operating characteristics of long-shunt and short-shunt compound motors.
- These motors are generally used where severe starting conditions are met and constant speed is required at the same time.

Direction of Rotation

The direction of rotation of a wound DC motor depends on the direction of the field and the direction of the current flow through the armature. If either the direction of the field current or the direction of the current flow through the armature of a wound DC motor is reversed, the rotation of the motor will reverse. If both of these two factors are reversed at the same time, however, the motor will continue rotating in the same direction.

For a series-wound DC motor, changing the polarity of either the armature or series field winding changes the direction of rotation. If you simply changed the polarity of the applied voltage, you would be changing the polarity of both series field and armature windings and the motor's rotation would remain the same.

Figure 5-24 shows the power and control circuit schematics for a typical DC reversing motor starter used to operate a series motor in the forward and reverse directions. In this application, reversing the polarity of the armature voltage changes the direction of rotation. The operation of the circuit can be summarized as follows:

- When the starter coil F is energized, the main F contacts close, connecting A1 to the positive side of the power supply and A2 to the negative side to operate the motor in the forward direction.
- When the starter coil R is energized, the main R contacts close, reversing the armature polarity so A2 is now positive and A1 is negative and the motor will now operate in the reverse direction.
- Notice that for both the forward and reverse directions, the polarity of the series field remains unchanged, with S1 being positive with respect to S2: only the armature's polarity is changed.
- The circuit is **electrically interlocked** by way of the normally closed (NC) R and F auxiliary control contacts. This prevents starter coils F and R from both being energized at the same time and in effect shorting out the motor armature circuit.
- If the reverse push button is pressed while the operating in the forward direction, the R starter coil cannot be energized as the circuit to the coil is opened by the NC F contact. In order to change the direction of rotation, the stop push button must first be pressed to de-energize the F starter coil and allow the NC F contacts to return to their closed position.

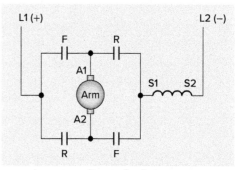

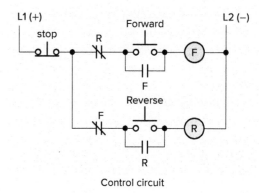

Figure 5-24 DC series motor reversing motor starter. This material and associated copyrights are proprietary to, and used with the permission of, Schneider Electric.

As in a DC series motor, the direction of rotation of a DC shunt and compound motor can be reversed by changing the

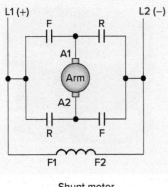

Shunt motor

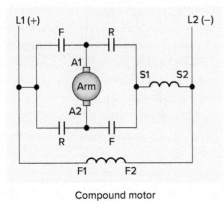

Compound motor

Figure 5-25 DC shunt and compound motor reversing.

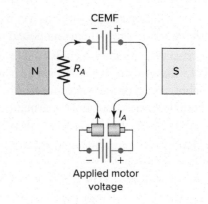

Applied motor voltage

Figure 5-26 Motor CEMF.

polarity of either the armature winding or the field winding. Figure 5-25 shows the power circuit schematics for typical DC shunt and compound motor reversing starters. The industry standard is to reverse the current through the armature while maintaining the current through the shunt and series field in the same direction. For the compound-wound motor, this ensures a cumulative connection (both fields aiding) for either direction of rotation.

Motor Counter Electromotive Force (CEMF)

As the armature rotates in a DC motor, the armature coils cut the magnetic field of the stator and induce a voltage, or electromotive force (EMF), in these coils. This occurs in a motor as a by-product of motor rotation and is sometimes referred to as the generator action of a motor. Because this induced voltage opposes the applied terminal voltage, it is called **counter electromotive force,** or CEMF. Counter EMF (sometimes referred to as back EMF) is a form of resistance that opposes and limits the flow of armature current, as illustrated in Figure 5-26.

The overall effect of the CEMF is that this voltage will be subtracted from the terminal voltage of the motor so that the armature motor winding will see a smaller voltage potential. Counter EMF is equal to the applied voltage minus the armature circuit $I_A R_A$ drop. The armature current, according to Ohm's law, is equal to:

$$I_A = \frac{V_{\text{MTR}} - \text{CEMF}}{R_A}$$

where I_A = armature current
V_{MTR} = motor terminal voltage
CEMF = counter electromotive force
R_A = armature-circuit resistance

EXAMPLE 5-1

Problem: The armature of a 250 V DC motor draws 15 A when operating at full load and has a resistance of 2 Ω. Determine the counter EMF produced by the armature when operating at full load.

Solution:

$$I_A = \frac{V_{\text{MTR}} - \text{CEMF}}{R_A}$$

$$\text{CEMF} = V_{\text{MTR}} - (I_A \times R_A)$$

$$= 250 \text{ V} - (15 \text{ A} \times 2 \text{ Ω})$$

$$= 250 - 30$$

$$= 220 \text{ V}$$

Counter EMF is directly proportional to the speed of the armature and the field strength. That is, the counter EMF increases or decreases if the speed is increased or decreased, respectively. The same is true if the field strength is increased or decreased. At the moment a motor starts, the armature is not rotating, so there is no CEMF generated in the armature. Full line voltage is applied across the armature, and it draws a relatively large amount of current. At this point, the only factor limiting current through the armature is the relatively low resistance of the windings. As the motor picks up speed, a counter electromotive force is generated in the armature, which opposes the applied terminal voltage and quickly reduces the amount of armature current.

When a motor reaches its full no-load speed, it is designed to be generating a CEMF nearly equal to the applied line voltage. Only enough current can flow to maintain this speed. When a load is applied to the motor, its speed will be decreased, which will reduce the CEMF, and more current will be drawn by the armature to drive the load. Thus, the load of a motor regulates the speed by affecting the CEMF and current flow.

Armature Reaction

The magnetic field produced by current flow through the armature conductors distorts and weakens the flux coming from the main field poles. This distortion and field weakening of the stator field of the motor are known as **armature reaction.** Figure 5-27 shows the position of the neutral plane under no-load and loaded motor operating conditions. As segment after segment of the rotating commutator pass under a brush, the brush short-circuits coil after coil in the armature. Note that armature coils A and B are positioned relative to the brushes so that at the instant each is short-circuited, it is moving parallel to the main field so that there is no voltage induced in them at this point. When operating under loaded conditions, due to armature reaction, the neutral plane is shifted backward, opposing the direction of rotation. As a result armature reaction affects the motor operation by:

- Shifting the neutral plane in a direction opposite to the direction of rotation of the armature.
- Reducing motor torque as a result of the weakening of the magnetic field.
- Arcing at the brushes due to short-circuiting of the voltage being induced in the coils undergoing commutation

When the load on the motor fluctuates, the neutral plane shifts back and forth between no-load and full-load positions. For small DC motors, the brushes are set in an intermediate position to produce acceptable commutation at all loads. In larger DC motors, **interpoles** (also called commutating poles) are placed between the main field poles, as illustrated in Figure 5-28, to minimize the effects of armature reaction. These narrow poles have a few turns of larger-gauge wire connected in series with the armature. The strength of the interpole field varies with the armature current. The magnetic field generated by the interpoles is designed to be equal to and opposite that produced by the armature reaction for all values of load current and improves commutation.

Speed Regulation

Motor **speed regulation** is a measure of a motor's ability to maintain its speed from no load to full load without a change in the applied voltage to the armature or fields. A motor has good speed regulation if the change between the no-load speed and full-load speed is small, with other conditions being constant. As an example, if the speed regulation is 3 percent for a motor rated 1,500 rpm with no load applied, then this means that the speed will drop by as much as 45 rpm (1,500 × 3%) with the motor fully loaded. The speed regulation of a direct current motor is proportional to the armature resistance and is generally expressed as a percentage of the motor base speed. DC motors that have a very low armature resistance will have a better

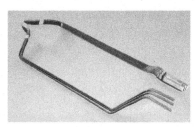

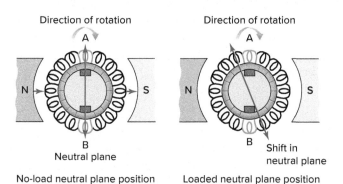

Figure 5-27 Position of the neutral plane under no-load and loaded motor operating conditions.
Photo courtesy of Rees Electric Company, www.ReesElectricCompany.com

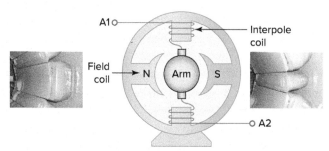

Figure 5-28 Interpoles are placed between the main field poles.
Photo courtesy of ERIKS UK, www.eriks.co.uk

speed regulation. Speed regulation is the ratio of the loss in speed, between no load and full load, to the full-load speed and is calculated as follows (the lower the percentage, the better the speed regulation):

Percent speed regulation

$$= \frac{\text{No-load speed} - \text{Full-load speed}}{\text{Full-load speed}} \times 100$$

EXAMPLE 5-2

Problem: A DC shunt motor is running with a measured no-load speed of 1775 rpm. When full load is applied, the speed drops slightly to 1725 rpm. Find the percentage speed regulation.

Solution:

Percent speed regulation

$$= \frac{\text{No-load speed} - \text{Full-load speed}}{\text{Full-load speed}} \times 100$$

$$= \frac{1775 \text{ rpm} - 1725 \text{ rpm}}{1725 \text{ rpm}} \times 100$$

$$= \frac{50}{1725} \times 100$$

$$= 2.9\%$$

Varying DC Motor Speed

The **base speed** listed on a DC motor's nameplate is an indication of how fast the motor will run with rated armature voltage and rated load amperes at rated field current (Figure 5-29). DC motors can be operated below base speed by reducing the amount of voltage applied to the armature and above base speed by reducing the field current. In addition, the maximum motor speed may also be listed on the nameplate. **Caution: Operating a motor above its rated maximum speed can cause damage to equipment and personnel. When only base speed is listed, check with the vendor before operating it above the specified speed.**

Perhaps the greatest advantage of DC motors is speed control. In armature-controlled adjustable-speed applications, the field is connected across a constant-voltage supply and the armature is connected across an independent adjustable-voltage source (Figure 5-30). By raising or lowering the armature voltage, the motor speed will rise or fall proportionally. For example, an unloaded motor might run at 1200 rpm with 250 V applied to the armature and 600 rpm with 125 V applied. Armature-controlled DC motors are capable of providing rated torque at any speed between zero and the base (rated) speed of the motor.

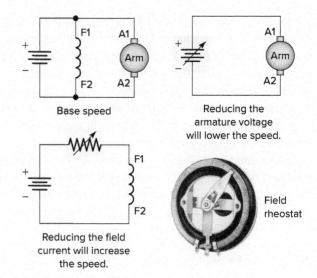

Figure 5-29 DC motor speed.
Photo courtesy of Jenkins Electric Company, www.jenkins.com

Horsepower varies in direct proportion to speed, and 100 percent rated horsepower is developed only at 100 percent rated motor speed with rated torque.

Shunt motors can be made to operate above base speed by field weakening.

- The motor is normally started with maximum field current to provide maximum flux for maximum starting torque.
- Decreasing the field current weakens the flux and causes the speed to rise.
- The reduction in field current results in less generated counter EMF and a greater armature current flow for a given motor load.
- One method for controlling field current is to insert a resistor in series with the field voltage source. This may be useful for trimming to an ideal motor speed for the application.
- Others methods use a variable-voltage field source.

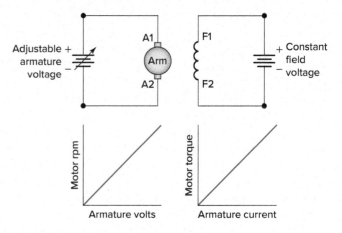

Figure 5-30 Armature-controlled DC motor.

Coordinated armature and field voltage control for extended speed range is illustrated in Figure 5-31. First the motor is armature voltage–controlled for constant-torque, variable-horsepower operation up to base speed. Once base speed is reached, field-weakening control is applied for constant-horsepower, variable-torque operation to the motor's maximum rated speed. **Caution: If a DC motor suffers a loss of field excitation current while operating, the motor will immediately begin to accelerate to the top speed that the loading will allow. This can result in the motor virtually flying apart if it is lightly loaded. For this reason some form of field loss protection must be provided in the motor control circuit that will automatically stop the motor in the event that current to the field circuit is lost or drops below a safe value.**

DC Motor Drives

In general, DC magnetic motor starters are intended to start and accelerate motors to normal speed and to provide protection against overloads. Unlike motor starters, **motor drives** are designed to provide, in addition to protection, precise control of the speed, torque, acceleration, deceleration, and direction of rotation of motors. In addition, many motor drive units are capable of high-speed communication with programmable logic controllers (PLCs) and other industrial controllers.

A motor drive is essentially an electronic device that uses different types of solid-state control techniques. Chapter 9, Motor Control Electronics, examines how these solid-state devices operate. Figure 5-32 shows the

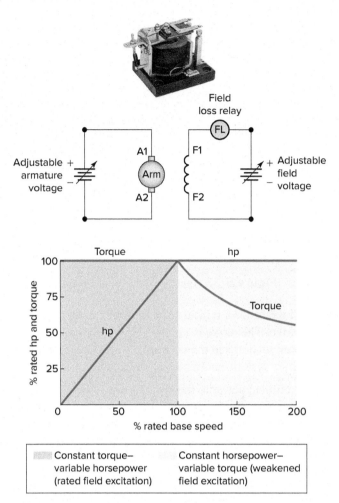

Figure 5-31 Armature- and field-controlled DC motor.
Photo courtesy of Jenkins Electric Company, www.jenkins.com

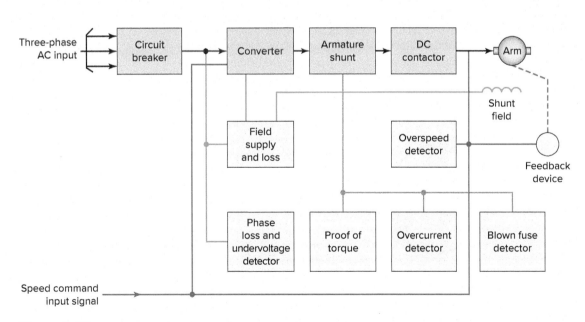

Figure 5-32 The block diagram for a typical DC motor drive.

Part 2 Direct Current Motors 119

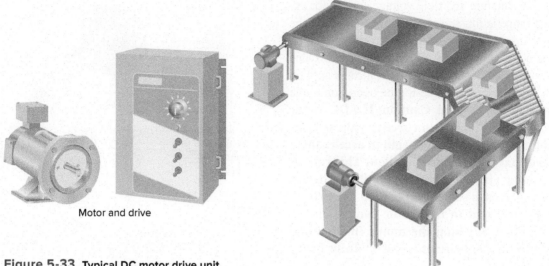

Figure 5-33 Typical DC motor drive unit.

block diagram for a typical DC electronic variable-speed motor drive. This drive is made up of two basic sections: the **power** section and the **control** section. The operation of the drive system can be summarized as follows:

- Controlled power to the DC motor is supplied from the power section, consisting of the circuit breaker, converter, armature shunt, and DC contactor.
- The converter rectifies the three-phase AC power, converting it to DC for the DC motor.
- Attaining precise control of the motor requires a means of evaluating the motor's performance and automatically compensating for any variations from the desired levels. This is the job of the control section, which is made up of the speed command input signal as well as various feedback and error signals that are used to control the output of the power section.

DC motor drives use a separately excited field because of the need to vary the armature voltage or the field current. When you vary the armature voltage, the motor produces full torque but the speed is varied. However, when the field current is varied, both the motor speed and the torque will vary. Figure 5-33 shows a typical DC motor drive unit used to provide very precise control over the operation of a conveyor system. In addition to managing motor speed and torque, it provides controlled acceleration and deceleration as well as forward and reverse motor operation.

Brushless DC Motors

As its name implies, **Brushless DC (BLDC)** motors do not use brushes. With brushed-type DC motors, the brushes deliver current through the commutator into the coils on the rotor. A BLDC has a rotor with permanent magnets and a stator with windings as illustrated in Figure 5-34:

- The electronic switching circuit function is similar to that of the mechanical commutator in a DC brushed motor.
- The stator coil windings (A, B, C) are DC powered through this electronic commutator that **sequentially** energizes the stator coils, generating a rotating magnetic field.

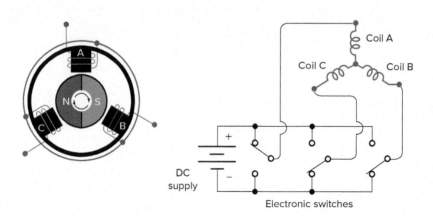

Figure 5-34 Brushless motor operation.

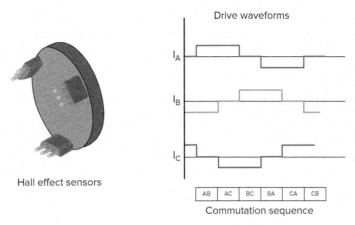

Figure 5-35 Hall effect sensor devices and drive waveform.

- The changing polarities of the rotating magnetic field react with the constant polarity of the permanent magnet rotor to produce the desired rotation.

In order to rotate the BLDC motor, the stator windings must be energized in sequence. It is important to know the rotor position in order to understand which winding will be energized following the energizing sequence. Rotor position is sensed using Hall-effect sensors embedded into the stator (Figure 5-35):

- Most BLDC motors have three Hall sensors inside the stator on the non-driving end of the motor.
- Whenever the rotor magnetic poles pass near the Hall sensors, they give a high or low signal indicating the N or S pole is passing near the sensors.
- Based on the combination of these three drive waveform sensor signals, the exact sequence of commutation can be determined.
- The Hall sensors are normally mounted on a printed circuit board and fixed to the enclosure cap on the non-driving end. This enables users to adjust the complete assembly of Hall sensors to align with the rotor magnets in order to achieve the best performance.

Characteristics of brushless DC motors include:

- Ability to develop high torque across their entire speed range. This feature is important for applications that require variable speed such as pumps and fans.
- Longevity and ease of maintenance due to a lack of a mechanical commutator. This reduction in the number of components results in fewer parts that wear out, break, need replacing, or require maintenance.
- Precision motor control, positioning, and mechanical displacement. This feature is utilized in operations such as servomotors and actuators for industrial robots.

Part 2 Review Questions

1. Give two reasons why DC motors are seldom the first motor of choice for some applications.
2. What special types of processes may warrant the use of a DC motor?
3. Explain the function of the commutator in the operation of a DC motor.
4. a. How is the direction of rotation of a permanent-magnet motor changed?
   b. How is the speed of a permanent-magnet motor controlled?
5. Summarize the torque and speed characteristics of a DC series motor.
6. Why should a DC series motor not be operated without some sort of a load coupled to it?
7. In what way is the shunt field winding of a shunt motor different from that of the series field winding of a series motor?
8. Compare the starting torque and load versus speed characteristics of the series motor to those of shunt wound motor.

9. How are the series and shunt field windings of the compound-wound DC motor connected relative to the armature?
10. In what way is a cumulative-compound motor connected?
11. Compare the torque and speed characteristics of a compound motor with those of the series and shunt motors.
12. How can the direction of rotation of a wound DC motor be changed?
13. Explain how counter EMF is produced in a DC motor.
14. A 5 hp, 230 V DC motor has an armature resistance of 0.1 Ω and a full-load armature current of 20 A. Determine
    a. the value of the armature current on starting.
    b. the value of the counter EMF with full load applied.
15. a. What is motor armature reaction?
    b. State three effects that armature reaction has on the operation of a DC motor.
16. Explain how interpoles minimize the effects of armature reaction.
17. a. A motor rated for 1750 rpm at no load has a 4 percent speed regulation. Calculate the speed of the motor with full load applied.
    b. In what way does a DC motor's armature resistance affect its speed regulation?
18. a. How is the base speed of a DC motor defined?
    b. How is the speed of a DC motor controlled below base speed?
    c. How is the speed of a DC motor controlled above base speed?
19. With armature voltage control of a DC shunt motor, what is the effect on the rated torque and horsepower when the armature voltage is increased?
20. With field current control of a shunt DC motor, what is the effect on the rated torque and horsepower when the armature voltage is increased?
21. Field loss protection must be provided for DC motors. Why?
22. List several control functions found on a DC motor drive that would not normally be provided by a traditional DC magnetic motor starter.
23. What is the function of the Hall effect sensors used in DC brushless motors?
24. List three characteristics of DC brushless motor applications.

PART 3 THREE-PHASE ALTERNATING CURRENT MOTORS

Rotating Magnetic Field

The main difference between AC and DC motors is that the magnetic field generated by the stator rotates in the case of AC motors. A **rotating magnetic field** is key to the operation of all AC motors. The principle is simple. A magnetic field in the stator is made to rotate electrically around and around in a circle. Another magnetic field in the rotor is made to follow the rotation of this field pattern by being attracted and repelled by the stator field. Because the rotor is free to turn, it follows the rotating magnetic field in the stator.

Figure 5-36 illustrates the concept of a rotating magnetic field as it applies to the stator of a three-phase AC motor. The operation can be summarized as follows:

- Three sets of windings are placed 120 mechanical degrees apart with each set connected to one phase of the three-phase power supply.
- When three-phase current passes through the stator windings, a rotating magnetic field effect is produced that travels around the inside of the stator core.
- Polarity of the rotating magnetic field is shown at six selected positions marked off at 60 degree intervals on the sine waves representing the current flowing in the three phases, A, B, and C.
- In the example shown, the magnetic field will rotate around the stator in a clockwise direction.
- Simply interchanging any two of the three-phase power input leads to the stator windings reverses the direction of rotation of the magnetic field.
- The number of poles is determined by how many times a phase winding appears. In this example, each winding appears twice, so this is a two-pole stator.

There are two ways to define AC motor speed. First is synchronous speed. The *synchronous speed* of an AC motor is the speed of the stator's magnetic field rotation. This is the motor's ideal theoretical, or mathematical, speed, since the rotor will always turn at a slightly

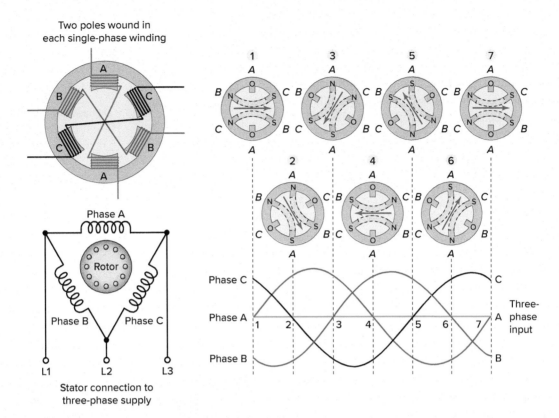

Figure 5-36 Rotating magnetic field.

slower rate. The other way motor speed is measured is called *actual speed*. This is the speed at which the shaft rotates. The nameplate of most AC motors lists the actual motor speed rather than the synchronous speed (Figure 5-37).

The speed of the rotating magnetic field varies directly with the frequency of the power supply and inversely with the number of poles constructed on the stator winding. This means the higher the frequency, the greater the speed and the greater the number of poles the slower the speed. Motors designed for 60 Hz use have synchronous speeds of 3,600, 1,800, 1,200, 900, 720, 600, 514, and 450 rpm.

The synchronous speed of an AC motor can be calculated by the formula:

$$S = \frac{120\,f}{P}$$

where

S = synchronous speed in rpm
f = frequency, Hz, of the power supply
P = number of poles wound in each of the single-phase windings

Example 5-3

Problem: Determine the synchronous speed of a four-pole AC motor connected to a 60 Hz electrical supply.

Solution:

$$S = \frac{120\,f}{P}$$

$$= \frac{120 \times 60}{4}$$

$$= 1{,}800 \text{ rpm}$$

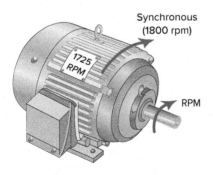

Figure 5-37 Synchronous and actual speed.

Part 3 Three-Phase Alternating Current Motors

Induction Motor

The AC **induction motor** is by far the most commonly used motor because it is relatively simple and can be built at less cost than other types. Induction motors are made in both three-phase and single-phase types. The induction motor is so named because no external voltage is applied to its rotor. There are no slip rings or any DC excitation supplied to the rotor. Instead, the AC current in the stator induces a voltage across an air gap and into the rotor winding to produce rotor current and associated magnetic field (Figure 5-38). The stator and rotor magnetic fields then interact and cause the rotor to turn.

A three-phase motor stator winding consists of three separate groups of coils, called **phases,** and designated A, B, and C. The phases are displaced from each other by 120 electrical degrees and contain the same number of coils, connected for the same number of poles. Poles refer to a coil or group of coils wound to produce a unit of magnetic polarity. The number of poles a stator is wound for will always be an even number and refers to the total number of north and south poles per phase. Figure 5-39 shows a typical connection of coils for a four-pole, three-phase Y-connected induction motor.

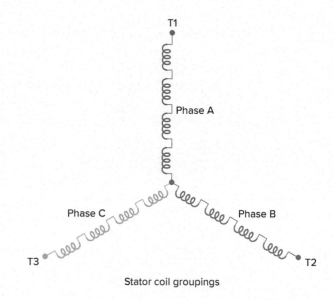

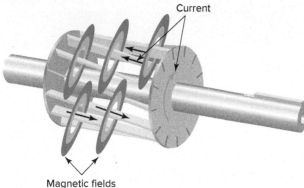

Figure 5-39 Stator coils for a Y-connected four-pole, three-phase inductor motor.
Photo courtesy of Swiger Coil LLC., www.swigercoil.com.

Squirrel-Cage Induction Motor

An induction motor rotor can be either wound rotor or a squirrel-cage rotor. The majority of commercial and industrial applications usually involve the use of a three-phase squirrel-cage induction motor. A typical squirrel-cage induction motor is shown in Figure 5-40.

- The rotor is constructed using a number of single bars short-circuited by end rings and arranged in a hamster-wheel or squirrel-cage configuration.
- When voltage is applied to the stator winding, a rotating magnetic field is established.
- This rotating magnetic field causes a voltage to be induced in the rotor, which, because the rotor bars are essentially single-turn coils, causes currents to flow in the rotor bars.
- These rotor currents establish their own magnetic field, which interacts with the stator magnetic field to produce a torque.

Figure 5-38 Induced rotor current.
Photo courtesy of Copper Development Association

124   Chapter 5   Electric Motors

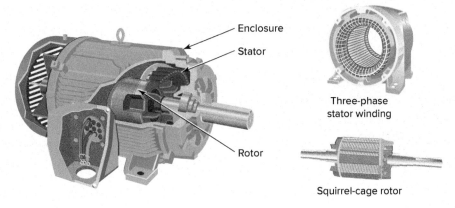

Figure 5-40 Squirrel-cage induction motor.

- The resultant production of torque spins the rotor in the same direction as the rotation of the magnetic field produced by the stator.
- The most common type of rotor has cast-aluminum conductors and short-circuiting end rings.

The resistance of the squirrel-cage rotor has an important effect on the operation of the motor. A high-resistance rotor develops a high starting torque at low starting current. A low-resistance rotor develops low slip and high efficiency at full load. Figure 5-41 shows how motor torque varies with rotor speed for three NEMA-type squirrel-cage induction motors:

NEMA Design B—Considered a standard type with normal starting torque, low starting current, and low slip at full load. Suitable for a broad variety of applications, such as fans and blowers, that require normal starting torque.

NEMA Design C—This type has higher than standard rotor resistance, which improves the rotor power factor at start, providing more starting torque. When loaded, however, this extra resistance causes a greater amount of slip. Used for equipment, such as a pump, that requires a high starting torque.

NEMA Design D—The even higher rotor resistance of this type produces a maximum amount of starting torque. This type is suitable for equipment with very high inertia starts such as cranes and hoists.

Operating characteristics of the squirrel-cage motor include the following:

- The motor normally operates at essentially constant speed, close to the synchronous speed.
- Large starting currents required by this motor can result in line voltage fluctuations.
- Interchanging any two of the three main power lines to the motor reverses the direction of rotation. Figure 5-42 shows the power circuit for reversing

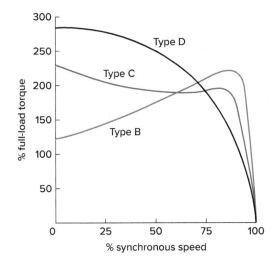

Figure 5-41 Typical squirrel-cage motor speed–torque characteristics.

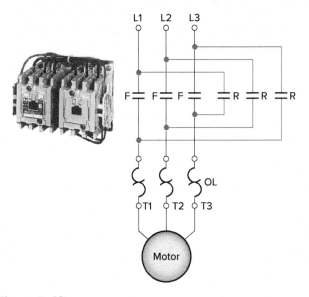

Figure 5-42 Power circuit for reversing a three-phase motor.
Photo courtesy of Eaton, www.eaton.com.

Part 3 Three-Phase Alternating Current Motors

a three-phase motor. The forward contacts F, when closed, connect L1, L2, and L3 to motor terminals T1, T2, and T3, respectively. The reverse contacts R, when closed, connect L1, L2, and L3 to motor terminals T3, T2, and T1, respectively, and the motor will now run in the opposite direction.

- Once started, the motor will continue to run, with a phase loss, as a single-phase motor. The current drawn from the remaining two lines will almost double, and the motor will overheat. The motor will not start from standstill if it has lost a phase.

The rotor does not revolve at synchronous speed, but tends to slip behind. Slip is what allows a motor to turn. If the rotor turned at the same speed at which the field rotates, there would be no relative motion between the rotor and the field and no voltage induced. Because the rotor slips with respect to the rotating magnetic field of the stator, voltage and current are induced in the rotor. The difference between the speed of the rotating magnetic field and the rotor in an induction motor is known as *slip* and is expressed as a percentage of the synchronous speed as follows:

$$\text{Percent slip} = \frac{\text{Synchronous speed} - \text{Actual speed}}{\text{Synchronous speed}} \times 100$$

The slip increases with load and is necessary to produce useful torque. The usual amount of slip in a 60 Hz, three-phase motor is 2 or 3 percent.

Example 5-4

Problem: Determine the percent slip of an induction motor having a synchronous speed of 1,800 rpm and a rated actual speed of 1,750 rpm.

Solution:

Percent slip

$$= \frac{\text{Synchronous speed} - \text{Actual speed}}{\text{Synchronous speed}} \times 100$$

$$= \frac{1,800 - 1,750}{1,800} \times 100$$

$$= 2.78\%$$

Loading of an induction motor is similar to that of a transformer in that the operation of both involves changing flux linkages with respect to a primary (stator) winding and secondary (rotor) winding.

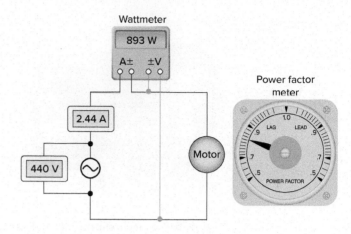

Figure 5-43 Motor power factor (PF).

- The no-load current is low and similar to the exciting current in a transformer.
- No-load current is composed of a magnetizing component that creates the revolving flux and a small active component that supplies the windage and friction losses in the rotor plus the iron losses in the stator.
- When the induction motor is under load, the rotor current develops a flux that opposes and, therefore, weakens the stator flux.
- This allows more current to flow in the stator windings, just as an increase in the current in the secondary of a transformer results in a corresponding increase in the primary current.

You may recall that **power factor (PF),** Figure 5-43, is defined as the ratio of the actual (or true) power (watts) to the apparent power (volt-amperes) and is a measure of how effectively the current drawn by a motor is converted into useful work. The motor exciting current and reactive power under load remain about the same as at no load. For this reason, whenever a motor is operating with no load, the power factor is very low in comparison to when it is operating at full load. At full load the PF ranges from 70 percent for small motors to 90 percent for larger motors. Induction motors operate at their peak efficiency if they are sized correctly for the load that they will drive. Oversized motors not only operate inefficiently, but they also carry a higher first cost than right-sized units.

The moment a motor is started, during the acceleration period, the motor draws a high inrush current. This inrush current is also called the **locked-rotor current.** Common induction motors, started at rated voltage, have locked-rotor starting currents of up to 6 times their nameplate full-load current. The locked-rotor current depends largely on

Figure 5-44 Power quality analyzer.
Photo courtesy of Fluke, www.fluke.com. Reproduced with Permission.

the type of rotor bar design and can be determined from the NEMA design code letters listed on the nameplate. High locked-rotor motor current can create voltage sags or dips in the power lines, which may cause objectionable light flicker and problems with other operating equipment (Figure 5-44). Also, a motor that draws excessive current under locked-rotor conditions is more likely to cause nuisance tripping of protection devices during motor start-ups.

A single-speed motor has one rated speed at which it runs when supplied with the nameplate voltage and frequency. A multispeed motor will run at more than one speed, depending on how the windings are connected to form a different number of magnetic poles. Two-speed, single-winding motors are called **consequent pole** motors. The low speed on a single-winding consequent pole motor is always one-half of the higher speed. If requirements dictate speeds of any other ratio, a two-winding motor must be used. With **separate winding** motors, a separate winding is installed in the motor for each desired speed.

Consequent pole single-winding motors have their stator windings arranged so that the number of poles can be changed by reversing some of the coil currents. Figure 5-45 shows a dual-speed three-phase squirrel-cage single-winding motor with six stator leads brought out. By making the designated connections to these leads, the windings can be connected in series delta or parallel wye. The series delta connection results in low speed and the parallel wye in high speed. The torque rating would be the same at both speeds. If the winding is such that the series delta connection gives the high speed and the parallel wye connection the low speed, the horsepower rating is the same at both speeds.

Single-speed AC induction motors are frequently supplied with multiple external leads for various voltage ratings in fixed-frequency applications. The multiple leads may be designed to allow either series to parallel reconnections, wye to delta reconnections, or combinations of these. Figure 5-46 shows typical connections for dual-voltage wye and delta series and parallel reconnections. These types of reconnections should not be confused with the reconnection of multispeed polyphase induction motors. In the case of multispeed motors, the reconnection results in a motor with a different number of magnetic poles and therefore a different synchronous speed at a given frequency.

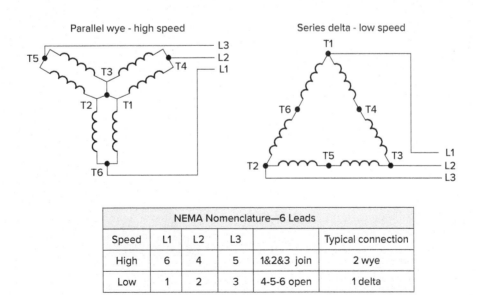

Figure 5-45 Dual-speed, three-phase squirrel-cage single-winding motor.

Part 3 Three-Phase Alternating Current Motors

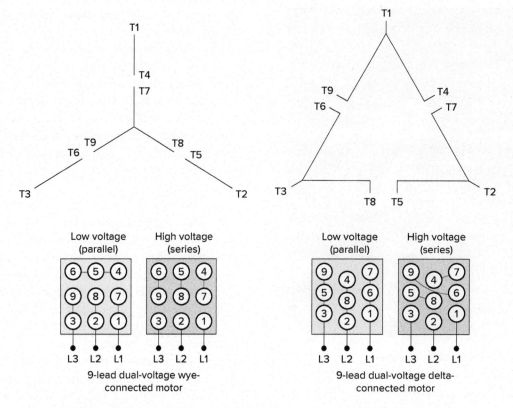

Figure 5-46 Typical connections for dual-voltage wye and delta series and parallel reconnections.

Wound-Rotor Induction Motor

The wound-rotor induction motor (sometimes called a **slip-ring** motor) is a variation on the standard cage induction motors. Wound-rotor motors have a three-phase winding wound on the rotor, which is terminated to slip rings as illustrated in Figure 5-47. The operation of the motor can be summarized as follows.

- The rotor slip rings connect to start-up resistors in order to provide current and speed control on start-up.
- When the motor is started, the frequency of current flowing through the rotor windings is nearly 60 Hz.
- Once up to full speed, the rotor current frequency drops down below 10 Hz to nearly a DC signal.
- The motor is normally started with full external resistance in the rotor circuit that is gradually reduced to zero, either manually or automatically.
- This results in a very high starting torque from zero speed to full speed at a relatively low starting current.
- With zero external resistance, the wound-rotor motor characteristics approach those of the squirrel-cage motor.
- Interchanging any two stator voltage supply leads reverses the direction of rotation.

A wound-rotor motor is used for constant-speed applications requiring a heavier starting torque than is obtainable with the squirrel-cage type. With a high-inertia load,

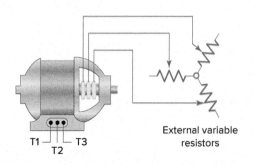

Figure 5-47 Wound-rotor induction motor.
Photo courtesy of TECO-Westinghouse Motor Co., http://www.tecowestinghouse.com/

a standard squirrel-cage induction motor may suffer rotor damage on starting due to the power dissipated by the rotor. With the wound rotor motor, the secondary resistors can be selected to provide the optimum torque curves and they can be sized to withstand the load energy without failure. Starting a high-inertia load with a standard squirrel-cage motor would require between 400 and 550 percent start current for up to 60 seconds. Starting the same machine with a wound-rotor motor (slip-ring motor) would require around 200 percent current for around 20 seconds. For this reason, wound rotor types are frequently used instead of the squirrel-cage types in larger sizes.

Wound-rotor motors are also used for variable-speed service. To use a wound-rotor motor as an adjustable-speed drive, the rotor control resistors must be rated for continuous current. If the motor is used only for a slow acceleration or high starting torque but then operates at its maximum speed for the duration of the work cycle, then the resistors will be removed from the circuit when the motor is at rated speed. In that case they will have been duty cycle–rated for starting duty only. Speed varies with this load, so that they should not be used where constant speed at each control setting is required, as for machine tools.

Three-Phase Synchronous Motor

All induction motors are **asynchronous** motors. The asynchronous nature of induction motor operation comes from the slip between the rotational speed of the stator field and somewhat slower speed of the rotor. The three-phase *synchronous motor* is a unique and specialized motor. As the name suggests, this motor runs at a constant speed from no load to full load in synchrony with the line frequency. As in squirrel-cage induction motors, the speed of a synchronous motor is determined by the number of pairs of poles and the line frequency.

A typical three-phase synchronous motor is shown in Figure 5-48. The operation of the motor can be summarized as follows.

- Three-phase AC voltage is applied to the stator windings and a rotating magnetic field is produced.
- DC voltage is applied to the rotor winding and a second magnetic field is produced.
- The rotor then acts like a magnet and is attracted by the rotating stator field.
- This attraction exerts a torque on the rotor and causes it to rotate at the synchronous speed of the rotating stator field.
- The rotor does not require the magnetic induction from the stator field for its excitation. As a result, the motor has zero slip compared to the induction motor, which requires slip in order to produce torque.

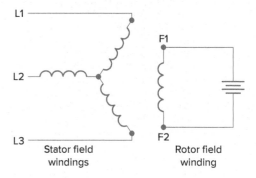

Figure 5-48 Three-phase synchronous motor.
Photo courtesy of ABB, www.abb.com

Synchronous motors are **not self-starting** and therefore require a method of bringing the rotor up to near synchronous speed before the rotor DC power is applied. Synchronous motors typically start as a normal squirrel-cage induction motor through use of special rotor amortisseur windings (Figure 5-49). Also, there are two basic methods of providing excitation current to the rotor. One method is to use an external DC source with current supplied to the windings through slip rings. The other method is to have the exciter mounted on the common shaft of the motor. This arrangement does not require the use of slip rings and brushes.

An electrical system's lagging power factor can be corrected by overexciting the rotor of a synchronous motor operating within the same system. This will produce a

Figure 5-49 Synchronous motor rotor with Amortisseur starting winding.
General Electric

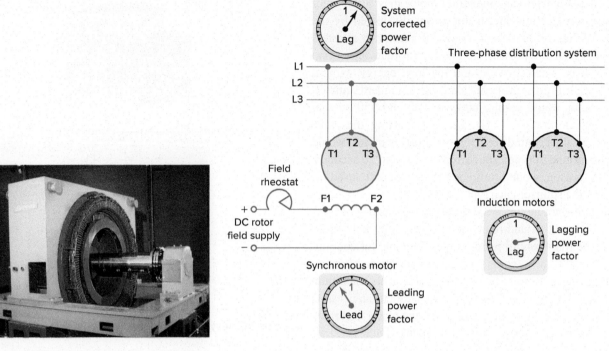

Figure 5-50 Synchronous motor used to correct power factor.
Photo courtesy of TECO-Westinghouse Motor Co, http://www.tecowestinghouse.com/

leading power factor, canceling out the lagging power factor of the inductive loads (Figure 5-50). An underexcited DC field will produce a lagging power factor and for this reason is seldom used. When the field is normally excited, the synchronous motor will run at a unity power factor. Three-phase synchronous motors can be used for power factor correction while at the same time performing a major function, such as operating a compressor. If mechanical power output is not needed, however, or can be provided in other cost-effective ways, the synchronous machine remains useful as a "nonmotor" means of controlling power factor. It does the same job as a bank of static capacitors. Such a machine is called a **synchronous condenser** or capacitor.

Part 3 Review Questions

1. A rotating magnetic field is the key to the operation of AC motors. Give a brief explanation of its principle of operation.
2. Compare synchronous speed and actual speed of an AC motor.
3. Calculate the synchronous speed of a six-pole AC motor operated from a standard voltage source.
4. Why is the induction motor so named?
5. Outline the operating principle of a three-phase squirrel-cage induction motor.
6. Explain what effect rotor resistance has on the operation of a squirrel-cage induction motor.
7. How is the direction of rotation of a squirrel-cage motor reversed?
8. If, while a three-phase induction motor is operating, power to one phase of its squirrel cage is lost, what will happen?
9. Define the term *slip* as it applies to an induction motor.
10. Calculate the percent slip of an induction motor having a synchronous speed of 3,600 rpm and a rated actual speed of 3,435 rpm.
11. What effect does loading have on the power factor of an AC motor?
12. What is the typical value of the locked-rotor motor current?
13. How is the speed of an induction motor determined?

14. Explain the difference between multispeed consequent pole and separate winding induction motors.
15. A wound-rotor induction motor is normally started with full external resistance in the rotor circuit that is gradually reduced to zero. How does this affect starting torque and current?
16. How is the direction of rotation of a wound-rotor induction motor changed?
17. When a wound-rotor motor is used as an adjustable-speed drive, rather than only for starting purposes, what must the duty cycle of the rotor resistors be rated for?
18. State two advantages of using three-phase synchronous motor drives in an industrial plant.

PART 4 SINGLE-PHASE ALTERNATING CURRENT MOTORS

Most home and business appliances operate on single-phase AC power. For this reason, single-phase AC motors are in widespread use. A single-phase induction motor is larger in size, for the same horsepower, than a three-phase motor. When running, the torque produced by a single-phase motor is pulsating and irregular, contributing to a much lower power factor and efficiency than that of a polyphase motor. Single-phase AC motors are generally available in the fractional to 10 hp range and all use a solid squirrel-cage rotor.

The single-phase induction motor operates on the principle of induction, just as does a three-phase motor. Unlike three-phase motors, they are not self-starting. Whereas a three-phase induction motor sets up a rotating field that can start the motor, a single-phase motor needs an **auxiliary** means of starting. Once a single-phase induction motor is running, it develops a rotating magnetic field. However, before the rotor begins to turn, the stator produces only a pulsating, stationary field.

A single-phase motor could be started by mechanically spinning the rotor, and then quickly applying power. However, normally these motors use some sort of automatic starting. Single-phase induction motors are classified by their start and run characteristics. The three basic types of single-phase induction motors are the split-phase, split-phase capacitor, and shaded-pole.

Split-Phase Motor

A single-phase split-phase induction motor uses a squirrel-cage rotor that is identical to that in a three-phase motor. Figure 5-51 shows the construction and wiring of a split-phase motor. To produce a rotating magnetic field, the single-phase current is split by two windings, the main **running** winding and an auxiliary **starting** winding, which is displaced in the stator 90 mechanical degrees from the running winding. The starting winding is connected in series with a switch, centrifugally or electrically operated, to disconnect it when the starting speed reaches about **75 percent** of full-load speed.

Phase displacement is accomplished by the difference in inductive reactance of the start and run windings as well as the physical displacement of the windings in the stator. The starting winding is wound on the top of the stator slots with fewer turns of smaller-diameter wire. The running winding has many turns of large-diameter wire wound in the bottom of the stator slots that give it a higher inductive reactance than the starting winding.

The way in which the two windings of a split-phase motor produce a rotating magnetic field is illustrated in Figure 5-52 and can be summarized as follows.

- When AC line voltage is applied, the current in the starting winding leads the current in the running winding by approximately 45 mechanical degrees.
- Since the magnetism produced by these currents follows the same wave pattern, the two sine waves can be thought of as the waveforms of the electromagnetism produced by the two windings.
- As the alternations in current (and magnetism) continue, the position of the north and south poles changes in what appears to be a clockwise rotation.

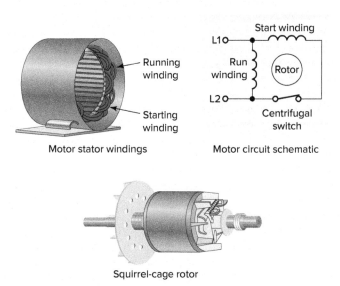

Figure 5-51 Split-phase induction motor.

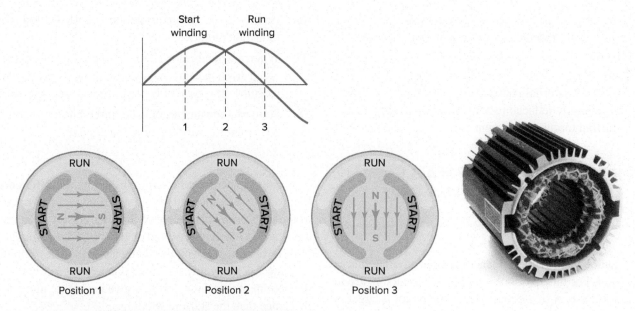

Figure 5-52 Rotating magnetic field of a split-phase motor.
Photo courtesy of Bodine Electric Company. www.bodine-electric.com

- At the same time the rotating field cuts the squirrel-cage conductors of the rotor and induces a current in them.
- This current creates magnetic poles in the rotor, which interact with the poles of the stator rotating magnetic field to produce motor torque.

Once the motor is running, the starting winding must be removed from the circuit. Since the starting winding is of a smaller gauge size, continuous current through it would cause the winding to burn out. Either a mechanical centrifugal or electronic solid-state switch may be used to automatically disconnect the starting winding from the circuit. The operation of a centrifugal-type switch is illustrated in Figure 5-53.

- It consists of a centrifugal mechanism, which rotates on the motor shaft and interacts with a fixed stationary switch whose contacts are connected in series with the start winding.
- When the motor approaches its normal operating speed, centrifugal force overcomes the spring force, allowing the contacts to open and disconnect the starting winding from the power source.
- The motor then continues operating solely on its running winding.
- Motors using such a centrifugal switch make a distinct clicking noise when starting and stopping as the centrifugal switch opens and closes.

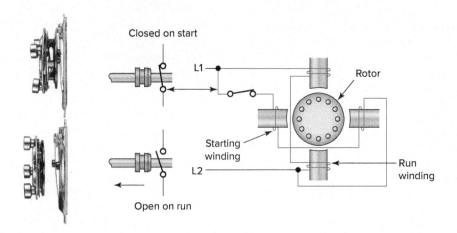

Figure 5-53 Centrifugal switch operation.
TORQ CORPORATION, 1956

132   Chapter 5   Electric Motors

The centrifugal switch can be a source of trouble if it fails to operate properly. Should the switch fail to **close** when the motor stops, the starting winding circuit will be open. As a result, when the motor circuit is again energized, the motor will not turn but will simply produce a low humming sound. Normally the starting winding is designed for operation across line voltage for only a short interval during starting. Failure of the centrifugal switch to **open** within a few seconds of starting may cause the starting winding to char or burn out.

The split-phase induction motor is the simplest and most common type of single-phase motor. Its simple design makes it typically less expensive than other single-phase motor types. Split-phase motors are considered to have low or moderate starting torque. Typical sizes range up to about ½ horsepower. Reversing the leads to either the start or run windings, but not to both, changes the direction of rotation of a split-phase motor. Popular applications of split-phase motors include fans, blowers, office machines, and tools such as small saws or drill presses where the load is applied after the motor has obtained its operating speed.

Dual-voltage split-phase motors have leads that allow external connection for different line voltages. Figure 5-54 shows a NEMA-standard single-phase motor with dual-voltage run windings. When the motor is operated at low voltage, the two run windings and the start winding are all connected in parallel. For high-voltage operation, the two run windings connect in series and the start winding is connected in parallel with one of the run windings.

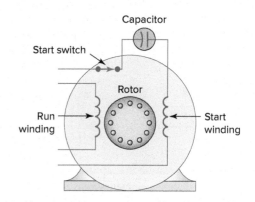

Figure 5-55 Capacitor start motor.
Photo courtesy of Leeson, www.leeson.com

Split-Phase Capacitor Motor

The **capacitor-start motor,** illustrated in Figure 5-55, is a modified split-phase motor. A capacitor connected in series with the starting winding creates a phase shift of approximately 80 degrees between the starting and running winding. This is substantially higher than the 45 degrees of a split-phase motor and results in a higher starting torque. Capacitor-start motors provide more than double the **starting torque** with one-third less starting current than the split-phase motor. Like the split-phase motor, the capacitor start motor also has a starting mechanism, either a mechanical centrifugal switch or solid-state electronic switch. This disconnects not only the start winding, but also the capacitor when the motor reaches about 75 percent of rated speed.

The capacitor-start motor is more expensive than a comparable split-phase design because of the additional cost of the start capacitor. However, the application range is much wider because of higher starting torque and lower starting current. The job of the capacitor is to improve the starting torque and not the power factor, as it is only in the circuit for a few seconds at the instant of starting. The capacitor can be a source of trouble if it becomes short-circuited or open-circuited. A **short-circuited** capacitor will cause an excessive amount of current to flow through the staring winding, while an **open** capacitor will cause the motor to not start.

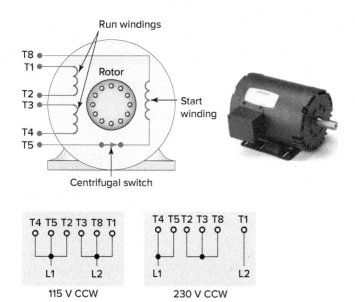

Figure 5-54 Dual-voltage split-phase motor connections.
Photo courtesy of Leeson, www.leeson.com

Part 4 Single-Phase Alternating Current Motors 133

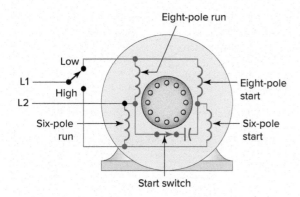

Figure 5-56 Two-speed capacitor-start motor wound.

Dual-speed capacitor-start motors have leads that allow external connection for low and high speeds. Figure 5-56 shows the wiring diagram for a typical two-speed capacitor-start motor wound with two sets of start and run windings. For low-speed 900-rpm operation, the eight-pole set of start and run windings connects to the source, and for high-speed 1,200 rpm, the six-pole set is used.

The **permanent-capacitor** motor has neither a centrifugal switch nor a capacitor strictly for starting. Instead, it has a run-type capacitor permanently connected in series with the start winding. This makes the start winding an auxiliary winding once the motor reaches running speed. Because the run capacitor must be designed for continuous use, it cannot provide the starting boost of the capacitor-start motor. Typical starting torques for permanent-capacitor motors are low, from 30 to 150 percent of rated load, so these motors are not suited for hard-to-start applications.

Permanent-capacitor motors are considered to be the most reliable of the single-phase motors, mostly because no starting switch is needed. The run and auxiliary windings are identical in this type of motor, allowing for the motor to be reversed by switching the capacitor from one winding to the other, as illustrated in Figure 5-57. Single-phase motors run in the direction in which they are started, so whichever winding has the capacitor connected to it will control the direction. Permanent split-capacitor motors have a wide variety of applications that include fans, blowers with low starting torque needs, and intermittent cycling uses such as adjusting mechanisms, gate operators, and garage door openers, many of which also need instant reversing. Since the capacitor is used at all times, it also provides improvement of motor power factor.

The **capacitor-start/capacitor-run motor,** shown in Figure 5-58, uses both start and run capacitors located in the housing connected to the top of the motor.

- When the motor is started, the two capacitors are connected in parallel to produce a large amount of capacitance and starting torque.
- Once the motor is up to speed, the start switch disconnects the start capacitor from the circuit.
- The motor start capacitor is typically an electrolytic type, while the run capacitor is an oil-filled type.
- The electrolytic type offers a large amount of capacitance when compared to its oil-filled counterpart.
- **It is important to note that these two capacitors are not interchangeable because an electrolytic capacitor used in an AC circuit for more than a few seconds will overheat.**

Capacitor start/capacitor-run motors operate at lower full-load currents and higher efficiency. Among other things, this means they run at lower temperature than other single-phase motor types of comparable horsepower. Their main disadvantage is a higher price, which is mostly the result of more capacitors, plus a starting switch. Capacitor start/capacitor-run motors are used over a wide range of single-phase applications, primarily starting hard loads that include woodworking machinery, air compressors, high-pressure water pumps, vacuum pumps, and other high-torque applications. They are available in sizes from ½ to 25 horsepower.

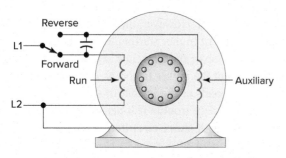

Figure 5-57 Reversible permanent-capacitor motor connection.

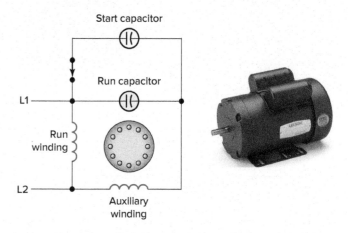

Figure 5-58 Capacitor-start/capacitor-run motor.
Photo courtesy of Leeson, www.leeson.com

Shaded-Pole Motor

Unlike other types of single-phase motors, shaded-pole motors have only one main winding and no start winding or switch. As in other induction motors, the rotating part is a squirrel-cage rotor. Starting is by means of a design that uses a continuous **copper loop** around a small portion of each motor pole, as illustrated in Figure 5-59.

- Currents in this copper loop delay the phase of magnetic flux in that part of the pole enough to provide a rotating field.
- This rotating field effect produces a very low starting torque compared with other classes of single-phase motors.
- Although the direction of rotation is not normally reversible, some shaded-pole motors are wound with two main windings that reverse the direction of the field.
- Slip in the shaded-pole motor is not a problem because the current in the stator is not controlled by a countervoltage determined by rotor speed as in other types of single-phase motors. Speed can, therefore, be controlled merely by varying voltage or through a multitap winding.

Shaded-pole motors are best suited to low-power household application, because the motors have low starting torque and efficiency ratings. Because of the weak starting torque, shaded-pole motors are built only in small sizes ranging from 1/20 to 1/6 hp. Applications for this type of motor include fans, can openers, blowers, and electric razors.

Universal Motor

The **universal motor,** shown in Figure 5-60, is constructed like a series-type DC motor with a wound series field (on the stator) and a wound armature (on the rotor). As in the DC series motor, its armature and field coils are connected in series. As the name implies, universal motors can be operated with either direct current or single-phase alternating current. The reason for this is that a DC motor will continue to turn in the same direction if the current through the armature and field are reversed at the

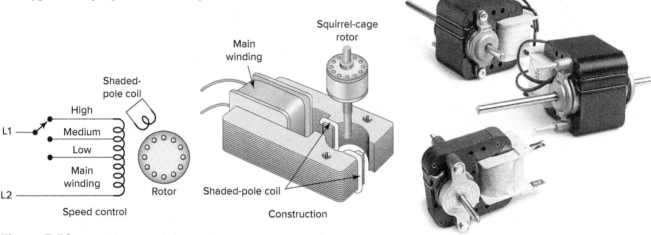

Figure 5-59 Shaded-pole motor.
Photo Courtesy of Mamco Corporation. www.mamcomotors.com

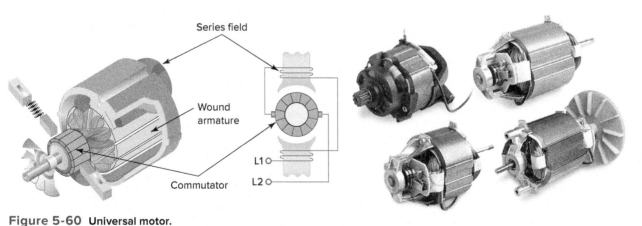

Figure 5-60 Universal motor.
Photo Courtesy of Mamco Corporation. www.mamcomotors.com

same time. This is exactly what happens when the motor is connected to an AC source. Universal motors are also known as AC series motors or AC commutator motors.

Although universal motors are designed to run on AC or DC, most are used for household appliances and portable hand tools that operate on single-phase AC power. Unlike other types of single-phase motors, universal motors can easily exceed one revolution per cycle of the main current. This makes them useful for appliances such as blenders, vacuum cleaners, and hair dryers, where high-speed operation is desired. The speed of the universal motor, like that of DC series motor, varies considerably from no load to full load, as can be observed when you apply varying pressure on a universal drill motor.

Both the speed and direction of rotation of a universal motor can be controlled, as illustrated in Figure 5-61.

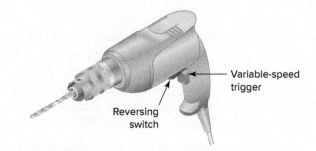

Figure 5-61 Motor speed and direction controls.

Reversing is accomplished just as in a DC series motor by reversing the current flow through the armature with respect to the series field. Varying the voltage that is applied to the motor controls the speed.

Part 4 Review Questions

1. What is the major difference between the starting requirements for a three-phase and a single-phase induction motor?
2. **a.** Outline the starting sequence for a split-phase induction motor.
   b. How is its direction of rotation reversed?
3. Dual-voltage split-phase motors have leads that allow external connection for different line voltages. How are the start and run windings connected for high and low line voltages?
4. What is the main advantage of capacitor motors over split-phase types?
5. Name the three types of capacitor motor designs.
6. Explain how the shaded-pole motor is started.
7. What type of DC motor is constructed like a universal motor?

PART 5 ALTERNATING CURRENT MOTOR DRIVES

AC drives, such as the one shown in Figure 5-62, connect to standard AC induction motors and have capabilities of adjustable speed, torque, and horsepower control similar to those of DC drives. Adjustable-speed drives have made AC squirrel-cage induction motors as controllable and efficient as their DC counterparts. AC induction motor speed depends on the number of motor poles and the frequency of the applied power. The number of poles on the stator of the motor could be increased or decreased, but this has limited usefulness. Although the AC frequency of the power source in the United States is fixed at 60 Hz, advances in power electronics make it practical to vary the frequency and resulting speed of an induction motor.

Variable-Frequency Drive

A **variable-frequency drive** (VFD) system, also known as a variable-speed drive system, generally consists of an AC motor, a controller, and an operator interface. Three-phase motors are usually preferred, but some types of single-phase motors can be used. Motors that are designed for fixed-speed main voltage operation are often used, but certain enhancements to the standard motor designs offer higher reliability and better VFD performance. A simplified diagram of a VFD controller is shown in Figure 5-63. The three major sections of the controller are as follows:

Converter—Rectifies the incoming three-phase AC power and converts it into DC.

DC filter (also known as the DC link or DC bus)—Provides a smooth, rectified DC voltage.

Inverter—Switches the DC on and off so rapidly that the motor receives a pulsating voltage that is similar to AC. The switching rate is controlled to vary the frequency of the simulated AC that is applied to the motor.

AC motor characteristics require the applied voltage to be proportionally adjusted by the drive whenever the frequency is changed. For example, if a motor is designed to

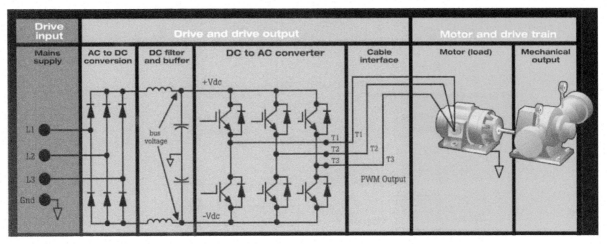

Figure 5-62 AC motor drive.
Photo courtesy of Rockwell Automation, www.rockwellautomation.com

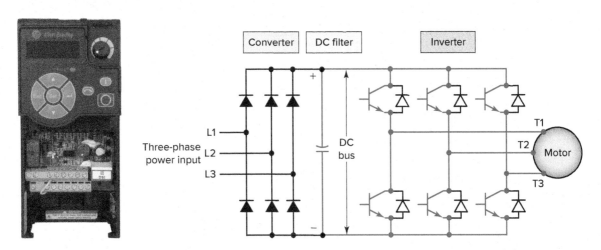

Figure 5-63 Variable-frequency drive controller.
Photo courtesy of Rockwell Automation, www.rockwellautomation.com

Part 5   Alternating Current Motor Drives   137

operate at 460 Volts at 60 Hz, the applied voltage must be reduced to 230 Volts when the frequency is reduced to 30 Hz, as illustrated in Figure 5-64. Thus, the ratio of volts per hertz must be regulated to a constant value (460/60 = 7.67 in this case). The most common method used for adjusting the motor voltage is called **pulse width modulation (PWM)**. With PWM voltage control, the inverter switches are used to divide the simulated sine-wave output waveform into a series of narrow voltage pulses and modulate the width of the pulses.

With a standard AC across-the-line motor starter, line voltage and frequency are applied to the motor and the speed is solely dependent on the number of motor stator poles (Figure 5-65). In comparison, an AC motor drive

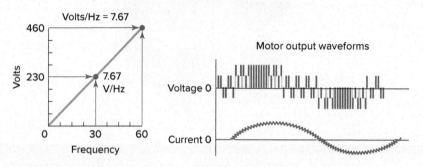

Figure 5-64 The ratio of volts per hertz is regulated to a constant value.

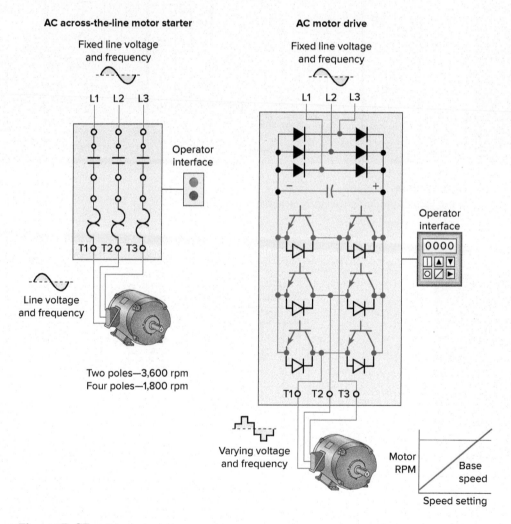

Figure 5-65 AC motor starter and drive control.

138   Chapter 5   Electric Motors

delivers a varying voltage and frequency to the motor, which determines its speed. The higher the frequency supplied to the motor, the faster it will run. Power applied to the motor through the drive can lower the speed of a motor below the nameplate base speed or increase the speed to synchronous speed and higher. Motor manufacturers list the maximum speed at which their motors can safely be operated.

Inverter Duty Motor

Inverter duty and **vector duty** describe a class of AC induction motors that are specifically designed for use with variable-frequency drives (Figure 5-66). The high switching frequencies and fast voltage rise times of AC motor drives can produce high voltage peaks in the windings of standard AC motors that exceed their insulation breakdown voltage. Also, operating motors for an extended time at low motor rpm reduces the flow of cooling air, which results in an increase in temperature. NEMA-rated inverter- or vector-duty motors use high-temperature insulating materials that can withstand higher voltage spikes and operating temperatures. This reduces the stress on the insulation system.

AC motors frequently drive variable loads such as pumps (Figure 5-67), hydraulic systems, and fans. In these applications, motor efficiency is often poor due to operation at low loads and can be improved by using a VFD in place of speed controllers such as belts and pulleys, throttle valves, fan dampers, and magnetic clutches. For example, a pump or fan, controlled by a variable-speed drive, running at half-speed consumes only one-eighth of the energy compared to one running at full speed, resulting in considerable energy savings.

Figure 5-67 In-line pump with integrated variable-frequency drive.
Taco, Inc., 2014

Figure 5-66 Inverter-duty AC induction motor.
Photo courtesy of Leeson, www.leeson.com

Part 5 Review Questions

1. List the three basic sections of an AC variable-frequency drive controller and state the function of each section.
2. An induction motor rated for 230 Volts at 60 Hz is to be operated by a VFD. When the frequency is reduced to 20 Hz, to what value must the voltage be reduced in order maintain the ratio of volts per hertz?
3. How does an AC drive vary the speed of an induction motor?
4. Inverter or vector duty AC induction motors are the types most often specified for use with variable-frequency drives. Why?

PART 6 MOTOR SELECTION

AC and DC motors come in many shapes and sizes. Some are standardized electric motors for general-purpose applications. Other electric motors are intended for specific tasks. In any case, electric motors should be selected to satisfy the requirements of the machines on which they are applied without exceeding rated electric motor temperature. The following are some of the important motor and load parameters that need to be considered as part of the selection process.

Mechanical Power Rating

The *mechanical power rating* of motors is expressed in either horsepower (hp) or watts (W): 1 hp = 746 W. Two important factors that determine mechanical power output are torque and speed. Torque and speed are related to horsepower by a basic formula, which states that:

$$\text{Horsepower} = \frac{\text{Torque} \times \text{Speed}}{\text{Constant}}$$

where

Torque is expressed in lb-ft.
Speed is expressed in rpm.

The value of the constant depends on the units that are used for torque. For this combination, the constant is 5,252.

The slower the motor operates, the more torque it must produce to deliver the same amount of horsepower. To withstand the greater torque, slow motors need stronger components than those of higher-speed motors of the same power rating. For this reason, slower motors are generally larger, heavier, and more expensive than faster motors of equivalent horsepower rating.

Current

Full-load amperes is the amount of amperes the motor can be expected to draw under full-load (torque) conditions and is also known as the *nameplate amperes*. The nameplate full-load current of the motor is used in determining the size of overload-sensing elements for the motor circuit.

Locked-rotor current is the amount of current the motor can be expected to draw under starting conditions when full voltage is applied and is also known as the *starting inrush current*.

Service-factor amperes is the amount of current the motor will draw when it is subjected to a percentage of overload equal to the service factor on the nameplate of the motor. For example, a service factor of 1.15 on the nameplate means the motor will handle 115 percent of normal running current indefinitely without damage.

Code Letter

NEMA **code letters** are assigned to motors for calculating the locked-rotor current based upon the kilovolt-amperes per nameplate horsepower. Overcurrent protection devices must be set above the locked-rotor current of the motor to prevent the overcurrent protection device from opening when the rotor of the motor is starting. The letters range in alphabetical order from A to V in increasing value of locked-rotor (LR) current.

Locked-Rotor Code, kVA/hp

A	.01–3.15	G	5.6–6.3
B	3.15–3.55	H	6.3–7.1
C	3.55–4.0	J	7.1–8.0
D	4.0–4.5	K	8.0–9.0
E	4.5–5.0	L	9.0–10.0
F	5.0–5.6	M	10.0–11.2

LR current (single-phase motors)

$$= \frac{\text{Code letter value} \times \text{hp} \times 1{,}000}{\text{Rated voltage}}$$

LR current (three-phase motors)

$$= \frac{\text{Code letter value} \times \text{hp} \times 577}{\text{Rated voltage}}$$

Design Letter

NEMA has defined four standard motor designs for AC motors, using the letters A, B, C, and D to meet specific requirements posed by different application loads. The **design letter** denotes the motor's performance characteristics relating to torque, starting current, and slip. Design B is the most common design. It has relatively high starting torque with reasonable starting currents. The other designs are used only on fairly specialized applications.

Efficiency

Motor **efficiency** is the ratio of mechanical power output to the electrical power input, usually expressed as a percentage. The power input to the motor is either transferred to the shaft as power output or is lost as heat through the body of the motor. Power losses associated with the operation of a motor include:

Core loss, which represents the energy required to magnetize the core material (known as hysteresis) and losses owing to the creation of small electric currents that flow in the core (known as eddy currents).

Stator and rotor resistance losses, which represent the I^2R heating loss due to current flow (I) through the resistance (R) of the stator and rotor windings, and are also known as *copper losses*.

Mechanical losses, which include friction in the motor bearings and the fan for air cooling.

Stray losses, which are the losses that remain after primary copper and secondary losses, core losses, and mechanical losses. The largest contributor to the stray losses is harmonic energy generated when the motor

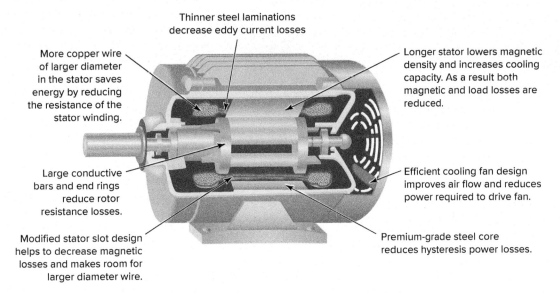

Figure 5-68 Typical energy-efficient motor.

operates under load. This energy is dissipated as currents in the copper windings, harmonic flux components in the iron parts, and leakage in the laminate core.

Energy-Efficient Motors

The efficiency of electric motors ranges between 75 and 98 percent. **Energy-efficient motors** use less energy because they are manufactured with higher-quality materials and techniques, as illustrated in Figure 5-68. To be considered energy-efficient, a motor's performance must equal or exceed the nominal full-load efficiency values provided by NEMA in publication MG-1.

Frame Size

Motors come in various **frame sizes** to match the requirements of the application. In general, the frame size gets larger with increasing horsepower or with decreasing speeds. In order to promote standardization in the motor industry, NEMA prescribes standard frame sizes for certain dimensions of standard motors. As an example, a motor with a frame size of 56 will always have a shaft height above the base of 3½ inches.

Frequency

This is the **frequency** of the line power supply for which an AC motor is designed to operate. Electric motors in North America are designed to operate on 60 Hz power, whereas most of the rest of the world uses 50 Hz. It is important to make sure equipment designed to operate on 50 Hz is properly designed or converted to provide good service life at 60 Hz. As an example, a three-phase change in frequency from 50 to 60 Hz can result in a 20 percent increase in rotor rpm.

Full-Load Speed

Full-load speed represents the approximate speed at which the motor will run when it is supplying full rated torque or horsepower. As an example, a typical four-pole motor running on 60 Hz might have a nameplate rating of 1,725 rpm at full load, while its synchronous speed is 1,800 rpm.

Load Requirements

Load requirements must be considered in selecting the correct motor for a given application. This is especially true in applications that require speed control. Important requirements a motor must meet in controlling a load are torque and horsepower in relation to speed.

Constant-torque loads—With a constant torque, the load is constant throughout the speed range, as illustrated in Figure 5-69. As speed increases, the torque required remains constant while the horsepower increases or decreases in proportion to the speed. Typical constant torque applications are conveyors, hoists, and traction devices. With such applications, as speed increases, the torque required remains constant while the horsepower increases or decreases in proportion to the speed. For example, a conveyor load requires about the same torque at 5 ft/min as it does at 50 ft/min. However, the horsepower requirement increases with speed.

Variable-torque loads—Variable torque is found in loads having characteristics that require low torque at low speed, and increasing values of torque as the speed is increased (Figure 5-70). Examples of loads that exhibit variable-torque characteristics are centrifugal fans, pumps, and blowers. When sizing motors for

Figure 5-69 Constant-torque load.
Photo courtesy of Hytrol, www.hytrol.com

Figure 5-70 Variable-torque load.
Photo courtesy of Preferred Utilities Manufacturing Corporation, www.preferred-mfg.com

variable-torque loads, it is important to provide adequate torque and horsepower at the maximum speed.

Constant-horsepower loads—Constant-horsepower loads require high torque at low speeds and low torque at high speeds, which results in constant horsepower at any speed (Figure 5-71). One example of this type of load would be a lathe. At low speeds, the machinist takes heavy cuts, using high levels of torque. At high speeds, the operator makes finishing passes that require much less torque. Other examples are drilling and milling machines.

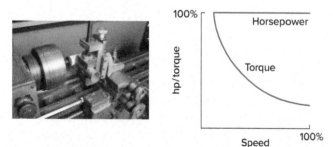

Figure 5-71 Constant-horsepower load.
Photo courtesy of Torchmate, www.torchmate.com.

High-inertia loads—Inertia is the tendency of an object that is at rest to stay at rest or an object that is moving to keep moving. A high-inertia load is one that is hard to start. A great deal of torque is needed to get the load up and running, but less torque is required to keep it operating. High-inertia loads are usually associated with machines using flywheels to supply most of the operating energy. Applications include large fans, blowers, punch presses, and commercial washing machines.

Motor Temperature Ratings

A motor's insulation system separates electrical components from each other, preventing short circuits, and thus winding burnout and failure. Insulation's major enemy is heat, so it's important to be familiar with the different motor temperature ratings in order to keep the motor operating within safe temperature limits.

Ambient temperature is the maximum safe room temperature surrounding the motor if it is going to be operated continuously at full load. When a motor is started, its temperature will begin to rise above that of the surrounding, or ambient, air. In most cases, the standardized ambient temperature rating is 40°C (104°F). While this standard rating represents a very warm room, special applications may require motors with a higher temperature capability such as 50°C or 60°C.

Temperature rise is the amount of temperature change that can be expected within the winding of the motor from its nonoperating (cool) condition to its temperature at full-load continuous operating condition. The heat causing the temperature rise is a result of electrical and mechanical losses, and a characteristic of a motor's design.

Hot-spot allowance must be made for the difference between the measured temperature of the winding and the actual temperature of the hottest spot within the winding, usually 5°C to 15°C, depending upon the type of motor construction. The sum of the temperature rise, the hot-spot allowance, and the ambient temperature must not exceed the temperature rating of the insulation.

Insulation class of a motor is designated by letter according to the temperature it is capable of withstanding without serious deterioration of its insulating properties.

Duty Cycle

The duty cycle refers to the length of time a motor is expected to operate under full load. Motor ratings according to duty are **continuous duty** and **intermittent duty.** Continuous duty–rated motors are rated to be run continuously without any damage or reduction in life of the motor. General-purpose motors will normally be rated for continuous duty. Intermittent-duty motors are rated for short operating periods and then must be allowed to stop and cool before restarting. For example, crane motors and hoists are often rated for intermittent duty.

Torque

Motor torque is the twisting force exerted by the shaft of a motor. The torque/speed curve of Figure 5-72 shows how a motor's torque production varies throughout the different phases of its operation.

Locked-rotor torque (LRT), also called **starting torque,** is produced by a motor when it is initially energized at full voltage. It is the amount of torque available to overcome the inertia of a motor at standstill. Many loads require a higher torque to start them moving than to keep them moving.

Pull-up torque (PUT) is the minimum torque generated by a motor as it accelerates from standstill to operating speed. If a motor is properly sized to the load, the pull-up torque is brief. If a motor's pull-up torque is less than that required by its application load, the motor will overheat

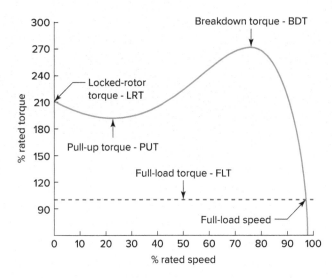

Figure 5-72 Motor torque/speed curve.

and stall. Some motors do not have a value of pull-up torque because the lowest point on the torque/speed curve may occur at the locked-rotor point. In this case pull-up torque is the same as the locked-rotor torque.

Breakdown torque (BDT), also called **pull-out torque,** is the maximum amount of torque a motor can attain without stalling. Typical induction motor breakdown torque varies from 200 to 300 percent of full-load torque. High breakdown torque is necessary for applications that may undergo frequent overloading. One such application is a conveyor belt. Often, conveyor belts have more product placed upon them than their rating allows. High breakdown torque enables the conveyor to continue operating under these conditions without causing heat damage to the motor.

Full-load torque (FLT) is produced by a motor functioning at a rated speed and horsepower. The operating life is significantly diminished in motors continually running at levels exceeding full-load torque.

Motor Enclosures

Motor enclosures are designed to provide adequate protection, depending on the environment in which the motor has to operate. The selection of the proper enclosure is vital to the successful safe operation of a motor. Using a motor enclosure inappropriate for the application can significantly affect motor performance and life. The two general classifications of motor enclosures are **open** and **totally enclosed,** examples of which are shown in Figure 5-73. An open motor has ventilating openings, which permit passage of external air over and around the motor windings. A totally enclosed motor is constructed to prevent the free exchange of air between the inside and outside of the frame, but not sufficiently enclosed to be termed airtight.

Open enclosure   Totally enclosed enclosure

Figure 5-73 Motor enclosures.
Photo courtesy of Leeson, www.leeson.com

Open and totally enclosed categories are further broken down by enclosure design, type of insulation, and/or cooling method. The most common of these types are:

Open drip-proof (ODP) motors are open motors in which all ventilating openings are so constructed that drops of liquid or solid particles falling on the motor at any angle from 0 to 15 degrees from vertical cannot enter the machine. This is the most common type and is designed for use in nonhazardous, relatively clean, industrial areas.

Totally enclosed, fan-cooled (TEFC) motors are enclosed motors equipped for external cooling by means of a fan integral with the motor, but external to the enclosed parts. They are designed for use in extremely wet, dirty, or dusty areas.

Totally enclosed, nonventilated (TENV) motors are enclosed motors generally limited to small sizes (usually under 5 hp) where the motor surface area is large enough to radiate and convey the heat to the outside air without an external fan or airflow. They are particularly effective in textile applications where a fan would regularly clog with lint.

Hazardous location motors are designed with enclosures suitable for environments in which explosive or ignitable vapors or dusts are present, or are likely to become present. These special motors are required to ensure that any internal fault in the motor will not ignite the vapor or dust. Every motor approved for hazardous locations carries a UL nameplate that indicates the motor is approved for that duty. This label identifies the motor as having been designed for operation in Class I or Class II locations. The class identifies the physical characteristics of the hazardous materials present at the location where the motor will be used. The two most common hazardous location motors are Class I explosion-proof and Class II dust ignition resistant.

Figure 5-74 Explosion-proof motor.
Photo courtesy of Siemens, www.siemens.com

Explosion-proof applies only to Class I environments, which are those that involve potentially explosive liquids, vapors, and gases (Figure 5-74). Dust ignition resistant motors are used in environments that contain combustible dusts such as coal, grain, or flour. Some motors may be approved for both Class I and II locations.

Metric Motors

When you need a replacement for a metric (IEC) motor installed on imported equipment, the most practical way to proceed is to get an exact metric replacement motor. When direct replacements are not available, the following may need to be considered:

- Metric motors (Figure 5-75) are rated in **kilowatts (kW)** rather than horsepower (hp). To convert from kilowatts to horsepower, multiply the kW rating of the motor by 1.34. For example, a 2 kW metric motor would be equal to approximately 2.7 hp and the closest NEMA equivalent would be 3 hp.

Available Ratings: D63 Frame – DF250 Frame – 230/460 Volt

B3 Foot Mount	.18 KW (1/4HP) – 37 KW (50HP)	3425 RPM
B3 Foot Mount	.18 KW (1/4HP) – 37 KW (50HP)	1700 RPM
B3 Foot Mount	.18 KW (1/4HP) – 37 KW (50HP)	1100 RPM

Figure 5-75 Metric motors.
Photo courtesy of Leeson, www.leeson.com

Speed, rpm				
	Frequency 50 Hz		Frequency 60 Hz	
Poles	Synchronous	Full load (Typical)	Synchronous	Full load (Typical)
2	3,000	2,850	3,600	3,450
4	1,500	1,425	1,800	1,725
6	1,000	950	1,200	1,150
8	750	700	900	850

- Metric motors may be rated for 50 rather than 60 Hz speed. The table shows a comparison of 50 and 60 Hz induction motor speeds.
- NEMA and IEC standards both use letter codes to indicate specific mechanical dimensions, plus number codes for general frame size. IEC motor frame sizes are given in metric dimensions, making it impossible to get complete interchangeability with NEMA frame sizes.
- Although there is some correlation between NEMA and IEC motor enclosures, it is not always possible to show a direct cross-reference from one standard to the other. Like NEMA, IEC has designations indicating the protection provided by a motor's enclosure. However, where the NEMA designation is in words, such as open drip proof or totally enclosed fan-cooled, IEC uses a two-digit index of protection (IP) designation. The first digit indicates how well protected the motor is against the entry of solid objects; the second digit refers to water entry.
- IEC winding insulation classes parallel those of NEMA, and in all but very rare cases use the same letter designations.
- NEMA and IEC duty cycle ratings are different. Where NEMA commonly designates continuous or intermittent duty, IEC uses eight duty cycle designations.
- **CE** is an acronym for the French phrase *Conformité Européene* and is similar to the UL or CSA marks of North America. However, unlike UL (Underwriters Laboratories) or CSA (Canadian Standards Association), which require independent laboratory testing, the motor manufacturer through "self-certifying" can apply the CE mark that its products are designed to the appropriate standards.

Part 6 Review Questions

1. What two factors determine the mechanical horsepower output of a motor?
2. Explain what each of the following motor current ratings represents: (a) full-load amperes; (b) locked-rotor current; (c) service factor amperes.
3. What does the code letter on the nameplate of a motor designate?
4. What NEMA design–type motor would be selected for driving a pump that requires a high starting torque at low starting current?
5. List four types of motor losses that affect the efficiency of a motor.
6. What motor specification defines the physical dimensions of a motor?
7. An imported machine motor rated for 50 Hz is operated at 60 Hz. What effect, if any, will this have on the speed of the motor? Why?
8. Would it be normally acceptable to replace a motor rated for NEMA A insulation with one rated for NEMA F insulation? Why?
9. Explain the basic load requirements for the following types of motor loads: (a) constant torque; (b) constant horsepower; (c) variable torque.
10. Explain what each of the following motor temperature ratings represents: (a) ambient temperature; (b) temperature rise; (c) hot spot temperature allowance.
11. What does the duty cycle rating of a motor refer to?
12. List the four types of torque associated with the operation of a motor.
13. What determines the type of motor enclosure selected for a particular application?
14. What type of motor enclosure would be best suited for extremely wet, dirty, or dusty areas?
15. Determine the equivalent NEMA horsepower rating for an 11 kW–rated metric motor.

PART 7 MOTOR INSTALLATION

Knowledge of proper installation techniques is vital to the effective operation of a motor. The following are some of the important motor installation procedures that need to be considered.

Foundation

A rigid foundation is essential for minimum vibration and proper alignment between motor and load. Concrete makes the best foundation, particularly for large motors and driven loads.

Mounting

Unless specified otherwise, motors can be mounted in any position or any angle. Mount motors securely to the mounting base of equipment or to a rigid, flat surface, preferably metallic. An adjustable motor base makes the installation, tensioning, and replacements of belts easier. Common types of motor mountings are shown in Figure 5-76 and include:

Rigid base, which is bolted, welded, or cast on the mainframe and allows the motor to be rigidly mounted on equipment.

Resilient base, which has isolation or resilient rings between motor mounting hubs and base to absorb vibration and noise. A conductor is imbedded in the ring to complete the circuit for grounding purposes.

NEMA C face mount, which has a machined face with a pilot on the shaft end that allows direct mounting with a pump or other direct-coupled equipment. Bolts pass through mounted part to threaded hole in the motor face.

Motor and Load Alignment

Misalignment between the motor shaft and the load shaft causes unnecessary vibration and failure due to mechanical problems. Premature bearing failure in the motor and/or the load can result from misalignment. Different types of alignment devices, such as the **laser alignment** kit shown in Figure 5-77, are used for motor and load alignment. Positioning a motor or placing a shim (thin piece of metal) under the feet of the motor is often part of the alignment process.

Direct-drive motors, as the name implies, supply torque and speed to the load directly. A motor coupling is used to mechanically connect axially located motor shafts with equipment shafts. Direct coupling of the motor shaft to the driven load results in a 1:1 speed ratio. For direct-coupled motors, the motor shaft must be centered with the load shaft to optimize operating efficiency. A flexible coupling permits the motor to operate the driven load while allowing for slight misalignments.

Coupling by means of gears or pulleys/belts may be used in cases where the application requires other than standard available speed. Variable speeds are possible by making available several gear ratios or pulleys with variable diameters. Matching a motor to a load involves transformation of power between shafts, often from a high-speed/low-torque drive shaft to a low-speed/high-torque

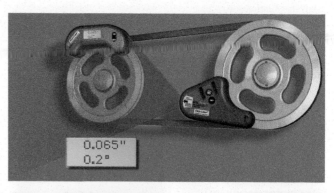

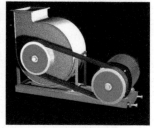

Figure 5-76 Common types of motor mountings.
Photo courtesy of Leeson, www.leeson.com

Figure 5-77 Laser alignment kit.
Photo courtesy of Damalini, www.damalini.com

load shaft. Multiple belts are often used together in order to increase carrying power. If the pulley wheels are different sizes, the smaller one will spin faster than the larger one. Changing pulley ratios does not change horsepower, only torque and speed. The following formula can be used to calculate speed and pulley sizes for belt-drive systems.

$$\frac{\text{Motor rpm}}{\text{Equipment rpm}} = \frac{\text{Equipment pulley diameter}}{\text{Motor pulley diameter}}$$

Example 5-5

Problem: You have a motor to drive a load (Figure 5-78). The motor operates at 1,725 rpm and has a pulley with a 2-inch diameter; the load must operate at 1,150 rpm. What size of pulley is needed for the load?

Solution:

$$\frac{\text{Motor rpm}}{\text{Equipment rpm}} = \frac{\text{Equipment pulley diameter}}{\text{Motor pulley diameter}}$$

$$\frac{1{,}725}{1{,}150} = \frac{\text{Equipment pulley diameter}}{2}$$

$$\frac{1{,}725 \times 2}{1{,}150} = 3\text{-inch pulley}$$

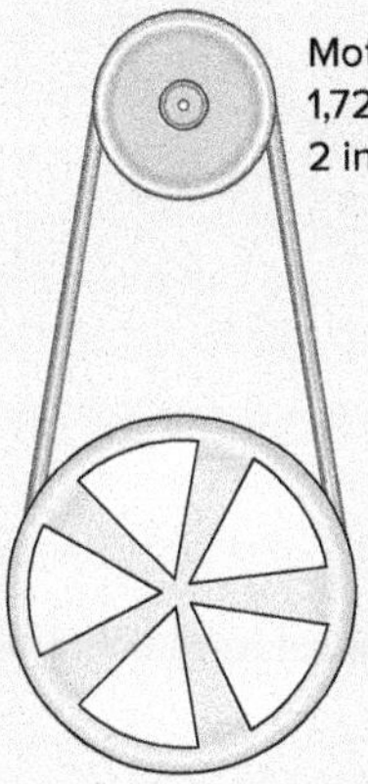

Motor drive pulley
1,725 rpm
2 inch diameter

Load pulley
1,150 rpm
diameter = ?

Figure 5-78 Example 5-5.

V belts are common belts used for power transmission. They have a flat bottom and tapered sides and transmit motion between two sheaves. When servicing a belt-drive system, the belts must be checked for proper tension and alignment as illustrated in Figure 5-79.

- The belt should be tight enough not to slip, but not so tight as to overload motor bearings.
- Belt deflection should be 1/64 inch per inch of span.
- A belt tension gauge is used to ensure proper specified belt tension.
- Misalignment is one of the most common causes of premature belt failure.

- Angular misalignment is misalignment caused by two shafts that are not parallel.
- Parallel misalignment is caused by two shafts that are parallel but not on the same axis.

Motor Bearings

The rotating shaft of a motor is suspended in the end bells by bearings that provide a relatively rigid support for the output shaft. Motors come equipped with different types of bearings properly lubricated to prevent metal-to-metal contact of the motor shaft (Figure 5-80). The lubricant used is usually either grease or oil. Most motors built today have sealed-bearing lubrication. This should be checked periodically to ensure the sealing has not been compromised and the bearing lubricant lost. For installations using older motors that require regular lubrication, this should be done on a scheduled basis, in conformance with the manufacturer's recommendations.

Sleeve bearings used on smaller light-duty motors consist of a bronze or brass cylinder, a wick, and a reservoir. The shaft of the motor rotates in the bronze or brass sleeve and is lubricated with oil from the reservoir by the wick, which transfers oil from the reservoir to the sleeve. Large motors (200 hp and over) are often equipped with large split-sleeve bearings that mount on the top and bottom half of the motor end shield. These bearings are usually poured with a material called a **babbitt** and then bored to size. Sleeve bearings are furnished with oil reservoirs, sight gauges, level gauges, and drain provisions.

Ball bearings are the most common type of bearing. They carry heavier loads and can withstand severe applications. In a ball bearing, the load is transmitted from the outer race to the ball, and from the ball to the inner race. Ball bearings come in three different styles: permanently lubricated, hand packed, and bearings that require lubrication through fitting. Not lubricating the bearings can damage a motor for obvious reasons; too much grease can overpack bearings and cause them to run hot, shortening their life. Excessive lubricant can find its way inside the motor where it collects dirt and causes insulation deterioration and overheating.

Roller bearings are used in large motors for belted loads. In these bearings, the roller is a cylinder, so this spreads the load out over a larger area, allowing the bearing to handle much greater loads than a ball bearing.

Thrust bearings consist of two thrust races and a set of rollers that are designed to handle higher than normal axial forces exerted on the shaft of the motors, as is the case with some fan and pump blade applications. Motors for vertically mounted motors typically use thrust bearings.

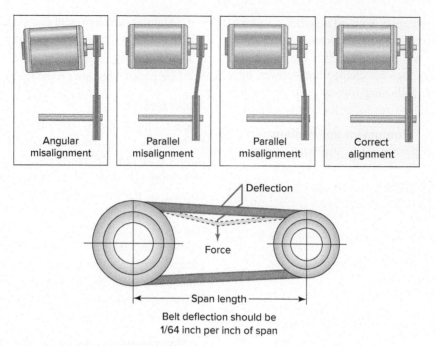

Figure 5-79 Servicing a belt-drive system.

(a) Split-sleeve bearing

(c) Roller bearing

(b) Ball bearing

(d) Thrust bearing

Figure 5-80 Motor bearings.
(a) Photo courtesy of Canadian Babbitt Bearings, www.cbb.ca; (b–d) Photo courtesy of The Timken Company

Electrical Connections

NEMA standards and **Article 430** of the NEC, as well as state and local codes, provide specific electrical and mechanical installation requirements and recommendations covering motors and motor controls. The motor must be connected to a power source corresponding to the voltage and frequency rating shown on the motor nameplate. After you've verified that the supply voltage requirements are correct, you then can make the motor terminal connections. Stator winding connections should be made as shown on the nameplate connection diagram or in accordance with the wiring diagram attached to the inside of the conduit box cover.

Grounding

Both your motor and the equipment or apparatus to which it is connected must be grounded as a precaution against the hazards of electrical shock and electrostatic discharge. This is done by using an **equipment-grounding conductor** that establishes a path or circuit for ground-fault current to facilitate overcurrent device operation. The equipment-grounding conductor may be a conductor (insulated or bare) run with the circuit conductors, or where metal raceways are used, the raceway may be the equipment-grounding conductor. The color green is reserved for an insulated grounding conductor. In addition to helping prevent electrical shock, grounding of an electronic motor drive also helps to reduce unwanted electrical noise that can interfere with the proper operation of the electronic motor drive circuits.

Electrical currents are induced onto the motor's rotor shaft and seek the least resistant path to ground—usually the motor bearings. Shaft voltages accumulate on the rotor until they exceed the dielectric capacity of the motor bearing lubricant; then the voltage discharges in a short pulse to ground through the bearing. The random and frequent discharging has an electric discharge machining (EDM) effect, causing pitting of the bearing's rolling elements and raceways that eventually can lead to bearing failure. This occurs more often in AC motors controlled by variable-frequency drives. For this reason, proper grounding is especially critical on the motor frame, between the motor and drive, and from the drive to earth. Grounding the motor shaft by installing a grounding device, such as shown in Figure 5-81, prevents bearing damage by dissipating shaft currents to ground.

Conductor Size

The size of the motor branch circuit conductors is determined in accordance with Article 430 of the NEC, based on the motor full-load current, and increased where required to limit voltage drop. Undersized wire between the motor and the power source will limit stating abilities and cause overheating of the motor.

Figure 5-81 Motor shaft grounding ring.
Photo courtesy of Electro Static Technology—an ITW Co., www.est-aegis.com.

Example 5-6

Problem: What size THW CU conductors are required for a single 15 hp, three-phase, 230 V squirrel-cage motor?

Solution:

Step 1 Determine the full-load current (FLC) rating of the motor to be used in determining the conductor size. NEC 430.6 requires that tables 430.247 through 430.250 be used to determine the FLC *and not* the nameplate rating. Table 430.250 deals with three-phase alternating current motors, and using this table, we find that for a 15 hp, 230 V, three-phase motor the FLC is 42 amperes.

Step 2 NEC 430.22 requires branch circuit conductors supplying a single motor to have an ampacity not less than 125 percent of the motor FLC. Therefore,

$$\text{Rated ampacity} = 42 \text{ A} \times 125\%$$
$$= 52.5 \text{ A}$$

Step 3 According to Table 310.15(B)(16) the conductor size required would be:

$$\text{6 AWG THW CU}$$

Voltage Levels and Balance

Motor voltages should be kept as close to the nameplate value as possible, with a maximum deviation of 5 percent. Although motors are designed to operate within 10 percent of nameplate voltage, large voltage variations can have negative effects on torque, slip, current, efficiency, power factor, temperature, and service life.

Unbalanced motor voltages applied to a polyphase induction motor may cause unbalanced currents, resulting in overheating of the motor's stator windings and rotor bars, shorter insulation life, and wasted energy in the form of heat. When three-phase line voltages are not equal in magnitude, they are said to be unbalanced. A voltage unbalance can magnify the percent current unbalance in the stator windings of a motor by as much as 6 to 10 times the percent voltage unbalance. Acceptable voltage unbalance is typically not more than 1 percent. When there is a 2 percent or greater voltage unbalance, steps must be taken to determine and rectify the source of the unbalance. In cases where the voltage unbalance exceeds 5 percent, it is not advisable to operate the motor at all. The voltage unbalance is calculated as follows:

Percent voltage unbalance =

$$\frac{\text{Maximum voltage deviation from the voltage average}}{\text{Average voltage}} \times 100$$

Example 5-7

Problem: What is the percent voltage unbalance for a three-phase supply voltage of 480 V, 435 V, and 445 V (Figure 5-82)?

Solution:

$$\text{Average voltage} = \frac{480 + 435 + 445}{3}$$

$$= \frac{1{,}360}{3}$$

$$= 453 \text{ V}$$

Maximum deviation from the average voltage = 480 − 453 = 27 V.

Percent voltage unbalance =

$$\frac{\text{Maximum voltage deviation from the voltage average}}{\text{Average voltage}} \times 100$$

$$= \frac{27}{453} \times 100$$

$$= 5.96\%$$

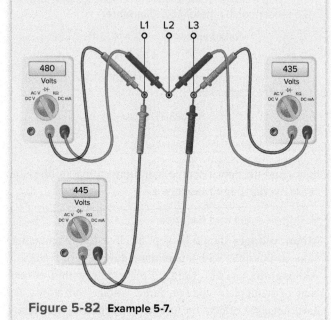

Figure 5-82 Example 5-7.

Built-in Thermal Protection

Overload relays mounted on the motor starter enclosure protect the motor by monitoring the motor current and resultant heat it creates inside the motor. They do not, however, monitor the actual amount of heat generated within the winding. Motors subject to such conditions as excessive starting cycles, high ambient motor temperatures, or inadequate ventilation conditions may experience rapid heat buildup that is not sensed by the overload relay. To minimize such risks, the use of motors with **thermal protectors** inside that sense motor winding temperature is advisable for most applications. These devices may be integrated into the control circuit to offer additional overload protection to the motor or connected in series with the motor windings on smaller single-phase motors, as illustrated in Figure 5-83. Basic types include:

Automatic reset: After the motor cools, this line-interrupting protector automatically restores power. It should not be used where unexpected restarting would be hazardous.

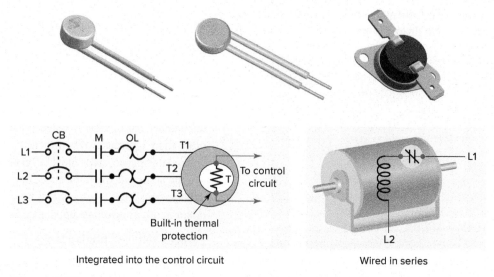

Figure 5-83 Built-in thermal motor protection.
Photos courtesy of Microtherm, www.microtherm.com.

Manual reset: This line-interrupting protector has an external button that must be pushed to restore power to the motor. Use where unexpected restarting would be hazardous, as on saws, conveyors, compressors, and other machinery.

Resistance temperature detectors: Precision-calibrated resistors are mounted in the motor and are used in conjunction with an instrument to detect high temperatures.

Part 7 Review Questions

1. List three popular types of motor mountings.
2. A motor with a 3-inch drive pulley operating at a speed of 3,600 rpm is belt-coupled to an equipment pulley 8 inches in diameter. Calculate the speed of the driven load.
3. List four basic types of bearings and give a typical application for each.
4. How can a motor be damaged by overlubricating a ball bearing?
5. Which NEC article deals specifically with requirements for electric motors?
6. Why is it desirable to ground the motor shaft in addition to the frame?
7. In what ways can undersized wiring between the motor and power source affect the operation of a motor?
8. What negative effects can unbalanced three-phase line voltages have on the operation of a motor?
9. In what type of application is it advisable to use manual-reset built-in thermal protectors?

PART 8 MOTOR MAINTENANCE AND TROUBLESHOOTING

Motor Maintenance

In general, motors are very reliable machines that require little maintenance. But while a typical electric motor might be a low-maintenance item, it still requires regular maintenance if it is to achieve the longest possible service life.

Schedule periodic inspections The key to minimizing motor problems is scheduled routine inspection and service. Keep records of all maintenance schedules and procedures performed. The frequency and procedures of routine service vary widely between applications. Motors should be inspected periodically for things such as shaft alignment, motor base tightness, and belt condition and tension.

Brush and commutator care For DC motors, remove the covers and perform checks on brush wear, spring tension, and commutator wear or scoring. Replace the brushes if there is any chance they will not last until the next inspection date. The commutator should be clean, smooth, and have a polished brown surface where the brushes ride. Observe the brushes while the motor is running. The brushes must ride on the commutator smoothly with little or no sparking and no brush chatter.

Testing winding insulation Twice yearly, test winding and winding-to-ground resistance to identify insulation problems. Motors that have been flooded or have low megger readings should be thoroughly cleaned and dried before being energized. The following are typical minimum motor insulation resistance values:

Rated Motor Voltage	Minimum Insulation Resistance
600 V and below	1.5 MΩ
2,300 V	3.5 MΩ
4,000 V	5.0 MΩ

Keep your motors clean Wipe, brush, vacuum, or blow accumulated dirt from the frame and air passages of the motor. Dirty motors run hot when thick dirt insulates the frame and clogged passages reduce cooling airflow. Heat reduces insulation life and eventually causes motor failure.

Keep your motors dry Motors that are used continuously are not prone to moisture problems. It is the intermittent use or standby motor that may have difficulties. Try to run the motor for at least a few hours each week to drive off moisture. Be careful that steam and water are not directed into open drip-proof motors.

Check lubrication Lubricate motors according to manufacturer specifications. Apply high-quality greases or oils carefully to prevent contamination by dirt or water.

Check for excessive heat, noise, and vibration Feel the motor frame and bearings for excessive heat or vibration. Listen for abnormal noise. All indicate a possible system failure. Promptly identify and eliminate the source of the heat, noise, or vibration.

Excessive starting is a prime cause of motor failures The high current flow during start-up contributes a great amount of heat to the motor. For motors 200 hp and below, the maximum acceleration time a motor connected to a high-inertia load can tolerate is about 20 seconds. The motor should not exceed more than about 150 **"start-seconds"** per day.

Troubleshooting Motors

Electric motor failures can be due to mechanical component failure or electrical circuit failure. Any type of electrical testing involves risk, and complacency can lead to injury! When working on any type of motor, to reduce the risk of injury be certain to:

- Disconnect power to the motor and complete lock-out and tag-out procedures before performing service or maintenance.
- Discharge all capacitors before servicing the motor.
- Always keep hands and clothing away from moving parts.
- Be sure required safety guards are in place before starting equipment.

Electrical contact accounts for one-fifth of all construction deaths. Never work on energized equipment unless this is absolutely necessary for examination, adjustment, servicing, or maintenance. When you find you must work on energized equipment, always wear the appropriate personal protective equipment and use appropriate tools and equipment. Use the "buddy rule" and never work on energized equipment alone. Always have a partner working with you, in case of emergency. Common devices used for troubleshooting motor operation (Figure 5-84) problems include:

- **Digital multimeter (DMM)**—most often used for precise voltage measurement, as well as circuit continuity and motor winding resistance. Multifunctional DMM types include capacitor capacitance testing and frequency measurement functions.
- **Clamp-on ammeter**—used for monitoring of motor current. The probe clamps around the outside of the conductor, thus avoiding having to open the circuit and connect the meter in series to measure the current.

Multimeter

Clamp-on ammeter

Megohmmeter

Infrared thermometer

Tachometer
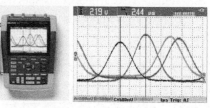
Portable oscilloscope

Figure 5-84 Instruments used for motor troubleshooting.
Photos courtesy of Fluke, www.fluke.com. Reproduced with Permission; e: Electromatic Equipment Co., Inc

- **Megohmmeter**—used for checking the motor's insulation resistance to ground resistance. Measures insulation resistance in millions of ohms.
- **Infrared (IR) thermometer**—used for noncontact motor surface temperature measurements. The laser sight will pinpoint high-temperature problems on any surface of the motor.
- **Tachometer**—used to check the speed of motors, from no load to full load.
- **Oscilloscope**—draws a graph of an electrical power or control voltage. Is commonly used for troubleshooting problems associated with electronic AC and DC motor drives.

The basic motor system consists of the power supply, controller, motor, and driven load. When a motor problem occurs, it is first necessary to find which of the parts of the system is at fault. Power supplies and controllers can fail as well as the motor itself. Mechanical loads can increase because of an increased size of the load the motor is driving, or failure of the bearings or coupling mechanisms can occur. Mechanical overloading is a prime cause of motor failure.

Troubleshooting Guides Once it has determined that the motor is at fault, you can proceed to locate the problem with the motor. A troubleshooting guide outlines a comprehensive variety of motor problems. Generally the categories are arranged according to symptoms, offering brief suggestions concerning what to look for when investigating motor failures and often providing advice on how to correct the problem once it has been identified. The following is an example of a troubleshooting guide that outlines fault symptoms common to most types of motors.

1. Symptom: The motor fails to start. Possible causes:

    Blown fuse or open-circuit breaker. Check the voltage at the input and output of the overcurrent protection device. If voltage is measured at the input but not at the output, the fuse is blown or the circuit breaker is open. Check the rating of the fuse or circuit breaker. It should be at least 125 percent of the motor's full-load current.

    Motor overload relay on starter tripped. Allow overload relay to cool and reset it. If the motor causes the overload relay to open after a short period, check for motor short circuits and grounds. Check the full-load current of the motor and compare it to the setting of the overload relay.

    Low voltage or no voltage applied to the motor. Check the voltage at the motor terminals. The voltage must be within 10 percent of the motor nameplate voltage. Determine the cause of the low voltage. Loose fuse clips and connections at the terminals of the disconnect switch or circuit breaker can result in low voltage at the motor.

    Mechanical overload. Rotate the motor shaft to see if a binding load is the problem. Check for frozen bearings. Check the air gap between the stator and the rotor. Reduce the load or try operating the motor with no load applied.

    Defective motor windings. Make resistance checks of the motor windings for opens and shorts in coil windings and coils shorted to ground. An ohmmeter reading of infinity across a set of coil windings means that there is an open somewhere—sometimes it is at one end of the coil and accessible for repair. A short circuit in only a few turns of a coil, while difficult to detect, will still result in a motor overheating. One way to test for a short-circuited coil winding is to compare its resistance reading with that of a known good identical coil.

    Burnt-out motor. If one or more of the motor windings looks blackened and smells burnt, it is most likely burnt out and needs to be replaced.

2. Symptom: The motor overheats. Possible causes:

    Load. A basic rule is that your motor should not get too hot to touch. Check ammeter reading against full-load current rating of motor. For a higher-than-normal current reading, reduce the load or replace motor with a larger sized one.

    Insufficient cooling. Remove any buildup of debris in or around the motor.

    Ambient temperature. Higher-than-normal ambient temperatures. Take steps to improve the motor's ventilation and/or lower the ambient temperature.

    Bearings and alignment. Bad bearings or poor coupling alignment can increase friction and heat.

    Source voltage. If the operating voltage is too high or too low, the motor will operate at a higher temperature. Correct voltage to within 10 percent of the motor's rating.

3. Symptom: Excessive motor noise and vibration. Possible causes:

    Bearings. With the motor stopped, try gently moving the shaft up and down to detect bearing wear. Use a stethoscope to check the bearings for noise. When the handle of a screwdriver is placed to the ear and the blade to the bearing housing, the screwdriver will amplify the noise, like a stethoscope. Replace worn or loose bearings. Replace dirty or worn-out oil or grease.

    Coupling mechanism. Check for bent shaft on motor or load. Straighten if necessary. Measure the alignment of the couplings. Realign if necessary.

    Loose hardware. Tighten all loose components on the motor and load. Check fasteners on the motor and load mounts. Motors with centrifugal mechanisms, brushes, slip rings, and commutators can cause noise due to wear and looseness of the mechanisms.

4. Symptom: Motor produces an electric shock when touched. Possible cause:

    Grounding. Broken or disconnected equipment grounding conductor. Motor winding short-circuited to frame. Check motor junction box for poor connections, damaged insulation, or leads making electrical connection with the frame.

5. **Symptom: Motor overload protector continually trips. Possible cause:**

Load. Load too high. Verify that the load is not jammed. Remove the load from the motor and measure the no-load current. It notably should be less than the full-load rating stamped on the nameplate.

Ambient temperature too high. Verify that the motor is getting air for proper cooling.

Overload protector may be defective. Replace the motor's protector with one of the correct rating.

Winding short-circuited or grounded. Inspect windings for defects and loose or cut wires that may cause a path to ground.

Troubleshooting Charts Troubleshooting charts may be used to quickly identify common problems and possible corrective courses of action. The following are examples that pertain to specific motor types.

Single–Phase Motors

Problem	Probable Cause and Course of Action
Split-phase motor hums, and it will run normally if started by hand.	Centrifugal switch is not operating properly. Disassemble the mechanism. Clean the contacts. Adjust spring tension. Replace switch.
Capacitor-start motor hums, and it will run normally if started by hand.	Centrifugal switch (same as for split-phase motor). Defective capacitor. Test capacitor. If defective replace.
Start capacitors continuously fail.	The motor is not coming up to speed quickly enough as a result of not being sized properly.
	The motor is being cycled too frequently. Capacitor manufacturers recommend no more than twenty 3-second starts per hour.
	Starting switch may be defective, preventing the motor from opening the start winding circuit.
Run capacitor fails.	Ambient temperature too high.
	Possible power surge to motor caused by high transient voltage. If a common problem, install a surge protector.
Universal motor sparks.	New brushes not properly seated. Seat brushes with fine sandpaper to fit contour of commutator.
	Worn or sticky brushes. Replace brushes or clean brush holder.
	Open- or short-circuited armature coils. Replace armature.

Three–Phase Motors

Problem	Probable Cause and Course of Action
Single-phasing—one phase of the three-phase system is lost. Motor will not start, but if in operation may continue to operate at increased current and diminished capacity. Unique high-pitched sound from motor.	A fuse is blown or one leg of a circuit breaker is open. Check each of the three-phase power lines for correct voltage.
Unbalanced three-phase voltage—the voltages of all phases of a three-phase power supply are not equal. A voltage imbalance of 3.5 percent between phases will cause a temperature rise of 25°C in the motor. Motor operates at a higher-than-normal temperature and reduced efficiency.	Blown fuse on power factor correction capacitor bank—find and replace fuse.
	Uneven single-phase loading—distribute single-phase loads more evenly on the three-phase circuit.
	Utility unbalanced voltages—if the incoming voltages are substantially unbalanced, contact the utility and ask them to correct the problem.
	Harmonic distortion—The presence of harmonic distortion in the applied voltage to a motor will increase motor temperature, which could result in insulation damage and possible failure.
	Locate the sources of the harmonics and use harmonic filters to control or reduce harmonics.
Wound-rotor induction motor fails to start or starts and runs erratically.	External rotor resistors—Look for failed components in the resistor bank when troubleshooting. Clean slip rings and check brushes for wear and proper pressure.
Synchronous motor experiences increased start-up times or erratic acceleration.	Damaged or defective amortisseur windings—Historical inrush testing that records the stator's current during start-up can greatly assist in determining if these windings have degraded over the life of the motor.

Direct Current Motors

Problem	Probable Cause and Course of Action
Excessive arcing at brushes.	Worn or sticky brushes. Replace brushes or clean brush holder.
	Incorrect brush position with respect to neutral plane. Rotate brush rigging to the correct position to aid in commutation.
	Overload. Measure current to the motor and compare to full-load current rating. If necessary, reduce motor load.
	Dirty commutator. The commutator surface should be clean and bright; slight scratches and discoloring can be removed with emery paper. Deep scratches/ridges require the commutator to be machined and mica-undercut.
	Armature faults. Test for open- and short-circuited windings in the armature and correct or replace motor.
	Field-winding faults. Test for short circuits, open circuits, and ground faults and correct or replace motor.
Rapid brush wear.	Wrong brush material, type, or grade. Replace with brushes recommended by manufacturer.
	Incorrect brush tension. Adjust brush tension so that the brush rides freely on the commutator. Replace brush springs if tension measured by a scale is insufficient.

Troubleshooting Ladder A troubleshooting ladder or tree may be used to guide you through the steps of the troubleshooting process. A troubleshooting ladder is sequential in nature, and its simplicity can often save time in arriving at the source of a motor problem. The following is a typical example of a troubleshooting ladder used to determine the cause of overheating of a three-phase squirrel-cage induction motor.

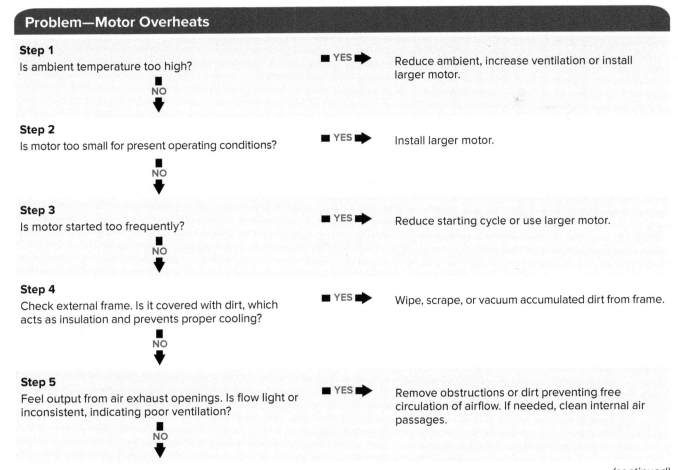

Problem—Motor Overheats

Step 1
Is ambient temperature too high? — YES → Reduce ambient, increase ventilation or install larger motor.
NO ↓

Step 2
Is motor too small for present operating conditions? — YES → Install larger motor.
NO ↓

Step 3
Is motor started too frequently? — YES → Reduce starting cycle or use larger motor.
NO ↓

Step 4
Check external frame. Is it covered with dirt, which acts as insulation and prevents proper cooling? — YES → Wipe, scrape, or vacuum accumulated dirt from frame.
NO ↓

Step 5
Feel output from air exhaust openings. Is flow light or inconsistent, indicating poor ventilation? — YES → Remove obstructions or dirt preventing free circulation of airflow. If needed, clean internal air passages.
NO ↓

(continued)

Problem—Motor Overheats (*Cont.*)

Step 6
Check input current while motor drives load. Is it excessive, indicating an overload? — **NO** → Go to Step 11

↓ YES

Step 7
Is the driven equipment overloaded? — **YES** → Reduce load or install larger motor.

↓ NO

Step 8
Are misalignments, or damaged components causing excessive friction in driven machine or power transmission system? — **YES** → Repair or replace bad components.

↓ NO

Step 9
Are motor bearings dry? — **YES** → Lubricate. Does motor still draw excessive current?

↓ NO   → YES

Step 10
Are damaged end bells, rubbing fan, bent shaft, or rubbing rotor causing excessive internal friction? — **YES** → Repair or replace motor.

↓ NO

Step 11
Are bad bearings causing excessive friction? — **YES** → Repair or replace bad bearings

↓ NO

Step 12
Check phase voltage. Does it vary between phases? — **YES** → Restore equal voltage on all phases.

↓ NO

Step 13
Is voltage more than 10 percent above or ten percent below nameplate? — **YES** → Restore proper voltage or install motor built for the voltage.

↓ NO

Step 14
Check stator. Are any coils grounded or short-circuited? — **YES** → Repair coils or replace motor.

Part 8 Review Questions

1. From a safety perspective, what is the first step to be taken before performing any type of motor maintenance?
2. Outline five common motor maintenance tasks that should be performed as part of a motor preventive maintenance program.
3. Outline how to test for each of the following suspected motor problems:
   a. Blown fuse or open circuit breaker.
   b. Low voltage applied to the motor.
   c. Defective motor windings.
4. List five possible causes of motor overheating.
5. The centrifugal switch of a spilt-phase motor fails and remains open at all times. How would this affect the operation of the motor?
6. The centrifugal switch of a capacitor start motor fails and remains closed at all times. How would this affect the operation of the motor?
7. List four possible causes of unbalanced voltages on the supply voltage of a three-phase motor circuit.
8. List five possible causes of excessive arcing at the brushes of a DC motor.

Troubleshooting Scenarios

1. Assume the tags used to identify the six motor leads of a compound-wound DC motor are suspected of being incorrectly marked or missing.
   a. Outline how an ohmmeter would be used to identify the armature, shunt field, and series field leads.
   b. What operating test could be made to ensure cumulative connection of the shunt and series field?
2. One of the three-phase line fuses to a squirrel-cage induction motor burns open while the motor is operating.
   a. Will the motor continue to rotate? Why?
   b. In what way might this operating condition damage the motor?
   c. Should the motor be able to restart on its own? Why?
3. A defective motor start capacitor rated for 130 µF and 125 V AC is replaced with one rated for 64 µF and 125 V AC. What would happen?
4. The speed of a motor is to be reduced by one-half by using two different size pulleys. What must the relative diameters of the motor drive and load pulley be?
5. A motor feels hot to the touch. Does this always indicate it is operating at too high a temperature? Explain.

Discussion Topics and Critical Thinking Questions

1. Explain how a squirrel-cage rotor produces a magnetic field.
2. List the different types of motor measurements that are used for troubleshooting motors.
3. Why does a single-phase motor have no starting torque if only a single winding is used?
4. How would you determine the running and starting winding of a single-phase motor from a visual inspection of the stator?
5. Arrange the following single-phase motors in the order of decreasing torque, with the highest torque first: split-phase, universal, shaded pole, capacitor.
6. How does slip affect motor speed?
7. Describe the major physical and electrical differences between the three major three-phase motor types.
8. Can a single-phase motor be operated from a three-phase power supply? Explain.
9. Assume you have to purchase a motor and load laser alignment kit. Search the Internet for suppliers and prepare a report on the features and operation of the one you would consider purchasing.
10. An energy-efficient motor produces the same shaft output power (hp), but uses less input power (kW) than a standard-efficiency motor. Visit the website of a motor manufacturer and compare the price and features of a standard-efficiency motor with that of an equivalent energy-efficient motor.
11. Explain why motors are more efficient at full load.

CHAPTER SIX

Contactors and Motor Starters

Photo courtesy of Rockwell Automation,
www.rockwellautomation.com

CHAPTER OBJECTIVES

This chapter will help you:

- List the basic uses of a contactor.
- Describe the requirements and operation of a capacitor switching contactor.
- Explain how arc suppression is applied to contacts.
- Describe the major factors in selecting the size of a contactor and type of enclosure.
- Explain the difference between a contactor and a motor starter.
- Explain the function and operation of motor overload relays.
- Compare NEMA and IEC types of contactors and starters.
- Understand the operation of a solid-state contactor and starter.

Motor starters and contactors are used for switching electric power circuits. Both use a small amount of control current to energize or de-energize the loads connected to them. This chapter examines how contactors and motor starters are used in the control of nonmotor and motor loads.

PART 1 MAGNETIC CONTACTOR

The National Electrical Manufacturers Association (NEMA) defines a **magnetic contactor** as a magnetically actuated device for repeatedly establishing or interrupting an electric power circuit. The magnetic contactor is similar in operation to the electromechanical relay. Both have one important feature in common: contacts operate when the coil is energized. Generally, unlike relays, contactors are designed to make and break electric power circuit loads in excess of 15 A without being damaged. Figure 6-1 shows a typical NEMA magnetic contactor used for switching

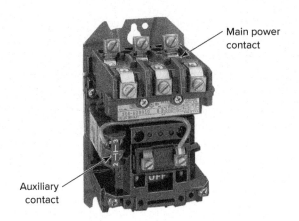

Figure 6-1 Typical magnetic contactor.
Photo courtesy of Rockwell Automation, www.rockwellautomation.com

AC motor loads for which overload protection is not required or provided separately. In addition to the three power contacts, one normally open auxiliary hold-in contact is provided to accommodate three-wire pushbutton control.

There are two circuits involved with the operation of a magnetic contactor: the control circuit and the power circuit. The **control** circuit is connected to the coil, and the **power** circuit is connected to the main power contacts. The operating principle of a three-pole magnetic contactor is illustrated in Figure 6-2.

- When voltage is applied to the terminals of the coil, the current flows through the coil, creating a magnetic field.
- The coil, in turn, magnetizes the stationary iron frame, turning it into an electromagnet.
- The electromagnet draws the armature toward it, pulling the movable and stationary contacts together.
- Power then flows through the contactor from the line side to the load side.
- Generally, a contactor is available in two-, three-, or four-pole contact configurations.

Switching Loads

Contactors are used in conjunction with pilot devices to automatically control high-current loads. The **pilot** device, with limited current handling capacity, is used to control current to the contactor coil, the contacts of which are used to switch heavier load currents. Figure 6-3 illustrates a contactor used with pilot devices to control the temperature and liquid level of a tank. In this application, the contactor coil connects with the level and temperature sensors to automatically open and close the power contacts to the solenoid and heating element loads.

Contactors may be used for switching motor loads when separate **overload protection** is provided. The most

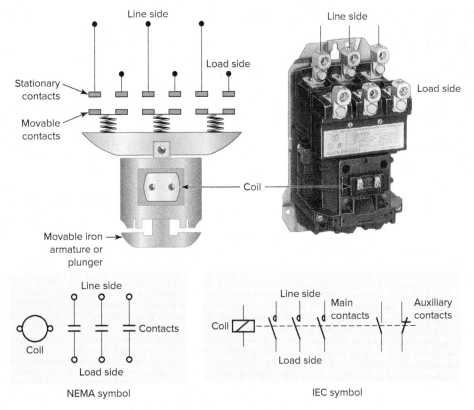

Figure 6-2 Three-pole magnetic contactor.
Photo courtesy of Rockwell Automation, www.rockwellautomation.com

Part 1   Magnetic Contactor

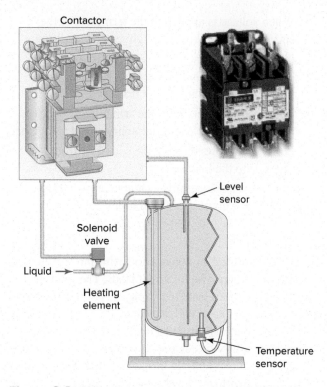

Figure 6-3 Contactor used in conjunction with pilot devices.
This material and associated copyrights are proprietary to, and used with the permission of, Schneider Electric.

common use for a contactor is in conjunction with an overload relay assembly in an AC motor starter. Figure 6-4 shows an IEC contactor used in combination with an overload relay module to switch a motor load. Two-wire or three-wire control may be used to switch the motor. The **two-wire** control circuit is commonly used in applications where the operation of a circuit is automatic. This may include such applications as pumps, electric heating, and air compressors where the pilot device starts the motor automatically as needed. **Three-wire** control is similar to the two-wire circuit except it has an extra set of contacts used to seal it in the circuit. The most common application for three-wire control is a motor controlled by momentary start/stop push buttons. In this case the push buttons must be pressed to energize or de-energize the contactor coil. Generally, start/stop push buttons are used to initiate and terminate the system processes.

High voltage may be handled by the contactor but kept entirely away from the operator, thus increasing the safety of an installation. When this is the case, a **step-down** control transformer is used to lower the AC voltage level required for the control circuit. Typically the secondary of the transformer is rated for 12, 24, or 120 V, while primary voltage may be 208, 230, 240, 460, 480, or 600 V. In all cases, the contactor coil voltage most be matched with that of the control circuit voltage.

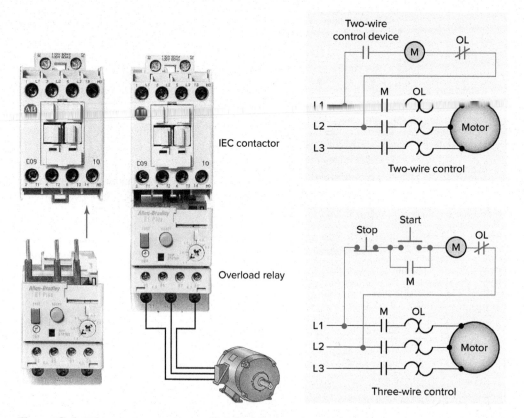

Figure 6-4 IEC contactor used in combination with an overload relay module to switch a motor load.
Photos courtesy of Rockwell Automation, www.rockwellautomation.com

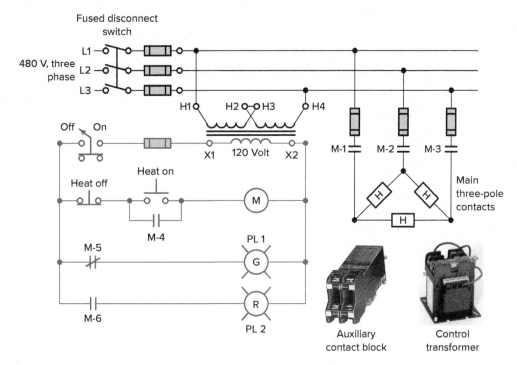

Figure 6-5 Heater circuit controlled by a magnetic contactor.
Photos courtesy of Rockwell Automation, www.rockwellautomation.com

The **auxiliary contacts** of a contactor have a much lower current rating than the main contacts and are used in control circuits for interlocking, holding, and status indication. Figure 6-5 shows the schematic circuit for a three-phase heater circuit controlled by a three-pole magnetic contactor and operated by a three-wire control circuit. The operation of the circuit can be summarized as follows:

- A control transformer is used to lower the 480 V line voltage to 120 V for control purposes.
- The three-wire control circuit is used to switch power to the heating elements.
- With the on/off switch closed, the heat on push button is depressed to energize coil M of the contactor.
- Main power contacts M-1, M-2, and M-3 close, energizing the heating elements at line voltage.
- Auxiliary contact M-4 closes to hold in the contactor coil by completing a circuit around the heat on push button.
- At the same time, auxiliary contact M-5 opens to switch off the green (off) pilot light and contact M-6 closes to switch on the red (on) pilot light.
- Depressing the heat off push button or opening the on/off switch will de-energize the coil, returning the circuit to its off state.

Definite-purpose contactors are specifically designed for applications such as air conditioning, refrigeration, resistive heating, data processing, and lighting. Lighting contactors provide effective control in applications such as office buildings, industrial plants, hospitals, stadiums, and airports. They can be used to handle the switching of tungsten (incandescent filament) or ballast (fluorescent and mercury arc) lamp loads, as well as other general non-motor loads. Contactors may be electrically held or mechanically held. With an **electrically held** contactor, the coil needs to be energized continuously all the time the main contacts are closed. **Mechanically held** contactors require only a pulse of coil current to change state. Once changed, a mechanical latch holds the main contacts in place so the control power can be removed, resulting in the contactor operation that is quieter, cooler, and more efficient. Figure 6-6 shows examples of mechanically and electrically held lighting contactors mounted in enclosures.

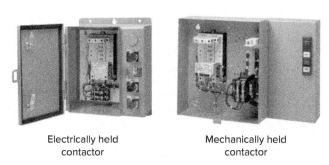

Figure 6-6 Electrically and mechanically held lighting contactors.
Photos courtesy of Eaton, www.eaton.com.

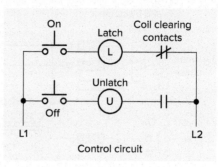
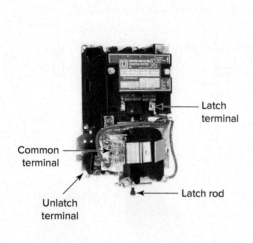
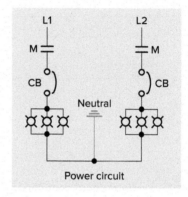

Figure 6-7 Mechanically held lighting contactor.
Photo courtesy of Rockwell Automation, www.rockwellautomation.com

Figure 6-7 shows a typical dual-coil mechanically held lighting contactor and associated circuit. Lighting circuits are single-phase and generally rated at 120 or 277 V. The operation of the circuit can be summarized as follows:

- When the on button is momentarily depressed, the latch coil is energized through the NC clearing contact.
- As a result, the contactor closes and latches mechanically to close the main contacts (M), lighting the bank of lamps, provided that the circuit breaker is closed.
- The coil clearing contacts change state (NC to NO and vice versa) alternately with a change in contactor latching position.
- To unlatch the contactor, thereby turning the lamps off, the off button is momentarily depressed, unlatching the contactor to open contacts M.
- Since the latch and unlatch coils are not designed for continuous duty, they are automatically disconnected by the coil clearing contacts to prevent accidental coil burnout should the push button remain closed.

Figure 6-8 shows a typical wiring diagram for a mechanically held lighting contactor with a single operating coil that is momentarily energized to either close or open the contactor. In this application the lighting contactor is being controlled from two remotely located (three-position, momentary, center off) control stations. A wide variety of automatic control devices such as programmable logic controllers (PLCs) and energy management systems can interface with the contactor as well.

Capacitor Switching Contactors

Capacitor switching contactors are specifically designed to handle the unique application requirements required to switch capacitors onto a power line for power factor correction. The IEC capacitor switching contactor shown in Figure 6-9 contains a damping resistor across each pole designed to minimize the effects of the large charging current.

- The damping resistors are controlled by separate auxiliary contacts interlocked with the main contacts in such a way as to allow the resistor to precharge the capacitors for a short period before the main contacts close.
- After the precharge, the main contacts close to conduct the continuous current.
- At the same time and the parallel damping resistors remain shunted or switched open.

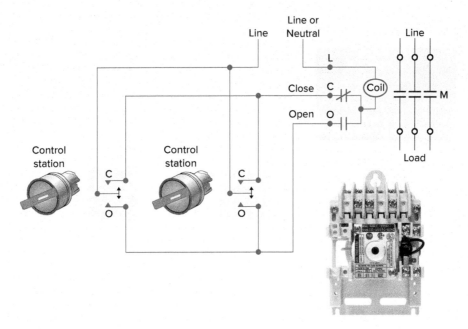

Figure 6-8 Single-operating-coil mechanically held lighting contactor.
Photo courtesy of Rockwell Automation, www.rockwellautomation.com.

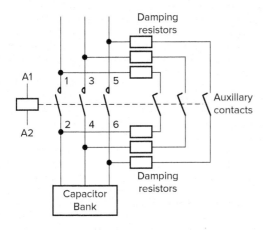

Figure 6-9 IEC capacitor switching contactor.

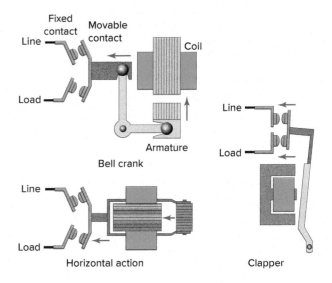

Figure 6-10 Typical operating mechanisms for magnetic contactors.

Contactor Assemblies

Figure 6-10 illustrates three typical operating mechanisms for magnetic contactors: bell-crank, horizontal-action, and clapper. Contactor operating mechanisms should be inspected periodically for proper functioning and freedom from sticking or binding. The magnetic circuit of the operating mechanism consists of soft steel with high permeability and low residual magnetism. The magnetic pull developed by the coil must be sufficient to close the armature against the forces of gravity and the contact spring.

The contactor coil is molded into an epoxy resin to increase moisture resistance and coil life. Its shape varies as a function of the type of contactor (Figure 6-11). A permanent air gap between the magnetic circuits in the closed state prevents the armature from being held in by residual magnetism.

If a coil exhibits evidence of overheating (cracked, melted, or burned insulation), it must be replaced. To measure the coil resistance, disconnect one of the coil leads and measure the resistance by setting the ohmmeter to its lowest resistance scale. A defective coil will read zero or infinity, indicating a short or opened coil, respectively.

Contactor coils have a number of insulated turns of wire designed to give the necessary ampere-turns to

Part 1 Magnetic Contactor

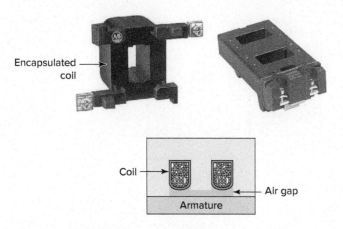

Figure 6-11 Contactor coil.
Photos courtesy of Rockwell Automation, www.rockwellautomation.com

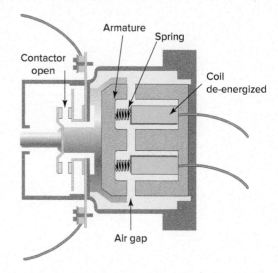

Figure 6-12 De-energized coil air gap.

operate on small currents. As contactors are used to control different line voltages, the voltage used to control the coil may vary. Therefore, when selecting coils, you must choose one that matches the available control voltage. The operational limit of the contactor is between 85 and 110 percent of the rated coil voltage. A coil voltage variation of greater than ±5 percent will maximize the contact wear and minimize life. The reason for this is that higher voltages will increase the speed of the electromagnet at closing. Lower voltages will decrease the speed at closing. Both these factors can lead to contact wear and erosion.

Magnetic coil voltage specifications include rated voltage, pickup voltage, hold-in voltage, and dropout voltage. **Rated** voltage refers to the coil supply voltage and must match that of the control circuit power source. **Pickup** voltage is the amount of voltage required to overcome the mechanical forces, like gravity and spring tension, trying to keep the contacts from closing. **Hold-in** voltage is the amount of voltage needed to maintain the contacts in their closed position after pickup voltage is reached (hold-in voltage is normally less than pickup voltage). All contactors that are electrically held in are sensitive to voltage dips occurring in the electrical supply. The **dropout** voltage is the amount of voltage below which the magnetic field becomes too weak to maintain the contacts in their closed position.

AC and DC contactor coils with the same voltage ratings are not normally interchangeable, the reason being that with a DC coil only the wire ohmic resistance limits the current flow, whereas with AC coils both resistance and reactance (impedance) limit the current flow. Direct current contactor coils have a large number of turns and a high ohmic resistance compared to their AC counterparts.

For a DC-operated coil, since current is limited by resistance only, the current flow through the coil upon closing is the same as the normal energized current flow. However, this is not the case when the coil is AC operated. With a de-energized AC coil, part of the magnetic path has an air gap because the armature is not pulled in (Figure 6-12). When the contactor closes, the armature closes the magnetic path, causing the inductive reactance of the coil to increase and the current to decrease. This results in a high current to close the contactor and low current to hold it. The inrush current for an AC coil may range from 5 to 20 times that of the sealed current.

When current in an inductive load, such as a contactor coil, is turned off, a very high **voltage spike** is generated. If they are not suppressed, these voltage spikes can reach several thousand volts and produce surges of damaging currents. This is especially true for applications requiring interface with solid-state components such as PLC modules. Figure 6-13 shows an *RC* suppression module wired in parallel (directly across) a contactor coil. The resistor and capacitor connected in series slows the rate of rise of the transient voltage when the push button is operated from the closed to open state.

Contactor coils operated from an AC power source experience changes in the magnetic field surrounding them. The attraction of an electromagnet operating on alternating current is pulsating and equals zero twice during each cycle. As the current goes through zero, the magnetic force decreases and tends to drop the armature out. When magnetism and force build up again, the armature is pulled back in. This motion of the armature, in and out, makes the contactor buzz or chatter, creating a **humming noise** and wear on the contactor's moving parts.

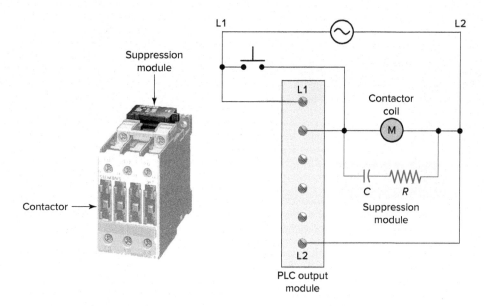

Figure 6-13 *RC* **suppression module.**
Photo courtesy of Siemens, www.siemens.com

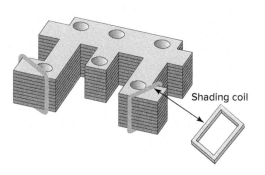

Figure 6-14 Shading coil.

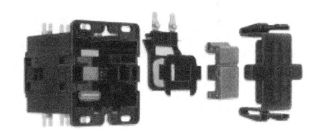

Figure 6-15 Contactor assembly alignment.
This material and associated copyrights are proprietary to, and used with the permission of, Schneider Electric.

The noise and wear of AC contactor assemblies can be prevented by the use of **shading coils** or **rings,** as illustrated in Figure 6-14. Unlike the contactor coil, shading coils are not electrically connected to the power source, but mounted to inductively couple with the contactor coil. The shading coil consists of a single turn of conducting material (generally copper or aluminum) mounted on the face of the magnetic assembly. It sets up an auxiliary magnetic attraction that is out of phase with the main field and of sufficient strength to hold the armature tight to the core even though the main magnetic field has reached zero on the sine wave. With well-designed shading coils, AC contactors can be made to operate very quietly. A broken or open shading coil will make its presence known; the contactor will immediately become extremely noisy.

The core and armature of an AC contactor assembly are made of laminated steel, whereas DC assemblies are solid. This is due to the fact that there are no eddy currents generated with continuous direct current applied. **Eddy currents** are small amounts of current flow induced in the core and armature material by the varying magnetic field produced by AC current flow through the contactor coil. Using a solid iron core would result in greater circulating currents, and for this reason the core of AC coils is made up of a stack of thin insulated laminations.

Misalignment or obstruction of the armature's ability to properly seat when energized causes increased current flow in an AC coil (Figure 6-15). This could occur as a result of pivot wear or binding, corrosion, or dirt buildup,

Part 1 Magnetic Contactor

Figure 6-16 Typical contactor replacement kit.
Photo courtesy of Rockwell Automation,
www.rockwellautomation.com

or pole face damage from impact over a long period of time. Depending on the amount of increased current, the coil may merely run hot, or it may burn out if the current increase is large enough and remains for a sufficient length of time. Improper alignment will create a slight hum coming from the contactor in the closed position. A louder hum will occur if the shading coil is broken because the electromagnet will cause the contactor to chatter.

Today, most contactor contacts are made of a low-resistance silver alloy (Figure 6-16). Silver contacts are used because they ensure a lower contact resistance than other less expensive materials. Depending on the size of the contactor, the main power contacts can be rated to control several hundred amperes. Most often silver inserts are brazed or welded on copper contacts (on the heel), so silver carries the current and copper carries the arc on interruption. Most manufacturers recommend that silver contacts *never* be filed. Silver contacts need not be cleaned because the black discoloration that appears is silver oxide, which is a relatively good conductor of electricity.

Contacts are subject to both electrical and mechanical wear as they establish and interrupt electric currents. In most cases, mechanical wear is minimal compared to electrical wear. Arcing when the contacts are establishing and interrupting currents causes electrical wear or erosion. Also, contacts will overheat if they transmit too much current, if they do not close quickly and firmly, or if they open too frequently. Any of these situations will cause significant deterioration of the contact surface and erratic operation of the contactor.

Contactor **auxiliary contact blocks** (Figure 6-17) are used for the operation of auxiliary control circuits. Auxiliary contacts have a current rating much lower than that of the main power contacts, but they are actuated by the same armature as the power contacts. One contactor block may have several auxiliary contacts, either normally open or normally closed.

Arc Suppression

One of the main reasons contacts wear is the electric arc that occurs when contacts are opened under load. As the contacts open, there will still be current flow between the

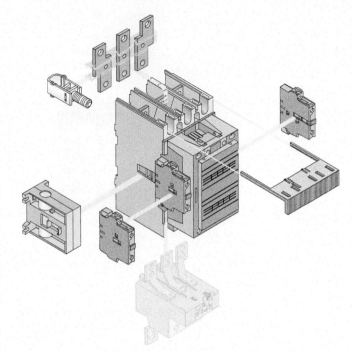

Figure 6-17 Contactor auxiliary contact blocks.

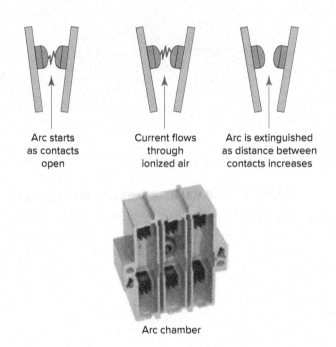

Figure 6-18 Electric arc occurs when contacts are opened under load.
Photo courtesy of Rockwell Automation,
www.rockwellautomation.com

contact surfaces if the voltage across the two points is high enough (Figure 6-18). The path for this continued flow is through the ionized air that creates the arc. As the distance between contacts increases, the resistance of the arc increases, the current decreases, and the voltage necessary to sustain the arc across the contacts increases.

Finally, a distance is reached at which full line voltage across the contacts is insufficient to maintain the arc. Arc current can create a substantial temperature rise on the surface of the contacts. This temperature rise may be high enough to cause the contact surfaces to become molten and emit vaporized metal into the gap between the contacts. Therefore, the sooner the arc is extinguished, the better; if allowed to continue, the hot arc will melt the contact surface. Most contactors contain some type of arc chamber to help extinguish the arc.

Factors that contribute significantly to contact arcing include:

- **The level of voltage and current being switched.** As circuit voltage and current increase, the gap between the opening contacts ionizes more rapidly into a conductive path.
- **Whether the voltage being switched is AC or DC.** Direct current arcs are considerably more difficult to extinguish than AC arcs. An AC arc is self-extinguishing; the arc will normally extinguish as the AC cycle passes through zero. In the case of a DC supply there is no current zero, as the current is always in one direction, so no natural arc extinction properties exist.
- **The type of load (resistive versus inductive).** With resistive loads, the duration of the arc is primarily determined by the speed at which the contacts separate. With inductive loads, the release of stored energy built up in the magnetic field serves to maintain the current and cause voltage spikes. Inductive loads in AC circuits are less of a problem than in DC circuits.
- **How quickly the contactor operates.** The faster the speed of contact separation, the quicker the arc will be extinguished.

Arcing may also occur on contactors when they are closing, for example, if the contacts come close enough together that a voltage breakdown occurs and the arc is able to bridge the open space between the contacts. Another way this can occur is if a rough edge of one contact touches the other first and melts, causing an ionized path that allows current to flow. In either case, the arc lasts until the contact surfaces are fully closed.

One major difference between AC and DC contactors is the electrical and mechanical requirements necessary for suppressing the arcs created in opening and closing contacts under load. To combat prolonged arcing in DC circuits, the contactor switching mechanism is constructed so that the contacts will separate rapidly and with enough air gap to extinguish the arc as soon as possible on opening.

Figure 6-19 DC contactor.
Photo courtesy of Hubbell Industrial Control, www.hubbell-icd.com.

DC contactors are **larger** than equivalently rated AC types to allow for the additional air gap (Figure 6-19). It is also necessary in closing DC contacts to move the contacts together as quickly as possible to avoid some of the same problems encountered in opening them. For this reason, the operating speeds of DC contactors are designed to be faster than those of AC contactors.

An **arc chute** or shield is a device designed to help confine, divide, and cool an arc, so that the arc is less likely to sustain itself. There is one arc chute for each set of contacts that is fitted above the moving and fixed contact (Figure 6-20). Arc chutes split the arc established at contactor tips while breaking the current to quench the arc. In addition, they also provide barriers between line voltages.

The arc chutes used in AC contactors are similar in construction to those used in DC contactors. However, in addition to arc chutes, most DC contactors employ magnetic blowout coils to assist with arc suppression. **Blowout coils** consist of heavy copper coils mounted above the contacts and connected in series with them (Figure 6-21). Current flow through the blowout coil sets up a magnetic field between the breaking contacts that "blows" out the arc. When an arc is formed, the arc sets up a magnetic field around itself. The magnetic field of the arc and the blowout coil repel each other. The net result is an upward push that makes the arc become longer and longer until it breaks and is extinguished.

Blowout coils seldom wear out or give trouble when operated within their voltage and current ratings. Arc chutes are constantly subjected to the intense heat of arcing and may eventually burn away, allowing the arc to

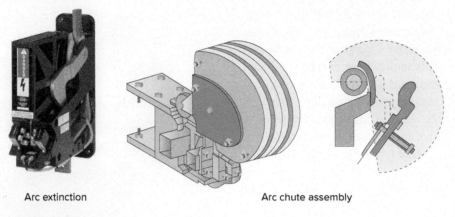

Arc extinction   Arc chute assembly

Figure 6-20 Arc chute.
Photo courtesy of Eaton, www.eaton.com.

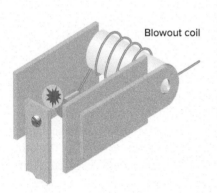

Figure 6-21 Magnetic blowout coil.
Photo courtesy of Monster Controls, www.monstercontrols.com

Sealed vacuum bottle

Figure 6-22 Vacuum contactor.
Photos courtesy of Rockwell Automation, www.rockwellautomation.com

short-circuit to the metal blowout pole pieces. Therefore, arc chutes should be inspected regularly and replaced before they burn through.

As part of a preventive maintenance program, large contactors should be checked periodically for contact wear, contact wipe, shunt terminal connections, free movement of the armature, blowout structure, blowout coil connections, coil structure, correct contact spring tension, and correct air gap. Normally the slight rubbing action and burning that occur during normal operation keep the contact surfaces clean for proper operation. Copper contacts, still used on some contactors, should be cleaned to reduce contact resistance. Worn contacts should always be replaced in pairs to ensure that complete and proper surface contact is maintained. High contact resistance produces overheating of contacts as well as a significant voltage drop across the contacts, resulting in less voltage being delivered to the load.

A **vacuum** contactor (Figure 6-22) switches power contacts inside a sealed vacuum bottle. The vacuum provides a better environment than free air for breaking the arc because without air to ionize, the arc extinguishes more quickly. Housed in vacuum bottles, the arc is isolated and the contacts are protected from dust and corrosion. Compared to conventional air contactors, they offer a significantly higher electrical endurance and are the preferred switching devices in applications with a high switching frequency, for heavy-duty starting, and for line voltages above 600 V.

DC power contactors are used to interrupt the current during shutdown of inverters in grid connected solar power systems. The direct current section of a typical photovoltaic system consists of parallel strings of solar panels connected in series. The presence of voltage in the range of 600 to 1,000 V DC and beyond allows the contactor to shut off the voltage other than by obscuring the solar panels. The main difference between contactors designed for AC power and those designed to switch DC is the enhanced ability to quench the arc so that it is not sustained any longer than necessary.

Part 1 Review Questions

1. What is the NEMA definition for a magnetic contactor?
2. Two circuits are involved in the operation of a magnetic contactor. Identify these circuits and the part of the contactor each connects to.
3. Give a brief description of how a magnetic contactor operates.
4. A micro switch, when activated, is used to switch current to a solenoid valve coil by way of a magnetic contactor. To which circuits of the contactor would each device be wired?
5. A magnetic contactor that has a coil rated for 24 V AC is fed from a 240 V AC power supply. What would most likely be used to lower the voltage level for the coil?
6. Compare the operation of electrically and mechanically held magnetic contactors.
7. List three types of operating mechanisms for magnetic contactors.
8. Why is the contactor coil molded into an epoxy resin?
9. What is the negative effect of operating a contactor coil above or below its rated voltage?
10. Which contactor coil rating refers to the amount of voltage below which the magnetic field becomes too weak to maintain the contacts in their closed position?
11. Explain why the inrush current to an AC contactor coil is much higher than its normal operating current.
12. Explain how a shading coil prevents an AC contactor from chattering.
13. Why are AC contactor assemblies made of laminated steel?
14. In what way can misalignment of the armature and core of an AC contactor cause a contactor coil to run hot?
15. Why do manufacturers recommend that discolored silver contacts not be filed?
16. In what way are the main power contacts of a contactor different from auxiliary contacts?
17. Why do contactors require some form of arc suppression?
18. Does the severity of contact arcing increase or decrease with each of the following changes?
    a. A decrease in the voltage level
    b. Use of an AC rather than a DC power source
    c. Change of load from resistive type to inductive type
    d. An increase in the speed of contact separation
19. Why is it harder to extinguish an arc on contacts passing direct current than on contacts passing alternating current?
20. Compare the design features of AC and DC contactors.
21. What is the function of an arc chute?
22. Explain the operation of the blowout coil used in DC contactors.
23. List six things to check as part of routine preventive maintenance for large contactors.
24. a. Explain the main advantage of using a vacuum contactor.
    b. List three common switching applications for vacuum contactors.
25. What is the function of the damping resistors used with a capacitor switching contactor?
26. What part of a grid-connected solar system would require a DC rated contactor to interrupt current flow during shut down of the system?

PART 2 CONTACTOR RATINGS, ENCLOSURES, AND SOLID-STATE TYPES

NEMA Ratings

The National Electric Manufacturers Association (NEMA) and the International Electrotechnical Commission (IEC) maintain guidelines for contactors. The NEMA standards for contactors differ from those of the IEC, and it is important to understand these differences.

A philosophy of the NEMA standards is to provide electrical interchangeability among manufacturers for a given **NEMA size.** Because the customer often orders a contactor by the current, motor horsepower, and voltage ratings, and may not know the application or duty cycle planned for the load, the NEMA contactor is designed by convention with sufficient reserve capacity to assure performance over a broad band of applications.

The continuous current rating and horsepower at the rated voltages categorize NEMA size ratings. NEMA contactor size guides for AC and DC contactors are shown in Figure 6-23. Because copper contacts are used on some contactors, the current rating for each size is an 8-hour open rating—the contactor must be operated at least once every 8 hours to prevent copper oxide from forming on the tips and causing excessive contact heating. For contactors with silver to silver-alloy contacts, the 8-hour rating is equivalent to a continuous rating. The NEMA current rating is for each main contact individually and not the contactor as a whole. As an example, a **Size 00** three-pole AC contactor rated at 9 A can be used for switching three separate 9 A loads simultaneously. Additional ratings for total horsepower are also listed. When selecting always ensure that the contactor ratings exceed the load to be controlled. NEMA contactor sizes are normally available in a variety of coil voltages.

As the NEMA size number classification increases, so does the current capacity and physical size of the contactor. Larger contacts are needed to carry and break the higher currents, and heavier mechanisms are required to open and close the contacts.

NEMA size 0

NEMA size 2

60 Hz AC contactor NEMA ratings 600 volts max	
NEMA size	Continuous amps
00	9
0	18
1	27
2	45
3	90
4	135
5	270
6	540
7	810
8	1215
9	2250

DC contactor NEMA ratings 600 volts max	
NEMA size	Continuous amps
1	25
2	50
3	100
4	150
5	300
6	600
7	900
8	1350
9	2500

Figure 6-23 NEMA contactor size guide.
Photos courtesy of Siemens, www.siemens.com

- **Inductive loads** such as industrial motors and transformers (low initial resistance until the transformer becomes magnetized or the motor reaches full speed; current lags behind voltage).
- **Capacitive loads** such as industrial capacitors for power factor correction (low initial resistance as capacitor charges; current leads voltage).

IEC Ratings

IEC contactors, compared to NEMA devices, generally are physically **downsized** to provide higher ratings in a smaller package (Figure 6-24). On average, IEC devices are 30 to 70 percent smaller than their NEMA counterparts. IEC contactors are not defined by standard sizes, unlike NEMA contactors. Instead, the IEC rating indicates that a manufacturer or laboratory has evaluated the contactor to meet the requirements of a number of defined "applications." With knowledge of the application

EXAMPLE 6-1

Problem: Use the table in Figure 6-23 to determine the NEMA size of an AC contactor required for a 480 V heating element load with a continuous current rating of 80 A.

Solution: According to the table, a size 2 contactor is rated for 45 A, while a size 3 is rated for 90 A. Since the load falls between these two values, the larger-size contactor must be used. The voltage requirement is satisfied because the controller can be used for all voltages up to 600 V.

Magnetic contactors are also rated for the type of load to be utilized or for actual applications. Load utilization categories include:

- **Nonlinear loads** such as tungsten lamps for lighting (large hot-to-cold resistance ratio, typically 10:1 or higher; current and voltage in phase).
- **Resistive loads** such as heating elements for furnaces and ovens (constant resistance; current and voltage in phase).

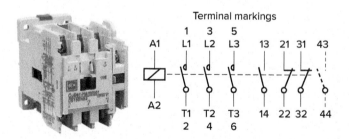

Figure 6-24 IEC-type contactor.
Photo courtesy of Automation Direct, www.automationdirect.com.

you can choose the appropriate contactor by defining the correct utilization category. This makes it possible to reduce contactor size, and therefore cost. The IEC rating system is broken down into different "utilization categories" that define the value of the current that the contactor must make, maintain, and break. The following category definitions are the most commonly used for IEC contactors:

AC Categories

AC-1: This applies to all AC loads where the power factor is at least 0.95. These are primarily noninductive or slightly inductive loads.

AC-3: This category applies to squirrel-cage motors where the breaking of the power contacts would occur while the motor is running. On closing, the contactor experiences an inrush, which is five to eight times the nominal motor current, and at this instant, the voltage at the terminals is approximately 20 percent of the line voltage.

AC-4: This applies to the starting and breaking of a squirrel-cage motor during an inch or plug reverse. On energization, the contactor closes on an inrush current approximately five to eight times the nominal current. On de-energization, the contactor breaks the same magnitude of nominal current at a voltage that can be equal to the supply voltage.

DC Categories

DC-1: This applies to all DC loads where the time constant (L/R) is less than or equal to 1 millisecond. These are primarily noninductive or slightly inductive loads.

DC-2: This applies to the breaking of shunt motors while they are running. On closing, the contactor makes the inrush current around 2.5 times the nominal rated current.

DC-3: This applies to the starting and breaking of a shunt motor during inching or plugging. The time constant is less than or equal to 2 milliseconds. On energization, the contactor sees current similar to that in category DC-2. On de-energization, the contactor will break around 2.5 times the starting current at a voltage that may be higher than the line voltage. This would occur when the speed of the motor is low because the back emf is low.

DC-5: This applies to the starting and breaking of a series motor during inching or plugging. The time constant is less than or equal to 7.5 milliseconds. On energization, the contactor sees about 2.5 times the nominal full-load current. On de-energization, the contactor breaks the same amount of current at a voltage that can be equal to the line voltage.

Contactor Enclosures

Enclosed magnetic contactors must be housed in an approved enclosure based on the environment in which they must operate to provide mechanical and electrical protection. Electrical codes mandate the type of enclosure to use. More severe environments require more substantial enclosures. Severe environmental factors to be considered include:

- Exposure to damaging fumes
- Operation in damp places
- Exposure to excessive dust
- Subject to vibration, shocks, and tilting
- Subject to high ambient air temperature

There are two general types of NEMA enclosures: nonhazardous-location enclosures and hazardous-location enclosures. **Nonhazardous**-location enclosures are further subdivided into the following categories:

- General-purpose (least costly)
- Watertight
- Oiltight
- Dust-tight

Hazardous-location enclosures are extremely costly, but they are necessary in some applications. Hazardous-location, explosion-proof enclosures involve forged or cast material and special seals with precision-fit tolerances. The explosion-proof enclosures are constructed so that an explosion inside will not escape the enclosure. If an internal explosion were to blow open the enclosure, a general-area explosion and fire could ensue. Hazardous-location enclosures are classified into two categories:

- Gaseous vapors (acetylene, hydrogen, gasoline, etc.)
- Combustible dusts (metal dust, coal dust, grain dust, etc.)

All industrial electrical and electronic enclosures must conform to standards published by NEMA to meet the needs of location conditions. Figure 6-25 shows typical NEMA enclosure types, including:

NEMA Type 1—general-purpose type, which is the least costly, and used in a location where unusual service conditions do not exist.

NEMA Type 4 and 4X—Watertight and dust-tight.

NEMA Type 12—Provides a degree of protection from noncorrosive dripping liquids, falling dirt, and dust.

NEMA Type 7 and 9—Designed for use in hazardous locations.

Although the enclosures are designed to provide protection in a variety of situations, the internal wiring and

Figure 6-25 Typical contactor enclosure types.
This material and associated copyrights are proprietary to, and used with the permission of, Schneider Electric.

Figure 6-26 Single-pole solid-state contactor.

NEMA enclosure-type number	IEC enclosure designation
1	IP10
2	IP11
3	IP54
3R	IP14
3S	IP54
4 and 4X	IP56
5	IP52
6 and 6P	IP67
12 and 12K	IP52
13	IP54

physical construction of the device remains the same. Consult the National Electrical Code (NEC) and local codes to determine the proper selection of an enclosure for a particular application.

The IEC provides a system for specifying the enclosures of electrical equipment on the basis of the degree of protection provided by the enclosure. Unlike NEMA, IEC does not specify degrees of protection for environmental conditions such as corrosion, rust, icing, oil, and coolants. For this reason, IEC enclosure classification designations cannot be exactly equated with those of NEMA. The table provides a guide for converting from NEMA enclosure-type numbers to IEC enclosure classification designations. The NEMA types meet or exceed the test requirements for the associated IEC classifications; for this reason the table should not be used to convert from IEC classifications to NEMA types, and the NEMA to IEC conversion should be verified by test.

Solid-State Contactor

Solid-state switching refers to interruption of power by nonmechanical electronic means. Figure 6-26 shows a single-pole AC solid-state contactor that uses electronic switching. In contrast to a magnetic contactor, an electronic contactor is absolutely silent, and its "contacts" never wear out. Solid-state or solid-state static contactors are recommended in applications that require a high switching frequency, such as heating circuits, dryers, single- and three-pole motors, and other industrial applications.

The most common high-power switching semiconductor used in solid-state contactors is the **silicon controlled rectifier (SCR).** An SCR is a three-terminal semiconductor device (anode, cathode, and gate) that acts like the power contact of a magnetic contactor. A gate signal, instead of an electromagnetic coil, is used to turn the device on, allowing current to pass from cathode to anode. Figure 6-27 shows three types of SCR construction styles designed for higher-current applications: the disk (also known as puck type), stud mount, and module. Flexible-lead stud-mounted SCRs have a gate wire, a flexible cathode lead, and a smaller cathode lead

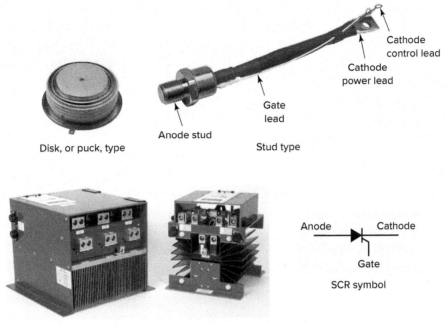

Figure 6-27 Silicon controlled rectifier (SCR) switching semiconductor.
Disk and stud type: Photos courtesy of Vishay Intertechnology. www.vishay.com.
Module type: Photo Control Concepts, Inc., www.ccipower.com.

that is used only for control purposes. The heat generated by the SCR must be dissipated; thus all contactors have some means to cool the SCR. Typically an aluminum **heat sink,** with fins to increase the surface area, is used to dissipate this energy to the air.

The SCR, like a contact, is in either the on state (closed contact) or the off state (open contact). SCRs are normally off switches that can be triggered on by a small current pulse into the gate electrode. Once turned on (or **triggered**), the component then stays in the conducting state even when the gate on signal is removed. It returns to the off (blocking) state only if the anode-to-cathode current falls below a certain minimum or if the direction of the current is reversed. In this respect, the SCR is analogous to a **latched** contactor circuit—once the SCR is triggered, it will stay on until its current decreases to zero.

The SCR testing circuit shown in Figure 6-28 is practical both as a diagnostic tool for checking suspected SCRs and as an aid to understanding how they operate. The operation of the circuit can be summarized as follows:

- A DC voltage source is used for powering the circuit, and two pushbutton switches are used to latch and unlatch the SCR.
- Momentarily closing the on push button connects the gate to the anode, allowing current to flow from the negative terminal of the battery, through the cathode-gate junction, through the switch, through the bulb, and back to the battery.
- This gate current should cause the SCR to latch on, allowing current to go directly from cathode to anode without further triggering through the gate.
- Momentarily opening the normally closed off push button interrupts the current flow to the SCR and bulb. The light turns off and remains off until the SCR is triggered back into conduction.
- If the bulb lights at all times, this is an indication that the SCR is shorted.
- If the bulb fails to light when the SCR is triggered into operation, this is an indication that the SCR is faulted open.

Since an SCR passes current in one direction only, **two SCRs** are necessary to switch single-phase AC power. The

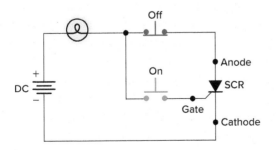

Figure 6-28 SCR testing circuit.

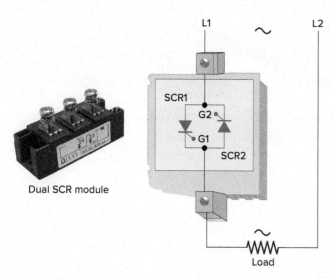

Figure 6-29 SCR connection for single-phase contactor.
Photo courtesy of Digi-Key Corporation, www.digikey.com

two SCRs are connected inverse-parallel (back-to-back), as shown in Figure 6-29: one to pass current during the positive half-cycle and the other during the negative half-cycle. Half the current is carried by each SCR, and sinusoidal AC current flows through the resistive load *R* when gates G1 and G2 are fired at 0 degrees and 180 degrees of the input, respectively.

Any device that makes use of a **coil** of wire can be classed as an **inductive** load. Common inductive loads include transformers, relay coils, motors, and solenoid coils. When current passes through a coil of wire, it stores electric energy in its magnetic field. Once this energy is created, it must be discharged when the current is removed. The discharge of this stored energy can create **voltage spikes** in the thousands of volts.

Inductive loads and voltage transients are both seen as problem areas in solid-state AC contactor control because they could falsely trigger an SCR into conduction. For this reason, for driving an inductive load, a **snubber circuit** is used to improve the switching behavior of the SCR. Figure 6-30 shows an electronic contactor, with a simple *RC* snubber circuit used to control an inductive transformer load. The snubber circuit consists of a resistor and capacitor wired in series with each other and placed in parallel with the SCRs. This arrangement suppresses any rapid rise in voltage across the SCR to a value that will not trigger it.

The abrupt switching of an SCR, particularly at higher current levels, can cause objectionable transients on the power line and create electromagnetic interference (EMI). By electrically switching an SCR on at the AC sine wave zero crossing point, it remains on through the half cycle of the sine wave and turns off at the next

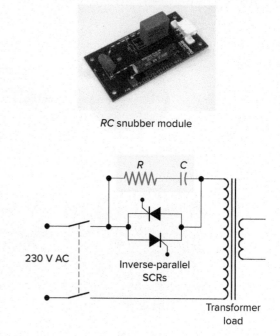

Figure 6-30 SCR snubber circuit.
Photo courtesy of Enerpro, www.enerpro-inc.com

zero crossing. In this scheme, known as **zero-fired control,** the SCR is turned on at or nearly at the zero crossing point so that no current is being switched under load. The result is virtually no power line disturbances or EMI generation.

Solid-state contactors are too **expensive** to build in very **high current** ratings, so magnetic contactors continue to dominate that application in industry today. Unlike magnetic contactors, solid-state types require very little input control power (coil current) to switch large loads. They can be switched ON and OFF by a lower power digital control signal as illustrated in Figure 6-31.

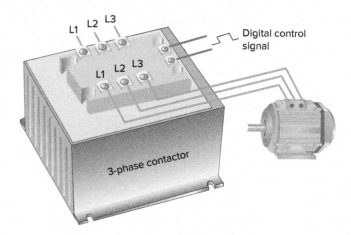

Figure 6-31 Solid-state contactor digital control.

174 **Chapter 6 Contactors and Motor Starters**

Part 2 Review Questions

1. Name the two major associations that maintain standard guidelines for contactors.
2. What three parameters are listed for each NEMA contactor size rating?
3. Use the NEMA contactor size guide to determine the NEMA size DC contactor required for a 240 V load with a continuous current rating of 80 A.
4. List four types of contactor load utilization categories.
5. Compare NEMA- and IEC-rated contactors with regard to
   a. physical size.
   b. the way in which they are rated.
6. Why are contactors mounted in an enclosure?
7. List the four categories of nonhazardous-location contactor enclosures.
8. Describe how explosion-proof contactor enclosures are constructed.
9. What does solid-state switching of contactors refer to?
10. For which type of switching operations are solid-state static contactors best suited? Why?
11. Answer the following with reference to high-powered SCR switching semiconductors used in solid-state contactors:
    a. Which circuit of the SCR is connected in series with the load, like the power contacts of a magnetic contactor?
    b. To which circuit of the SCR is the control signal that triggers the device into conduction applied?
    c. In what way is the operation of an SCR analogous to that of a latched contactor circuit?
    d. What effect can inductive loads and voltage transients have on the normal operation of an SCR?
    e. Describe the method commonly used to dissipate the heat generated by an SCR.
12. What type of current applications are solid-state contactors limited to? Why?

PART 3 MOTOR STARTERS

Magnetic Motor Starters

The basic use for the magnetic contactor is for switching power in resistance heating elements, lighting, magnetic brakes, and heavy industrial solenoids. Contactors can also be used to switch motors if **separate overload** protection is supplied. In its most basic form, a magnetic motor starter (Figure 6-32) is a contactor with an overload protective device, known as an overload relay (OL), physically and electrically attached. The overload protective device protects the motor from overheating and burning up. Normally magnetic starters come equipped

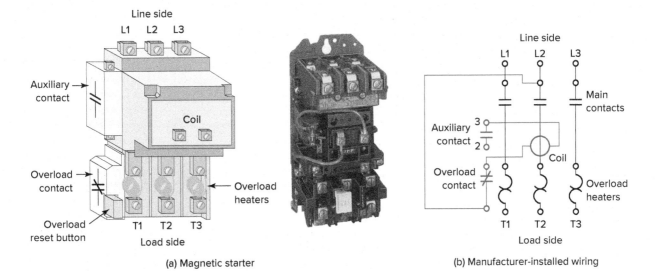

Figure 6-32 Magnetic motor starter.
Photo courtesy of Rockwell Automation, www.rockwellautomation.com

with some manufacturer-installed control wiring, which may include:

- A wire connected from the overload relay contacts to the starter coil
- A wire connected from the other side of the starter coil to the holding contacts
- A wire connected from L2 to the other side of the overload relay contacts (note that this wire must be removed when a control transformer is used)

In their simplest and most widely used form, magnetic motor starters consist of a two-, three-, or four-pole magnetic contactor and an overload relay mounted in a suitable enclosure (Figure 6-33). Enclosures are essentially boxes that **"enclose"** motor control devices such as contactors, motor starters, and push buttons. They may be of general-purpose sheet-metal construction, dust-tight, water-tight, or explosion-resisting, or whatever may be required by the installation to protect motor control equipment and people. Start and stop push buttons may be mounted in the cover of the enclosure.

A separately mounted start/stop pushbutton station may also be used, in which case only the reset button would be mounted in the cover, as illustrated in Figure 6-34. Starters are also built in skeleton form, without enclosure, for mounting in a motor control center or control panel on a machine. The control circuit of a magnetic motor starter is very simple. It involves only energizing the starter coil when the start button is pressed and de-energizing it when the stop button is pressed or when the overload relay trips.

Motor Overcurrent Protection

Motor branch circuits can be broken down into several major NEC requirements for motor installations, as illustrated in Figure 6-35. They include:

- Motor circuit and controller disconnecting means
- Motor branch short-circuit and ground-fault protection
- Motor controller and overload protection
- Sometimes motor disconnecting means, often referred to as the "at the motor" disconnecting means

When an AC motor is first energized, a high inrush of current occurs. During the initial half-cycle, this inrush

Figure 6-33 Magnetic motor starter enclosure.
Photo courtesy of Rockwell Automation,
www.rockwellautomation.com

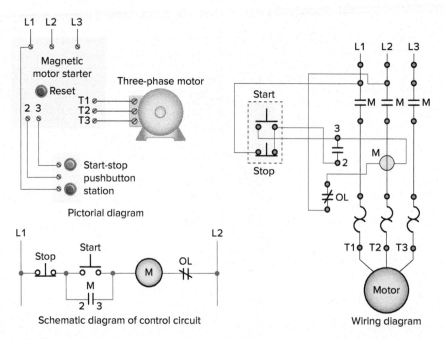

Figure 6-34 Magnetic motor starter with separately mounted start-stop pushbutton station.

Chapter 6 Contactors and Motor Starters

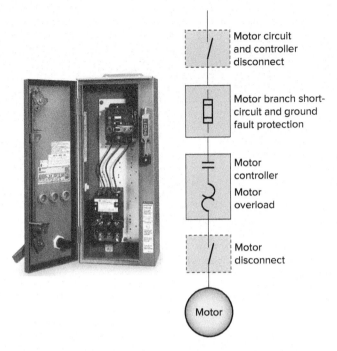

Figure 6-35 Major functional blocks for motor operation.
This material and associated copyrights are proprietary to, and used with the permission of, Schneider Electric.

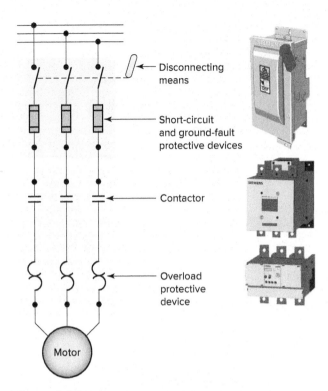

Figure 6-36 Motor overcurrent protection.
Photos courtesy of Siemens, www.siemens.com

current is often 20 times the normal full-load current. After the first half-cycle, the motor begins to rotate and the starting current subsides to four to eight times the normal current for several seconds. As a motor reaches running speed, the current subsides to its normal running level. Because of the inrush current, motors require special overload protective devices that can withstand the temporary overloads associated with starting currents and yet protect the motor from sustained overloads.

Motor starting characteristics make motor protection requirements different from those for other types of loads. When providing overcurrent protection for most circuits, we use a fuse or circuit breaker that combines overcurrent protection with short-circuit and ground-fault protection. Motor overcurrent protection is normally provided by separating the overload protection devices from the short-circuit and ground-fault protection devices, as illustrated in Figure 6-36. In addition, the NEC also requires a disconnect means.

Motor overcurrent protection can be summarized as follows:

- **Short-circuit and ground-fault motor protection.** Branch and feeder fuses and circuit breakers protect motor circuits against the very high current of a short circuit or a ground fault. Fuses and circuit breakers connected to motor circuits must be capable of ignoring the initial high inrush current and allow the motor to draw excessive current during start-up and acceleration.

- **Overload protection.** Overload devices are intended to protect motors, motor control apparatus, and motor-branch circuit conductors against excessive heating due to motor overloads and failure to start. Motor overload may include conditions such as a motor operating with an excessive load or a motor operating with low line voltages or, on a three-phase motor, a loss of a phase. The motor overload devices are most often integrated into the motor starter.

The basic difference between a contactor and motor starter is the addition of overload relays as shown in Figure 6-37. Contactor use is restricted to fixed lighting loads, electric furnaces, and other resistive loads that have set current values. Motors are subject to high starting currents and periods of load, no-load, short duration overload, and so on. They must have protective devices with the flexibility required of the motor and driven equipment. The purpose of overload protection is to protect the motor windings from excessive heat resulting from motor overloading. The motor windings will not be damaged when overloaded for a short period of time. If the overload should persist, however, the sustained increase in current should cause the overload relay to operate, shutting off the motor.

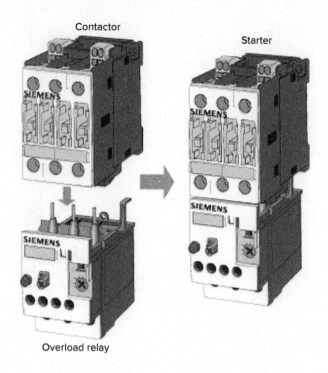

Figure 6-37 The basic difference between a contactor and motor starter is the addition of overload relays.
Photo courtesy of Siemens, www.siemens.com

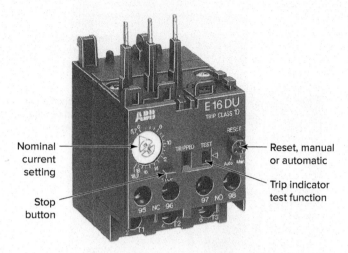

Figure 6-38 Overload relay trip indicator.
Photo courtesy of ABB, www.abb.com

Motor Overload Relays

Overload relays are designed to meet the special protective needs of motor control circuits. Overload relays:

- Allow harmless temporary overloads (such as motor starting) without disrupting the circuit.
- Will trip and open a circuit if current is high enough to cause motor damage over a period of time.
- Can be reset once the overload is removed.

Overload relays are rated by a *trip class*, which defines the length of time it will take for the relay to trip in an overload condition. The most common trip classes are Class 10, Class 20, and Class 30. A Class 10 overload relay, for example, has to trip the motor off line in 10 seconds or less at 600 percent of the full-load amperes (which is usually sufficient time for the motor to reach full speed). The class designation is an important consideration in applying OL relays in motor-control circuits. For example, a high-inertia industrial load may require a Class 30 overload relay that trips in 30 seconds rather than a Class 10 or 20. The overload relay itself will have markings to indicate which class it belongs to.

Normally overload protection devices have a **trip indicator** built into the unit to indicate to the operator that an overload has taken place. Overload relays can have either a manual or an automatic reset. A **manual** reset requires operator intervention, such as pressing a button, to restart the motor. An **automatic** reset allows the motor to restart automatically, usually after a cooling-off period; this allows the motor time to cool. After an overload relay has tripped, the cause of the overload should be investigated. Motor damage can occur if repeated resets are attempted without correcting the cause of the overload relay tripping. Figure 6-38 shows a three-pole Class 10 overload relay that features a manual or automatic resetting mode selection. The nominal current setting allows the relay to be set to the full-load current shown on the motor rating plate and can be adjusted to the desired trip point.

External-overload protection devices, which are mounted in the starter, attempt to monitor the heating and cooling of a motor by sensing the **current** flowing to it. The current drawn by the motor is a reasonably accurate measure of the load on the motor and thus of its heating. Overload relays can be classified as thermal, magnetic, or electronic.

Thermal Overload Relays A thermal overload relay uses a **heater** connected in series with the motor supply. Current flowing from the motor contactor to the motor passes through the motor overload heaters (one per phase), which are mounted in the overload relay block. Each thermal overload relay (Figure 6-39) consists of the main overload block, which houses the contacts, a tripping mechanism with reset button, and interchangeable heaters sized for the motor being protected. The amount of heat produced increases with supply current. If an overload

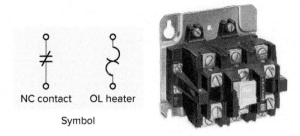

Figure 6-39 Thermal overload relay.
Photo courtesy of Rockwell Automation,
www.rockwellautomation.com

occurs, the heat produced causes a set of contacts to open, interrupting the circuit. Installing a different heater for the required trip point changes the tripping current. This type of protection is very effective because the heater closely approximates the actual heating within the windings of the motor and has a thermal "memory" to prevent immediate reset and restarting.

Thermal overload relays can be further subdivided into two types: melting alloy and bimetallic. The **melting alloy type,** illustrated in Figure 6-40, utilizes the principle of heating solder to its melting point. It consists of a heater coil, eutectic alloy, and mechanism to activate a tripping device when an overload occurs. The term **eutectic** means the mixture melts and solidifies at a single temperature. The eutectic alloy in the heater element is a material that goes from a solid to liquid state without going through an intermediate putty stage. The operation of the device can be summarized as follows:

- When the motor current exceeds the rated value, the temperature will rise to a point where the alloy melts; the ratchet wheel is then free to rotate, and the contact pawl moves upward under spring pressure, allowing the control circuit OL contacts to open.
- After the heater element cools, the ratchet wheel will again be held stationary and the overload contacts can be reset.
- The thermal overload relay uses an **integral heater element,** the trip setting of which is adjustable to match the motor's full-load amperes.

The **bimetallic type** of thermal overload relay illustrated in Figure 6-41 uses a bimetallic strip made up of two pieces of dissimilar metal that are permanently joined by lamination. The operation of the device can be summarized as follows:

- Heating the bimetallic strip causes it to bend because the dissimilar metals expand and contract at different rates.
- Overload heating elements connected in series with the motor circuit heat the bimetal tripping elements in accordance with the motor load current.
- The movement/deflection of the bimetallic strip is used as a means of operating the trip mechanism and opening the normally closed overload contacts.

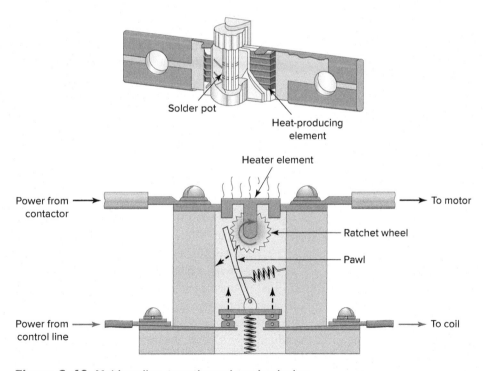

Figure 6-40 Melting alloy–type thermal overload relay.

Part 3 Motor Starters

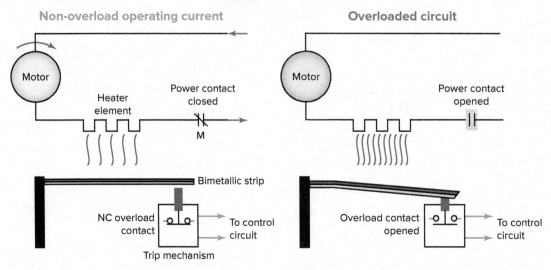

Figure 6-41 Bimetallic type of thermal overload relay.
Photo courtesy of Rockwell Automation, www.rockwellautomation.com

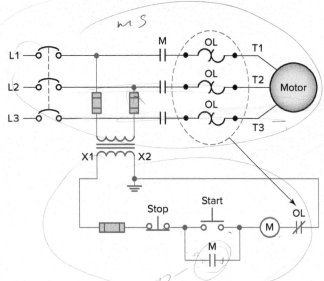

Figure 6-42 Thermal overload relay circuit operation.

With a thermal overload relay, the same current that goes to the motor coils (causing the motor to heat) also passes through the thermal elements of the overload relays. The thermal element is connected mechanically to an NC overload (OL) contact (Figure 6-42). When an excessive current flows through the thermal element for a long enough time period, the contact is tripped open. This contact is connected in series with the control coil of the starter. The starter coil is de-energized when the contact opens. In turn, the starter's main power contacts open to disconnect the motor from the line.

Damage due to overloads accounts for most motor failures. Selecting the proper heater size for thermal OL relays is critical for ensuring maximum motor protection. Overload heaters for continuous-duty motors are selected from manufacturer's tables or charts, similar to that illustrated in Figure 6-43, and based on compliance with Section 430.32(A1) of the NEC. Selection tables normally list OL heaters according to the **full-load amperage (FLA)** of the motor, and the motor **starter size.** The full load ampere (FLA) rating is found on the nameplate of a motor and differs from the full load current (FLC) rating given in the NEC tables. The lists show the range of motor currents with which they should be used. These may be in increments of from 3 to 15 percent of FLA. The smaller the increment, the closer the selection can be to match the motor to its actual work.

When the overload heater element is rated according to the motor FLA, the calculations required by the NEC to determine the necessary level of protection have already been completed. For example, an OL heater rated at 10 A

180   Chapter 6   Contactors and Motor Starters

Heater Type No.	Full-Load Amps.							
	Size 00	Size 0	Size 1	Size 1P	Size 2	Size 3	Size 4	Size 5
W10	0.19	0.19	0.19					
W11	0.21	0.21	0.21					
W12	0.23	0.23	0.23					
W13	0.25	0.25	0.25					
W14	0.28	0.28	0.28					
W15	0.31	0.31	0.31					
			4.08					
W43	4.52	4.52	4.52					226
W44	4.98	4.98	4.98					249
W45	5.51	5.51	5.51	5.80				276
W46	6.07	6.07	6.07					
W47	6.68	6.68	6.68					

Figure 6-43 Typical motor overload heater selection chart.

in the selection table is intended for use with a motor that has a 10 A FLA. Typically, it is assumed that the motor has a service factor of 1.15 or greater and a temperature rise not over 40°C, which allows the motor to be protected up to **125 percent** of the nameplate FLA rating. NEMA standards permit classifying OL heater elements in this manner, but require the manufacturer to provide conversion factors for selecting devices to protect motors that have a service factor less than 1.15 or a temperature rise over 40°C (104°F).

Thermal overload relays react to heat, regardless of the origin of the heat. **Ambient temperatures** can greatly affect the tripping time of a thermal overload relay. Cooler temperatures increase tripping times, while warmer temperatures decrease tripping times. Ambient compensated bimetal overload relays are designed to overcome this problem. A compensating bimetal strip is used along with the primary bimetal. As the ambient temperature changes, both bimetals will bend equally and the overload relay will not trip.

Electronic Overload Relays Unlike electromechanical overload relays that pass motor current through heating elements to provide an indirect simulation of motor heating, an electronic overload relay measures motor current directly through a *current transformer*. It uses a signal from the current transformer, as illustrated in Figure 6-44, along with precision solid-state measurement components to provide a more accurate indication of the motor's thermal condition. Electronic circuitry calculates the average temperature within the motor by monitoring its starting and running currents. When a motor overload occurs, the control circuit operates to open the NC overload relay contacts.

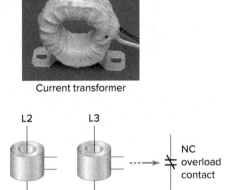

Figure 6-44 Electronic overload relays use current transformers to sense motor current.
Photo courtesy of Hammond Mfg. Co., www.hammondmfg,com.

Figure 6-45 shows an **electronic** overload relay designed to be mounted in a two-component (contactor and OL relay) starter assembly. Heaters are not used; full-load current is set with a dial. This particular electronic overload relay is adjustable for a full-load motor amperes of from 1 to 5 A. This wide 5:1 adjustment range results in the need for half as many catalog numbers as the bimetallic alternative in order to cover the same current range. A separate phase loss detection circuit incorporated into the overload relay allows it to respond quickly to phase loss conditions. The self-enclosed latching trip relay contains a set of isolated NC and NO contacts that provide trip and reset functions for control circuits. Whenever an overload motor condition is sensed, these contacts change state and trigger a control circuit that interrupts current flow to the

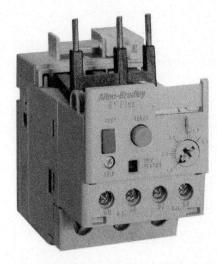

Figure 6-45 Electronic solid-state overload relay.
Photo courtesy of Rockwell Automation, www.rockwellautomation.com

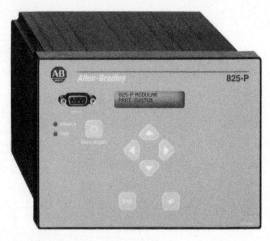

Figure 6-46 Microprocessor-based type modular overload relay.
Photo courtesy of Rockwell Automation, www.rockwellautomation.com

motor. The low energy consumption of the electronic design minimizes temperature rise issues inside control cabinets. Dual-in-line package (DIP) switch settings allow the selection of trip class (10, 15, 20, or 30) and the reset mode (manual or automatic).

Advantages of solid-state electronic overload relays over thermal-overload types include the following:

- No buying, stocking, installing, or replacing of heater coils
- Reduction in the heat generated by the starter
- Energy savings (up to 24 W per starter) through the elimination of heater coils
- Insensitivity to temperature changes in the surrounding environment
- High repeat trip accuracy (±2 percent)
- Easily adjustable to a wide range of full-load motor currents

Although both bimetal and eutectic trip mechanisms are still used, solid-state overload relays are more popular for the majority of newer motor control installations. Despite the differences between NEMA and IEC motor controls, the two types have a major similarity—the solid-state overload relay. There is little difference between solid-state overload relays used for either type. In some applications, the same solid-state overload relay can be used in NEMA and in IEC units, leaving the contactor and enclosure the main differences between the two.

Another form of the electronic overload relay is the **microprocessor-based** modular overload relay, shown in Figure 6-46. They provide better motor control management by eliminating unnecessary trips and isolating faults.

In addition to motor overload protection, other protective features may include overtemperature, instantaneous overcurrent, ground fault, phase loss/phase reversal/phase unbalance (both voltage and current), overvoltage, and undervoltage protection. Some units can tabulate the number of starts per programmed unit of time and lock out the starting sequence, preventing inadvertent excessive cycling.

Dual-Element Fuses Dual-element (time-delay) fuses, when properly sized, provide both overload and fault protection. This type of fuse contains dual fuse elements with both thermal and instantaneous trip features that allow the high motor starting current to flow for a short time without blowing the fuse. Figure 6-47 shows the construction of a dual-element fuse. The operation of a dual-element fuse can be summarized as follows:

- Under sustained overload conditions, the trigger spring fractures the calibrated fusing alloy of the overload element, releasing the connector. The insets in Figure 6-47 represent a model of the overload element before and after.
- A short-circuit fault causes the restricted portions of the short-circuit element to vaporize, and arcing commences. The special granular, arc-quenching filler material quenches the arcs, creating an insulating barrier that forces the current flow to zero.

NEMA and IEC Symbols

As the electrical standards adopted by various nations vary, the **markings** and **symbols** used to describe electrical control products vary as well. Today's motor

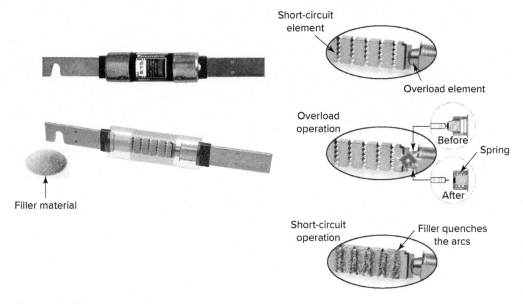

Figure 6-47 Dual-element fuse.
Courtesy of Cooper Industries, Copyright ©2011 Cooper Industries.

control panel may contain products from various parts of the world. The need to recognize and understand these symbols is crucial to reading different types of motor control diagrams. Figure 6-48 shows a comparison of typical NEMA and IEC symbols commonly found in motor control schematic/wiring diagrams.

Figure 6-49 shows a comparison of NEMA and IEC symbols found in a typical motor control schematic for a combination starter with a fused disconnect switch. IEC devices have power terminals marked 1, 3, 5 and 2, 4, 6 corresponding in function to L1, L2, L3 and T1, T2, T3 on NEMA controllers. The terminals of the IEC auxiliary contacts are marked with a two-digit number. Terminals that belong together are marked with the same location digit (first digit). The second digits (called the function digits) identify the function of each contact. For example, on the motor control IEC schematic:

- The numbers 13 and 14 represent an auxiliary contact.
- The number 1 identifies that this is the first contact in the sequence.
- The numbers 3 and 4 identify this is as normally open contact.
- The numbers 21 and 22 represent another auxiliary contact.
- The number 2 identifies that this is the second contact in the sequence.
- The numbers 1 and 2 identify this as a normally closed contact.

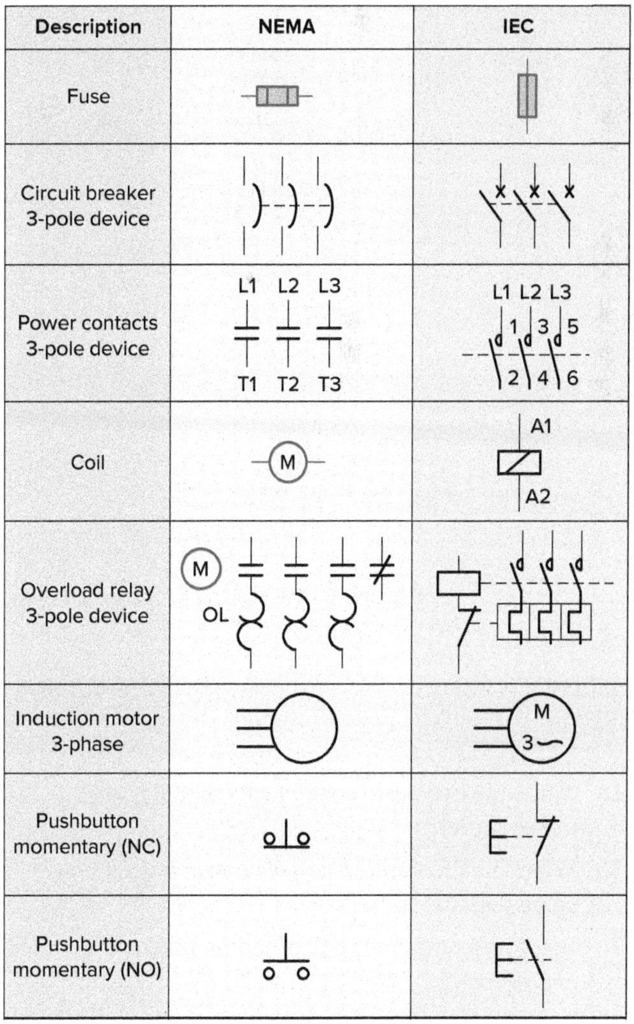

Figure 6-48 Comparison of NEMA and IEC symbols.

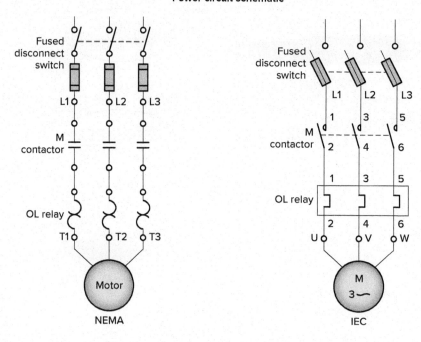

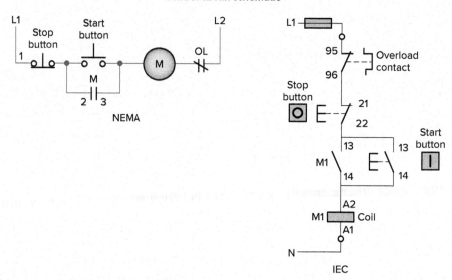

Figure 6-49 Typical NEMA and IEC symbols found in a motor control schematic.

Part 3 Review Questions

1. Name the two basic components of a magnetic motor starter.
2. What manufacturer-installed control wiring may come with the starter?
3. List four common types of motor starter enclosures.
4. Identify four major NEC requirements for motor installations.
5. Explain how motor overload protection devices function to protect the motor.
6. In what manner is motor overcurrent protection normally provided?
7. Outline three important operating characteristics of overload relays.
8. Assume an overload relay is rated for trip Class 20. What exactly does this mean?

9. Compare the operation of the manual reset and automatic reset overload relays.
10. List three ways in which motor overload relays are classified.
11. How do thermal overload relays provide an indirect monitoring of motor heating?
12. Compare how the tripping device is activated in a melting alloy (eutectic)–and a bimetal-type thermal overload relay.
13. List four major factors to be considered when selecting the proper heater size for a motor thermal overload relay.
14. How is motor current sensed in an electronic overload relay?
15. List five advantages of electronic overload relays over thermal overload relay types.
16. Explain the operating principle of a dual-element (time-delay) fuse.

Troubleshooting Scenarios

1. Identify possible causes or things to investigate for each of the following reported problems with a magnetic contactor or motor starter.
   a. Noisy coil assembly
   b. Coil failure
   c. Excessive wear on electromagnet
   d. Overheating of the blowout coil
   e. Pitted, worn, or broken arc chutes
   f. Failure to pick up
   g. Short contact life
   h. Broken flexible shunt
   i. Failure to drop out
   j. Insulation failure
   k. Failure to break arc
   l. An overload that trips on low current
   m. Failure to trip (motor burnout)
   n. Failure to reset

2. A magnetic contactor coil rated for 24 V AC is incorrectly replaced with an identical physical size coil rated for 24 V DC. How might this affect the operation of the contactor or starter?

3. Solid-state SCR switching contactor modules may become faulted as short circuits or open circuits. Discuss symptoms that might be associated with each type of failure.

4. One of three starter thermal overload heater elements has become open-circuited because of overheating and is to be replaced. Why is it recommended that the set of three rather than the single heater element be replaced?

5. List the things to investigate in determining the cause of excessive tripping of a motor overload device.

Discussion Topics and Critical Thinking Questions

1. The key to understanding conductor and motor protection is to know the meaning of ground fault, short-circuit fault, and overload. Demonstrate your understanding of these terms by citing motor circuit examples of each.

2. Explain how electronic overload relays protect against single phasing.

3. Why are IEC contactors and starters much smaller in size than their NEMA counterparts?

4. Identify the different types of contacts found on a magnetic motor starter and describe the function each performs.

5. Search the Internet for motor starter enclosures and identify the NEMA type that would be suitable for each of the following environments:
   a. In a paint booth
   b. In a boiler room
   c. At a grain feed mill
   d. Inside a plant for lathe controls

6. Inherent motor protection devices are located in the motor housing or mounted directly to the motor and accurately sense the heat being generated by the motor. Draw the schematic for a standard three-wire control circuit showing this type of device integrated into the control circuit.

CHAPTER SEVEN

Relays

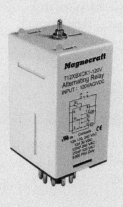
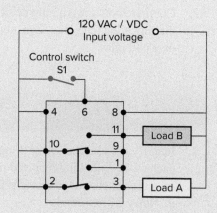

Magnecraft photo courtesy of Schneider Electric USA, Inc., www.serelays.com

CHAPTER OBJECTIVES

This chapter will help you:

- Compare electromagnetic, solid-state, timing, and latching relays in terms of construction and operation.
- Identify relay symbols used on schematic diagrams.
- Identify different types of relay applications.
- Explain how relays are rated.
- Describe the operation of on-delay and off-delay timer relays.
- Summarize the use of relays as control elements in motor circuits.
- Explain the operation and application of interposing and analog-switching relays.
- Compare timed and instantaneous relay contacts.

Many motor applications in industry and in process control require relays as critical control elements. Relays are used primarily as switching devices in a circuit. This chapter explains the operation of different types of relays and the advantages and limitations of each type. Relay specifications are also presented to show how to determine the correct relay type for different applications.

PART 1 ELECTROMECHANICAL CONTROL RELAYS

Relay Operation

An **electromechanical relay** (EMR) is best defined as a switch that is operated by an electromagnet. The relay turns a load circuit on or off by energizing an electromagnet, which opens or closes contacts connected in series with a load. A relay is made up of two circuits: the coil input or **control** circuit and the contact output or **load** circuit, as illustrated in Figure 7-1. Relays are used to control small loads of 15 A or less. In motor circuits, electromechanical relays are often used to control coils in motor contactors and starters.

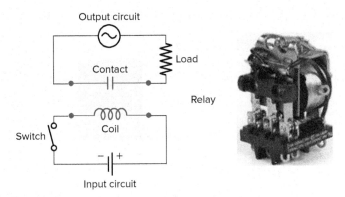

Figure 7-1 Electromechanical control relay.
Photo courtesy of Tyco Electronics, www.tycoelectronics.com

Other applications include switching of solenoids, pilot lights, audible alarms, and small motors (⅛ hp or less).

Operation of a relay is very similar to that of a contactor. The main difference between a control relay and a contactor is the size and number of contacts. Control relay contacts are relatively small because they need to handle only the small currents used in control circuits. The small size of control relay contacts allows control relays to contain multiple isolated contacts.

A relay will usually have only one coil, but it may have any number of different contacts. Electromechanical relays contain both stationary and moving contacts, as illustrated in Figure 7-2. The moving contacts are attached to the armature. Contacts are referred to as **normally open** (NO) and **normally closed** (NC). When the coil is energized, it produces an electromagnetic field. Action of this field, in turn, causes the armature to move, closing the NO contacts and opening the NC contacts. The distance that the plunger moves is generally short—about ¼ inch or less. A letter is used in most diagrams to designate the coil. The letter M frequently indicates a motor starter, while CR is used for control relays. The associated contacts will have the same identifying letters.

Normally open contacts are open when no current flows through the coil but closed as soon as the coil conducts a current or is energized. Normally closed contacts are closed when the coil is de-energized and open when the coil is energized. Each contact is normally drawn as it would appear with the **coil de-energized.** Some control relays have some provision for changing contacts from normally open to normally closed types, or vice versa. The provisions range from a simple flip-over contact to removing the contacts and relocating with spring location changes.

Relays are used to control several switching operations by a single, separate current. One relay coil/armature assembly may be used to actuate more than one set of contacts. Those contacts may be normally open, normally closed, or any combination of the two. A simple example of this type of application is the relay control with two pilot lights illustrated in Figure 7-3. The operation of the circuit can be summarized as follows:

- With the switch open, coil CR1 is de-energized.
- The circuit to the green pilot light is completed through normally closed contact CR1-2, so this light will be on.

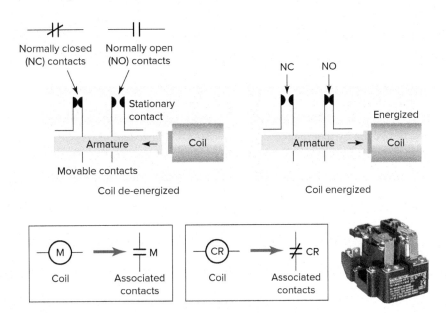

Figure 7-2 Relay coil and contacts.
Photo courtesy of Eaton, www.eaton.com.

Part 1 Electromechanical Control Relays 187

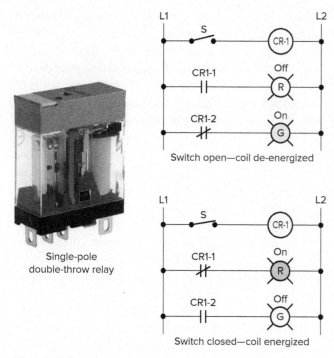

Figure 7-3 Relay switching operation.
Photo courtesy of Digi-Key Corporation, www.digikey.com

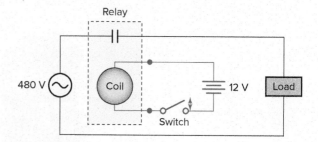

Figure 7-4 Relay used to control a high-voltage circuit with a low-voltage circuit.

- At the same time, the circuit to the red pilot light is opened through normally open contact CR1-1, so this light will be off.
- With the switch closed, the coil is energized.
- The normally open contact CR1-1 closes to switch the red pilot light on.
- At the same time, the normally closed CR1-2 opens to switch the green pilot light off.

Relay Applications

Relays are extremely useful when we need to control a large amount of current and/or voltage with a small electrical signal. The relay coil, which produces the magnetic field, may consume only a fraction of a watt of power, while the contacts closed or opened by that magnetic field may be able to conduct hundreds of times that amount of power to a load.

You can use a relay to control a **high-voltage load** circuit with a low-voltage control circuit as illustrated in the circuit of Figure 7-4. This is possible because the coil and contacts of the relay are electrically insulated from each other. The relay's coil is energized by the low-voltage (12 V) source, while the contact interrupts the high-voltage (480 V) circuit. Closing and opening the switch energizes and de-energizes the coil. This, in turn, closes and opens the contacts to switch the load on and off.

You can also use a relay to control a **high-current load** circuit with a low-current control circuit. This is possible because the current that can be handled by the contacts can be much greater than what is required to operate the relay coil. Relay coils are capable of being controlled by low-current signals from integrated circuits and transistors, as illustrated in Figure 7-5. The operation of the circuit can be summarized as follows:

- The electronic control signal switches the transistor on or off, which in turn causes the relay coil to energize or de-energize.
- The current in the transistor control circuit and relay coil is quite small in comparison to that of the solenoid load.
- Transistors and integrated circuits (ICs, or chips) must be protected from the brief high-voltage spike produced when the relay coil is switched off.
- In this circuit a diode is connected across the relay coil to provide this protection.
- Note that the diode is connected backward so that it will normally not conduct. Conduction occurs only when the relay coil is switched off; at this moment current tries to continue flowing through the coil and it is harmlessly diverted through the diode.

Relay Styles and Specifications

Control relays are available in a variety of styles and types. One popular type is the general-purpose "ice cube" relay, so named because of its size and shape and the clear plastic enclosure surrounding the contacts. Although the contacts are nonreplaceable, this relay is designed to plug into a socket, making replacement fast and simple in the event of failure. An eight-pin plug-in-style ice cube relay is shown in Figure 7-6. This relay contains two separate single-pole double-throw contacts. Because the relay plugs into a socket, the wiring is connected to the socket, not the relay. The numbering on the socket base designates a terminal with the corresponding pin position. Care must be taken not to confuse the base numbers with the wire reference numbers used to label control wires.

Relay options that aid in troubleshooting are also available. An on/off indicator is installed to indicate the state

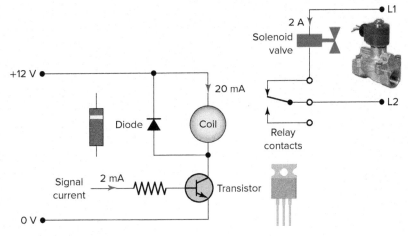

Figure 7-5 Using a relay to control a high-current load circuit with a low-current control circuit.

Top photo courtesy of GC Valves; bottom photo courtesy Fairchild Semiconductor, www.fairchildsemi.com

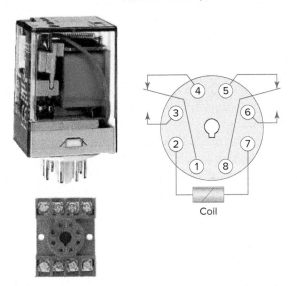

Figure 7-6 Plug-in style ice cube relay.

Photo courtesy of Rockwell Automation, www.rockwellautomation.com

Figure 7-7 Relay manual override push-to-test button.

Magnecraft photo courtesy of Schneider Electric USA, Inc., www.serelays.com

Figure 7-8 Typical DIN rail.

Magnecraft photos courtesy of Schneider Electric USA, Inc., www.serelays.com

(energized or de-energized) of the relay coil. A manual override button (Figure 7-7), mechanically connected to the contact assembly, may be used to move the contacts into their energized position for testing purposes. Use caution when exercising this feature because the circuit controlling the coil is bypassed and loads may be energized or de-energized without warning.

A **DIN rail** (Figure 7-8) is a metal rail often used for mounting electrical devices, such as relays, inside control panels. The main advantage of a DIN rail mounting system is that it offers quick replacement and panel space-saving solutions for common relay control applications.

Like contactors, relay coils and contacts have separate ratings. Relay coils are usually rated for type of operating current (DC or AC), normal operating voltage or current, permissible coil voltage variation (pickup and dropout), resistance, and power. Coil voltages of 12 V DC, 24 V DC, 24 V AC, and 120 V AC are most common. Sensitive relay

Part 1 Electromechanical Control Relays

coils that require as little as 4 mA at 5 V DC are used in relay circuits operated by transistor or integrated circuit chips.

Relays are available in a wide range of switching configurations. Figure 7-9 illustrates common relay contact switching arrangements. Like switch contacts, relay contacts are classified by their number of poles, throws, and breaks.

- The number of **poles** indicates the number of completely isolated circuits that a relay contact can switch. The single-pole contact can conduct current through only one circuit at a time while a double-pole contact can conduct current through two circuits simultaneously.
- A **throw** is the number of closed contact positions per pole (single or double). The single-throw contact can control current in only one circuit while the double-throw contact can control two circuits.
- The term **break** designates the number of points in a set of contacts where the current will be interrupted during opening of the contacts. All relay contacts are constructed as single break or double break. Single-break contacts have lower current ratings because they break the current at only one point.

In general, relay contact ratings are rated in terms of the maximum amount of current the contacts are capable of handling at a specified voltage level and type (AC or DC). Current ratings specified may include:

- Inrush or make-contact capacity
- Normal or continuous carrying capacity
- Opening or break capacity

The load-carrying capacity of contacts is normally given as a current value for a **resistive** load. Lamp filaments are resistive but change in value by a large factor from their cold state to their operating state resistance. This effect is so great that the inrush current can be expected to be 10 to 15 times greater than the steady-state value. Normal practice is to derate contacts to 20 percent of their resistive load capabilities for a lamp load. **Inductive** loads, such as transformers, act as energy storage devices and can cause excessive contact arcing when the relay breaks the circuit. For inductive-type loads, contacts are normally derated to 50 percent of their resistive load capacity.

Relay contacts often have two ratings: AC and DC. These ratings indicate how much power can be switched through the contacts. One way to determine the maximum power capacity of relay contacts is to multiply the rated volts times the rated amperes. This will give you the total watts a relay can switch. For instance, a 5 A relay rated at 125 V AC can also switch 2.5 A at 250 V AC. Similarly, a 5 A relay rated at 24 V DC can switch 2.5 A at 48 V DC, or even 10 A at 12 V DC.

Interposing Relay

The term **interposing relay** refers to a type of relay that enables the energy in a high-power circuit to be switched by a low-power control signal. The interposing relay circuit of Figure 7-10 is an example of a low current temperature

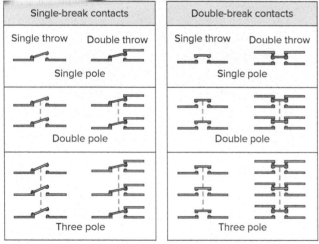

Figure 7-9 Common relay contact switching arrangements.
Photo courtesy of Omron Automation and Safety, www.Omron247.com

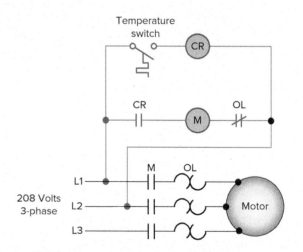

Figure 7-10 Interposing relay current control.

switch used to switch the higher current required by motor starter coil.

- The voltages applied to the interposing relay CR coil and motor starter coil M are the same.
- However, the ampere rating of the temperature switch is lower than that required to operate the starter coil.
- A relay is interposed between the temperature switch and starter coil.
- The current drawn by the relay coil is within the current rating of the temperature switch, and the relay contact has an adequate current rating for the current drawn by the starter coil.

The interposing relay circuit of Figure 7-11 is an example of a condition that exists in which the voltage rating of the temperature switch is too low to permit its direct connection into the motor starter power circuit.

- The coil of the interposing relay CR and the temperature switch are wired to the 24-volt source of power compatible with the rating of the temperature switch.

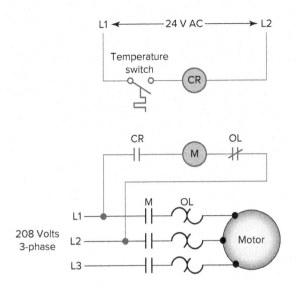

Figure 7-11 Interposing relay voltage control.

- The relay contact CR, with its higher 208 volt rating, is then used in the operation of the starter coil M.

Part 1 Review Questions

1. What exactly is an electromechanical control relay?
2. A relay involves two circuits. Name the two circuits and explain how they interact with each other.
3. Compare control relays with contactors.
4. Describe the switching action of normally open and normally closed relay contacts.
5. Outline three basic ways in which control relays are put to use in electric and electronic circuits.
6. An eight-pin octal-base ice-cube-style relay is to be wired into a control circuit that requires a set of NO and NC contacts electrically isolated from one another. State the number of the pin connections you would use for each contact.
7. How many breaks can relay contacts have?
8. What does SPDT stand for?
9. List three types of current ratings that may be specified for relay contacts.
10. The load-carrying capacity of contacts is normally given as a current value for a resistive load. Name two types of load devices that require this value to be derated.
11. How many amperes of current can a relay contact rated for 10 A at 250 V AC safely switch at 125 V AC?
12. For what type of applications are interposing relays used?

PART 2 SOLID-STATE RELAYS

Operation

A **solid-state relay** (SSR) is an electronic switch that, unlike an electromechanical relay, contains no moving parts. Although EMRs and solid-state relays are designed to perform similar functions, each accomplishes the final results in different ways. Unlike electromechanical relays, SSRs do not have actual coils and contacts. Instead, they use semiconductor switching devices such as bipolar transistors, MOSFETs, silicon-controlled rectifiers (SCRs), or triacs mounted on a printed circuit board. All SSRs are constructed to operate as two separate sections: input and output. The input side receives a voltage signal from the control circuit and the output side switches the load.

SSRs are manufactured in a variety of configurations that include both **"hockey-puck"** and **"ice-cube"** types (Figure 7-12). Most often, a square or rectangle will be used on the schematic to represent the relay. The internal circuitry will not be shown, and only the input and output connections to the box will be given.

Like electromechanical relays, solid-state relays provide electrical isolation between the input control circuit and the switched load circuit. A common method used to provide isolation is to have the input section illuminate a light-emitting diode (LED) that activates a photodetector device connected to the output section. The photodetector device triggers the output side, actuating the load. Relays that use this method of coupling the two circuits are said to be **optoisolated.**

Solid-state relays are constructed with different main switching devices depending on the type of load being switched. If the relay is designed to control an **AC load,** a triac is commonly used as the main switching semiconductor. A simplified diagram of an optically coupled solid-state relay used to switch an AC load is shown in Figure 7-13. The operation of the circuit can be summarized as follows:

- A current flow is established through the LED connected to the input when conditions call for the relay to be actuated.
- The LED conducts and shines light on the phototransistor.
- The phototransistor conducts switching on the triac and AC power to the load.
- The output is isolated from the input by the simple LED and phototransistor arrangement.
- Since a light beam is used as the control medium, no voltage spikes or electrical noise produced on the load side of the relay can be transmitted to the control side of the relay.

Solid-state relays intended for use with **DC loads** have a power transistor rather than a triac connected to the load circuit as shown in Figure 7-14. The operation of the circuit can be summarized as follows:

- The LED section of the relay acts like the coil of the electromechanical relay and requires a DC voltage for its operation.
- When the input voltage turns the LED on, the photodetector connected to the base of the transistor turns the transistor on, allowing current flow to the load.
- The transistor section of the optocoupler inside the SSR is equivalent to the contacts in a relay.
- Because solid-state relays have no moving parts, their switching response time is many times faster than that of electromechanical relays. For this reason, when loads are to be switched continually and quickly, the SSR is the relay of choice.

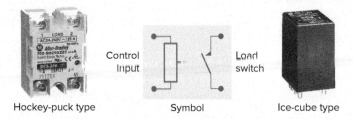

Figure 7-12 Typical solid-state relay (SSR).
Photos courtesy of Rockwell Automation, www.rockwellautomation.com

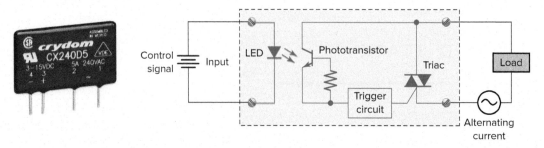

Figure 7-13 Optically coupled SSR used for AC loads.
Photo courtesy of Custom Sensors & Technologies, www.cstsensors.com

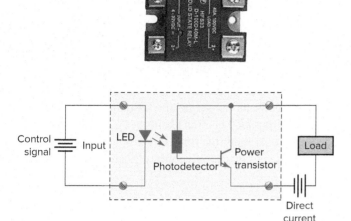

Figure 7-14 Optically coupled SSR used for DC loads.
Photo courtesy of Futurlec, www.futurlec.com

Specifications

Applying the specified amount of pickup voltage activates the SSR input control circuit of an SSR. Most SSRs have a variable input voltage range, such as 5 V DC to 24 V DC. This voltage range makes the SSR compatible with a variety of electronic input devices. Output voltage ratings range from 5 V DC up to 480 V AC. Although most SSRs are designed for a rated output current of under 10 A, relays mounted on heat sinks are capable of controlling up to 40 A.

The majority of SSRs are single-pole devices, as multipole relays pose a greater power dissipation problem. When multiple poles are required, a multipole solid-state module can be used. Another solution is to wire several SSR control circuits in parallel, as illustrated in Figure 7-15, to provide the equivalent function as a multipole electromagnetic relay. In this application, three single-pole solid-state relays are used to switch current to a three-phase load. The input section may receive a signal from a variety of sources such as device contacts or sensor signals. When the control circuit contact closes, all three relays actuate to complete the current path to the load.

The standard single-pole SSR configuration works fine with two-wire control; however, when it becomes necessary for it to be used in a three-wire control scheme, the problem of the holding circuit arises. An additional relay can be wired in parallel to the SSR to act as the holding contact. Another solution is to use a DC control circuit with a silicon-controlled rectifier (SCR) for latching the load. Figure 7-16 shows a three-wire motor control circuit utilizing a solid-state relay and an SCR. The operation of the circuit can be summarized as follows:

- The SCR will not allow current flow from anode (A) to cathode (K) until current is applied to the gate (G).
- When the start push button is pressed, current flows through the gate, which triggers the anode-to-cathode section of the SCR and relay control circuit into conduction.
- The SCR remains latched on after the start push button is released, and the circuit must be opened to stop the anode-to-cathode current flow. This is accomplished by pressing the stop push button.

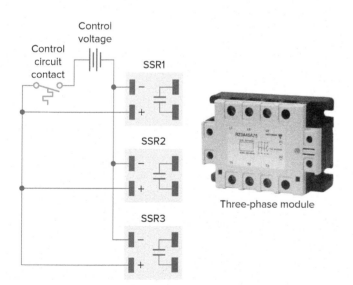

Figure 7-15 Multiple-pole solid-state relay connections.
Photo courtesy of Carlo Gavazzi, www.GavazziOnline.com

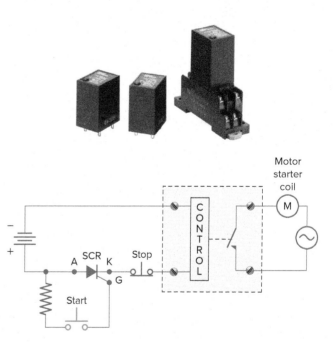

Figure 7-16 Three-wire control utilizing a solid-state relay and an SCR.
Photo courtesy of Omron Automation and Safety, www.Omron247.com

Part 2 Solid-State Relays 193

Switching Methods

SSRs operate with several different switching methods. The type of load is an important factor in the selection of the switching method.

- **Zero-switching relay.** A **zero-switching relay** is designed to turn on an AC load when the control voltage is applied and the voltage at the load passes through zero. The relay turns off the load when the control voltage is removed and the current in the load crosses zero. This allows resistive loads such as lamp filaments to last longer because they are not subjected to high-voltage transients from switching AC voltage and current when the sine wave is at a peak. Figure 7-17 shows a simplified diagram of a zero-switching SSR.

- **Peak-switching relay.** A **peak-switching relay** is an SSR that turns the load on when the control voltage is present and the voltage at the load is at its peak. The relay turns off when the control voltage is removed and the current in the load crosses zero. Peak switching is preferred when the output circuit is mostly inductive or capacitive and the voltage and current are approximately 90 degrees out of phase. In this case, when the voltage is at or near its peak value, the current will be at or near its zero value.

- **Instant-on relay. Instant-on relays** are typically specified when the controlled load is a combination of resistance and reactance. In this case the voltage and current phase angle vary, so there is no advantage to disconnecting the load at any specific time on the sine wave.

- **Analog-switching relay.** An **analog-switching relay** has an infinite number of output voltages within the relay's rated range. This relay contains a built-in synchronizing circuit that controls the amount of output voltage as a function of the input voltage. For example, in a closed-loop temperature control system, if there is a small temperature difference between the actual and set temperature input, the output load heating element is given low power output. At the same time, large input temperature differences gives high power output.

Solid-state relays have several advantages over electromechanical types:

- The SSR is more reliable and has a longer life because it has no moving parts.
- It is compatible with transistor and IC circuitry and does not generate as much electromagnetic interference.
- The SSR is more resistant to shock and vibration, has a much faster response time, and does not exhibit contact bounce.

Solid-state relays contain semiconductors that are more susceptible to damage from voltage and current spikes. Other issues related to solid-state relay applications include:

- **Thermal dissipation.** One of the major considerations when using an SSR is the amount of heat it generates. This heat is a result of the power loss due to the operating voltage drop created across the switching semiconductor. Solid-state relays controlling loads rated at more than 5 amps require a heat sink (Figure 7-18) for reliable operation. The size

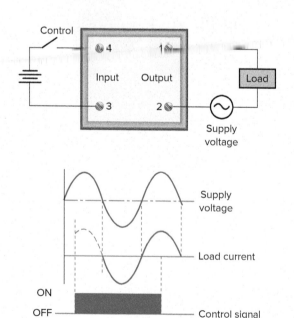

Figure 7-17 Zero-switching SSR.

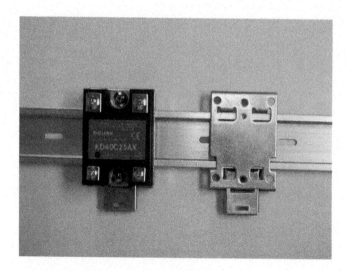

Figure 7-18 Solid-state relay heat sink.
Sun West Technical Sales, Inc.

and thermal rating of the heat sink increases as the load current carried by the SSR increases, or as the operational ambient temperature increases.

- **Leakage current.** A solid-state relay in the blocked or output off state does not have an infinite impedance across its terminals. As a result, a small amount of current, called *leakage current,* comes through the load in the off state. Although the leakage current is normally less than 1 milliamp, it may present a problem in applications controlling very low loads (small solenoid valves, etc.) because this current may be sufficient to maintain the load supplied once the relay is switched off.

- **Higher cost.** An electromechanical relay consists, basically, of an input coil and output contact. In a solid-state relay, these components are replaced with various types of semiconductors, as illustrated in Figure 7-19. In general, solid-state relays cost more than comparable electromechanical types. If your application is one in which multiple loads need to be switched, the cost disadvantage of the SSR can become much more pronounced.

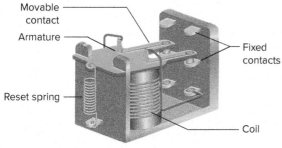

Electromechanical relay (EMR)

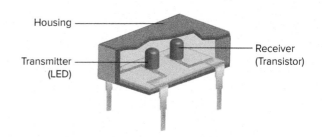

Solid-state relay (SSR)

Figure 7-19 Electromechanical versus solid-state relay construction.

Part 2 Review Questions

1. What is the fundamental difference between an electromechanical and a solid-state relay?
2. A common method employed in SSRs to provide isolation between input and output circuits is optoisolation. Give a brief explanation of how this works.
3. State the type of SSR main switching semiconductor used to control
   a. AC loads.
   b. DC loads.
4. A given SSR has an input control voltage rating of 3 to 32 V DC. What does this imply as far as actuation of the relay is concerned?
5. Why are the majority of solid-state relays constructed as single-pole devices?
6. List four common switching modes for SSRs.
7. Explain the advantage gained by using a zero-switching relay to control a resistive load.
8. List three advantages SSRs have compared to electromechanical types.
9. Why do solid-state relays generate more heat during normal operation than electromechanical types?
10. Explain the operation of an analog-switching relay.

PART 3 TIMING RELAYS

Timing relays are a variation of the standard instantaneous control relay in which a fixed or adjustable **time delay** occurs after a change in the control signal before the switching action occurs. Typical types of timing relays are shown in Figure 7-20. Timers allow a multitude of operations in a control circuit to be automatically started and stopped at different time intervals. The use of timers can eliminate the labor-intensive process of trying to manually control each step of a process.

Motor-Driven Timers

Timer functions include timing a cycle of operation, delaying the starting or stopping of an operation, and

Solid-state timing relay　　Pneumatic timing relay　　Plug-in timing relay

Figure 7-20 Timing relays.
Photos courtesy of Rockwell Automation,
www.rockwellautomation.com

controlling time intervals within an operation. **Motor-driven timers** are used to time a cycle of operations. Synchronous clock types, such as shown in Figure 7-21, use a small electric clock motor driven from the AC power line to maintain sync with standard time. A mechanical connection to the clock mechanism controls the contacts. The operation of the device can be summarized as follows:

- The motor turns the mechanism and actuates normally open or normally closed contacts.
- Adjustable on/off tabs set along the clock's timing wheel trip the contact open or closed.
- The timer motor is supplied with continuous power. If power is lost, the timing will be delayed by an amount of time the power was off, and the correct time must be manually reset.
- These types of timers are best suited for applications such as lighting and water sprinkler control where precise timing is not critical.

Dashpot Timers

Dashpot timers manage their timing function by controlling fluid flow or airflow through a small orifice. The pneumatic (air) timing relay shown in Figure 7-22 uses an air-bellows system to achieve its timing cycle. The operation of the device can be summarized as follows:

- The bellows design allows air to enter through an orifice at a predetermined rate to provide time-delay increments.
- As soon as the coil is energized or de-energized, the timing process begins and the rate of airflow determines the length of the time delay.
- Smaller orifice openings restrict the flow rate, resulting in longer time delays.
- Pneumatic timers have relatively small adjustable range settings. The timing range for the timer shown is adjustable from 0.05 to 180 seconds with an accuracy of approximately ± 10 percent.

Solid-State Timing Relays

Solid-state timing relays use electronic circuitry to provide their timing functions. The two broad categories of solid-state timers are analog and digital. Different methods are used to control the time-delay period. Some use a resistor/capacitor (*RC*) charge and discharge circuit to obtain the time base, while others use quartz clocks as the time base. These electronics-based timers are much more **accurate** than their dashpot counterparts and can control timing functions ranging from a fraction of a second to hundreds of hours. In order to maintain their timing operations, solid-state timers are normally constantly powered. Some are equipped with batteries or internal memory to retain their settings during power failures.

 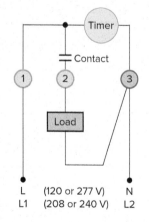

Figure 7-21 Synchronous clock timer.
NSi Industries, LLC, 2014. www.nsiindustries.com

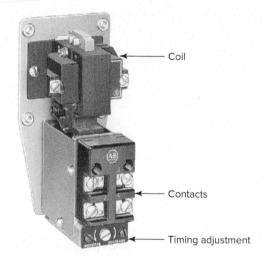

Figure 7-22 Pneumatic timer.
Photo courtesy of Rockwell Automation,
www.rockwellautomation.com.

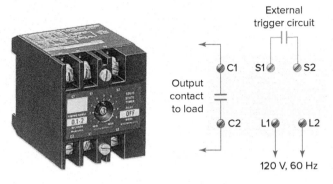

Figure 7-23 Solid-state timing relay connections.
Photo courtesy of Rockwell Automation,
www.rockwellautomation.com

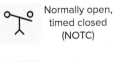
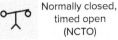

Figure 7-24 On-delay timer contacts.
Photo courtesy of Tyco Electronics, www.tycoelectronics.com

The timing functions of dashpot timers are initiated when the electromagnetic coil is energized or de-energized. In comparison, solid-state timing functions are initiated when the electronic circuit of the timer is energized or a triggering signal is received or removed. Electronic timers are available in a variety of rated input operating voltages. Figure 7-23 shows a typical solid-state timing relay. The operation of the device can be summarized as follows:

- Connections provided include timed contacts (C1, C2), voltage input (L1, L2), and external trigger switch (S1, S2).
- A timing-delay period of from 0.1 to 2 seconds is set by the adjustment of an internal potentiometer located on the front panel of the timer.
- The timer is energized continuously, and timing is initiated when the external trigger circuit is closed.
- The timed contact is convertible between on-delay and off-delay.

Timing Functions

There are four basic timing functions: on-delay, off-delay, one-shot, and recycle.

On-Delay Timer The on-delay timer is sometimes referred to as **DOE**, which stands for delay on energize. The time delay of the contacts begins once the timer is switched on; hence the term **on-delay timing.** Figure 7-24 shows the NEMA symbols for the on-delay timer normally open (NO) and normally closed (NC) contacts. The operation of the timed contacts can be summarized as follows:

- Once initiated, DOE timed contacts change state after a set time period has passed.
- After that time has passed, all normally open timed contacts close and all normally closed contacts open.
- Once the timed contacts change state, they will remain in this position until the power is removed from the coil or electronic circuit.

The circuit shown in Figure 7-25 illustrates the timing function of an on-delay timing relay. In this example, a simple dashpot timer with a time delay setting of 10 seconds can be assumed. The same operation applies to electronic timers that perform a similar function. The operation of the circuit can be summarized as follows:

- When the switch is closed, power is applied to the coil but the contacts are delayed from changing position.
- With the switch still closed, after the 10-second timing period the normally open contacts (TR1-1) close to energize load 1 and the normally closed contacts (TR1-2) open to de-energize load 2.
- If the switch is then opened, the coil de-energizes immediately, returning both timed contacts to their normal state, switching load 1 on and load 2 off.

Figure 7-26 shows an application for an on-delay timer that uses a NOTC contact. This circuit is used as a warning signal when moving equipment, such as a conveyor motor, is about to be started. The operation of the circuit can be summarized as follows:

- Coil CR is energized when the start pushbutton PB1 is momentarily actuated.
- As a result, contact CR closes to seal in the CR coil, energize timer coil TD, and establish the circuit to sound the horn.
- After a 10-second time-delay, normally open timer contact TD closes to energize the motor starter coil M to start the motor and turn off the horn.

Off-Delay Timer The off-delay timer is sometimes referred to as **DODE,** which stands for delay on de-energize. The operation of the off-delay timer is the exact opposite of that of the on-delay timer. When power is applied to the coil or electronic circuit, the timed contacts will change state immediately. When power is removed, however, there is a time delay before the timed contacts change to

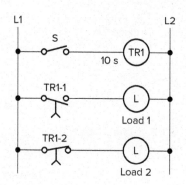

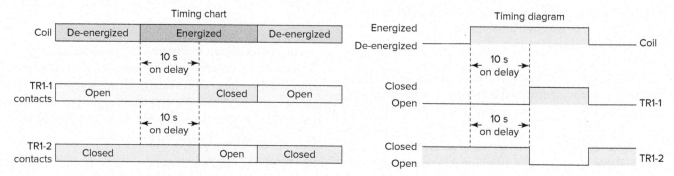

Figure 7-25 On-delay relay timer circuit.

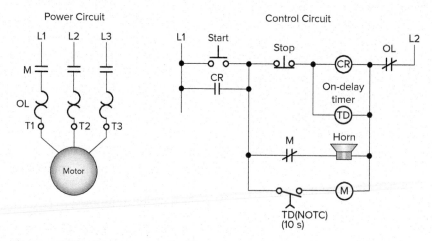

Figure 7-26 Conveyor motor warning signal.

their normal de-energized positions. Figure 7-27 shows the standard NEMA symbols and illustrates the timing function of an off-delay timing relay.

Figure 7-28 shows the wiring diagram for the automatic pumping down of a sump using a level sensor switch and plug-in cube-type off-delay timer. A solid-state timing circuit drives an internal electromechanical relay within the timer. The operation of the circuit can be summarized as follows:

- When the level rises to point A, the level sensor contact closes to energize the relay timer coil and close the NO contacts to the pump motor starter.

- This immediately turns the pump on to initiate the pumping action.
- When the height of the vessel level decreases, the sensor contacts open and timing begins.
- The pump continues to run and empty the tank for the length of the time-delay period.
- At time-out the relay coil de-energizes and the normally open relay contact reopens, turning the pump off.
- The timer has a built-in time adjustment potentiometer that is adjusted to empty the tank to a desired level before the pump shuts off.

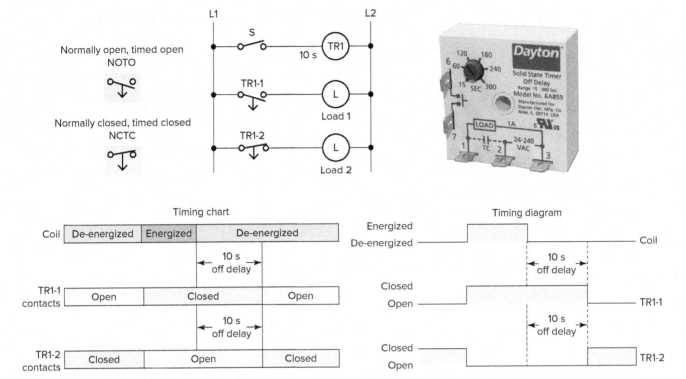

Figure 7-27 Off-delay relay timer.
Photo courtesy of W.W. Grainger, www.grainger.com

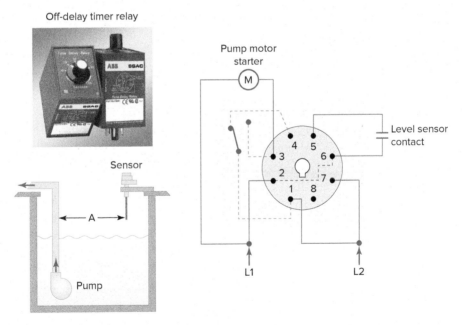

Figure 7-28 Off-delay timer automatic pumping circuit.
Photo courtesy of ABB, www.abb.com

Some types of timer relays come with a combination of timed and instantaneous contacts in both NO and NC configurations. The **instantaneous contacts** are controlled directly by the timer coil, as in a general-purpose control relay. The control circuit of Figure 7-29 is that of an off-delay timer with both instantaneous and timed contacts. The operation of the circuit is summarized as follows:

- When power is first applied (limit switch LS open), motor starter coil M1 is energized and the green pilot light is on.

Part 3   Timing Relays   199

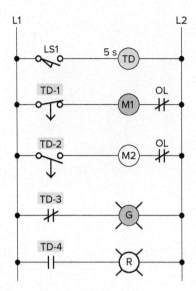

Figure 7-29 Off-delay timer with both instantaneous and timed contacts.

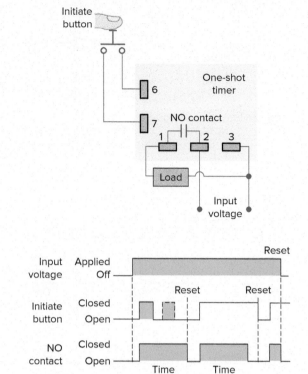

Figure 7-30 One-shot timer.

- At the same time, motor starter coil M2 is de-energized, and the red pilot light is off.
- When limit switch LS1 closes, off-delay timer coil TD energizes.
- As a result, timed contact TD-1 opens to de-energize motor starter coil M1, timed contact TD-2 closes to energize motor starter coil M2, instantaneous contact TD-3 opens to switch the green light off, and instantaneous contact TD-4 closes to switch the red light on. The circuit remains in this state as long as limit switch LS1 is closed.
- When limit switch LS1 is opened, the off-delay timer coil TD de-energizes and the time-delay period is started.
- Instantaneous contact TD-3 closes to switch the green light on, and instantaneous contact TD-4 opens to switch the red light off.
- After a 5-second time delay, timed contact TD-1 closes to energize motor starter M1 and timed contact TD-2 opens to de-energize motor starter M2.

One-Shot Timer With a one-shot timer, momentary or continuous closure of the initiate circuit results in a single timed **pulse** being delivered to the output. The one-shot causes this action to happen only once, and then must be reinitiated if it is to continue to operate. The circuit of Figure 7-30 illustrates the wiring and timing function of a typical one-shot timer. The operation of the circuit can be summarized as follows:

- Input voltage must be applied before and during timing.
- Upon momentary or maintained closure of the initiate button, the output load is energized.
- The load remains energized for the duration of the time-delay period and then is return to its normal de-energized state ready to be initiated for another cycle of operation.
- Opening or reclosing the initiate button during timing has no effect on the time delay. Reset occurs when the time delay is complete and the initiate button is open.
- If power is interrupted to a one-shot timer during the time-delay operation, the time delay is canceled. When power is restored to the timer, the time-delay function will not begin again until the one-shot has been reinitiated.
- One-shot timers do not have dedicated contact symbols. Instead, the standard NO and NC contact symbols are used and referenced to the timer that controls them.

At times, it's required to have a momentary input trigger a device or operation for a preset period of time. Figure 7-31 illustrates an application of a one-shot used to energize a solenoid valve for 10 seconds:

- The proximity sensor is positioned to detect passing of a product.

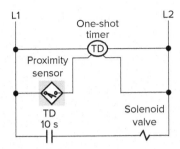

Figure 7-31 One-shot used to energize a solenoid valve.

- When the product passes, the proximity sensor the sensor's contact provides a trigger signal that changes from open, to closed, and back to open.
- The timer's output contact closes causing the solenoid valve to be activated and the one-shot time delay relay to begin its timing cycle.
- The output contacts remain closed and TD continues to time for 10 seconds.
- Only when TD has timed out does its output contact open.
- After time-out, the one-shot is ready to be retriggered for the next cycle.

Recycle Timer The contacts of a recycle timer **alternate** between the on and off states when the timer is initiated. Solid-state circuits within the device drive an internal electromagnetic relay. The operation of the recycle timers shown in Figure 7-32 can be summarized as follows:

- Upon application of the input voltage, the first delay period (TD1) begins and the output remains de-energized, or off.
- At the end of the first delay, or off period, the relay coil will energize and the second delay (TD2), or on period, begins.

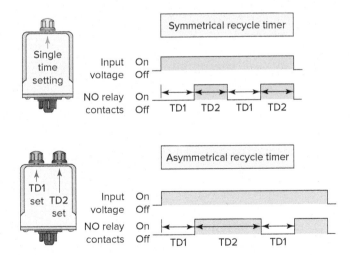

Figure 7-32 Recycle timers.

- When the second delay period ends, the relay de-energizes.
- This recycling sequence will continue until input voltage is removed.
- In some recycling timers, the on time may be configured for the first delay. Removing the input voltage resets the output and time delays, and returns the sequence to the first delay.
- Recycle timers are available in two configurations: symmetrical and asymmetrical.
- In symmetrical timing, the on and off periods are equal. The length of the timing period is adjustable, but the time between the on and off operations remains constant. Flashers are an example of symmetrical timing.
- Asymmetrical timers allow independent adjustments for the on and off periods. They come equipped with individual on and off time adjustment knobs and use standard NO and NC contact symbols referenced to the timer that controls them.

Figure 7-33 illustrates an application of a symmetrical recycle timer flasher circuit used to operate the high-temperature pilot light on and off.

- The time setting of the timer is one second.
- The timer is trigger into operation upon closure of the high-temperature switch.
- The timer's output contact closes and opens at one second intervals to operate the pilot light.
- The pilot continues to flash on and off as long as a high-temperature state exists.

Multifunction and PLC Timers

Multifunction Timer The term *multifunction timer* refers to timers that perform more than one timing function. Multifunction timers are more versatile in that they can perform many different timing functions and therefore are

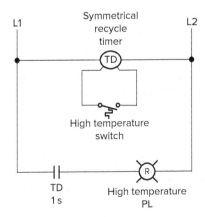

Figure 7-33 Symmetrical recycle timer flasher circuit.

Part 3 Timing Relays 201

Figure 7-34 Multifunction digital timer.
Photo courtesy of Omron Automation and Safety, www.Omron247.com

more common. Figure 7-34 shows a multifunction digital timer that is capable of performing all of the basic timing functions.

PLC Timers **Programmable logic controllers (PLCs)** can be programmed to operate like conventional timing relays. The PLC timer instruction can be used to activate or deactivate a device after a preset interval of time. One advantage of the PLC timer is that its timer accuracy and repeatability are extremely high. The most common types of PLC timer instructions are the on-delay timer (TON), off-delay timer (TOF), and retentive timer on (RTO).

Figure 7-35 illustrates how an Allen-Bradley Pico programmable logic controller is wired and programmed to implement an on-delay timer function. This application calls for the pilot light to turn on any time the pressure switch closes for a sustained period of 5 seconds or more. The procedure followed can be summarized as follows:

- The pressure switch is hard wired to input I3 and the pilot light output to Q1 according to the wiring diagram.

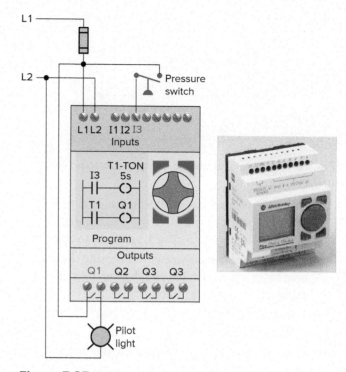

Figure 7-35 PLC programmed on-delay timer.
Photo courtesy of Rockwell Automation, www.rockwellautomation.com

- Next, the ladder logic program is entered, using the front keypad and LCD display.
- When the pressure switch contacts close, the programmed timing coil T1 energizes, initiating the time-delay period.
- After 5 seconds have passed, the programmed timer contact T1 closes to energize output relay coil Q1 and turn the pilot light on.
- Opening of the pressure switch contacts at any time resets the timed value to zero and turns off the pilot light.

Part 3 Review Questions

1. In what way is a timing relay different from a standard control relay?
2. Explain how contacts are closed and opened in a synchronous clock timer.
3. What types of applications are synchronous clock timers best suited for?
4. Assume power is lost and later returned to a synchronous clock timer. In what way would this affect its operation?
5. Explain how timing is achieved in a dashpot timer.
6. Compare the manner in which the instantaneous and timed contacts of a dashpot timer operate.
7. Compare the accuracy and timing range of solid-state and dashpot timers.
8. Dashpot timers rely on an electromagnetic coil to initiate their timing functions. How is this accomplished with solid-state relays?
9. List four basic types of timing functions.

10. State what the timer abbreviations DOE and DODE stand for.
11. Outline the switching operation of the NOTC and NCTO contacts of an on-delay timer.
12. Outline the switching operation of the NOTO and NCTC contacts of an off-delay timer.
13. The normally open contacts of a one-shot timer are used to control a solenoid valve. Explain what happens when the timing function is momentarily initiated.
14. Assume power is lost and later returned to a one-shot timer. In what way will this affect its operation?
15. Explain the switching operation of the timed contacts of a recycle timer.
16. Compare the manner in which the timed contacts of a symmetrical and an asymmetrical recycle timer can be set to operate.
17. To what general classification of timers do multifunction timers belong?
18. List the three most common PLC timer instructions.
19. How are the instantaneous contacts of a timer relay controlled?

PART 4 LATCHING RELAYS

Latching relays typically use a mechanical latch or permanent magnet to hold the contacts in their last energized position without the need for continued application of coil power. They are especially useful in applications where power must be conserved, such as a battery-operated device, or where it is desirable to have a relay stay in one position if power is interrupted.

Mechanical Latching Relays

Mechanical latching relays use a **locking mechanism** to hold their contacts in their last set position until commanded to change state, usually by energizing a second coil. Figure 7-36 shows a two-coil mechanical latching relay. The latch coil requires only a single pulse of current to set the latch and hold the relay in the latched position. Similarly, the unlatch or release coil is momentarily energized to disengage the mechanical latch and return the relay to the unlatched position.

Figure 7-37 illustrates the operation of a two-coil mechanical latching relay circuit. There is no normal position for the contacts of a latching relay. The contact is shown with the relay in the unlatched condition—that is, as if the unlatch coil were the last one energized. The operation of the circuit can be summarized as follows:

- In the unlatched state, the circuit to the pilot light is open, so the light is off.
- When the on button is momentarily actuated, the latch coil is energized to set the relay to its latched position.
- The contacts close, completing the circuit to the pilot light, so the light is switched on.
- Note that the relay coil does not have to be continuously energized to hold the contacts closed and keep the light on. The only way to switch the lamp off is to actuate the off button, which will energize the unlatch coil and return the contacts to their open, unlatched state.

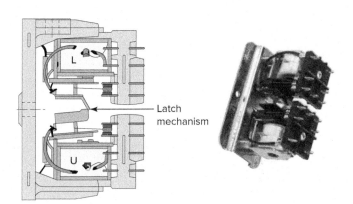

Figure 7-36 Two-coil mechanical latching relay.
Photo courtesy of Relay Service Company, www.relayserviceco.com.

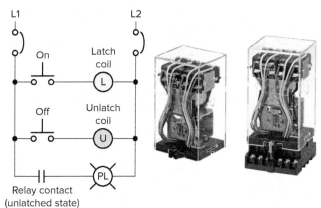

Figure 7-37 Operation of a two-coil latching relay circuit.
Photo courtesy of Omron Automation and Safety, www.Omron247.com.

- In cases of power loss, the relay will remain in its original latched or unlatched state when power is restored. This arrangement is sometimes referred to as a memory relay.

Magnetic Latching Relays

Magnetic latching relays are typically single-coil relays designed to be **polarity** sensitive. When voltage is momentarily applied to the coil with a predetermined polarity, the relay will latch. A permanent magnet is used to hold the contacts in the latch position without the need for continued power to the coil. When the polarity is reversed, and current momentarily applied to the coil, the armature will push away from the coil, overcoming the holding effect of the permanent magnet, causing the contacts to unlatch or reset. Figure 7-38 shows a single coil magnetic latching relay used with an 11-pin plug-in socket. The direction of current flow through the coil determines the position of the relay contacts. Repeated pulses from the same input have no effect. The double pole double throw (DPDT) relay contacts can handle control circuit loads and are shown in the relay reset position.

Latching Relay Applications

The latching relay has several advantages in electrical circuit design. For example, it is common in a control circuit to have to remember when a particular event takes place and not permit certain functions once this event occurs. Running out of a part on an assembly line may signal the shutdown of the process by momentarily energizing the unlatch coil. The latch coil would then have to be momentarily energized before further operations could occur.

Another application for a latching relay involves power failure. Circuit continuity during power failures is often important in automatic processing equipment, where a sequence

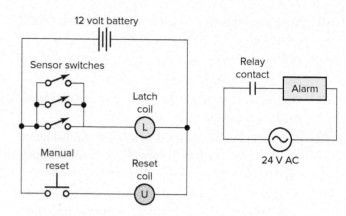

Figure 7-39 Battery-operated latching alarm circuit.

of operations must continue from the point of interruption after power is resumed rather than return to the beginning of the sequence. In applications similar to this, it is important not to have the relay control any devices that could create a safety hazard if they were to restart after a power interruption.

Latching relays are useful in applications where power must be conserved, such as a battery-operated device. Figure 7-39 shows a simplified diagram for a battery-operated latching alarm circuit. The circuit uses a latching relay to conserve power. Regardless of whether the circuit is reset or latched, there is no current drain on the battery. Momentarily closing any normally open sensor switch will cause the relay to latch, closing the contact to power the alarm circuit. The manual reset button must be depressed with all sensors in their normally open state to reset the circuit.

Alternating Relays

Alternating relays (also known as impulse relays) are a form of latching relay that **transfers** the contacts with each pulse. They are used in special applications where the optimization of load usage is required by equalizing the run time of two loads. Figure 7-40 shows a plug-in alternating

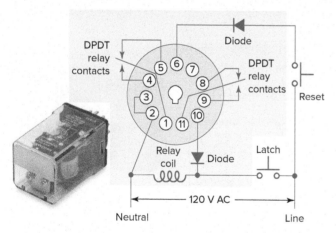

Figure 7-38 Single-coil magnetic latching relay.
Photo courtesy of Automation Direct, www.automationdirect.com.

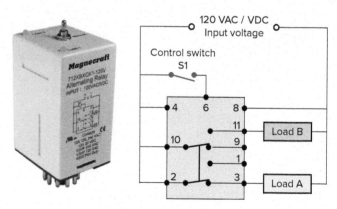

Figure 7-40 Alternating or impulse relay.
Magnecraft photo courtesy of Schneider Electric USA, Inc., www.serelays.com

204   Chapter 7   Relays

relay made up of a magnetic latch relay that is operated by a solid-state steering circuit. The operation of the circuit can be summarized as follows:

- An external control switch such as a float switch, manual switch, timing relay, pressure switch, or other isolated contact initiates the alternating action.
- The input voltage must be applied at all times, and the S1 control switch voltage must be from the same supply as the unit's input voltage (no other external voltage should be connected to it).
- Each time the control switch S1 is opened, the output contacts will change status. LEDs indicate the status of the internal relay and which load is selected to operate.
- Loss of input voltage resets the unit; load A becomes the lead load for the next operation.
- To terminate alternating operation and cause only the selected load to operate, the toggle switch located on the top of the relay is shifted to position A to lock load A or to position B to lock load B. This feature allows users to select one of the two loads or alternate between the two.

In certain pumping applications, two identical pumps are used for the same job. A standby unit is made available in case the first pump fails. However, a completely idle pump might deteriorate and provide no safety margin. Alternating relays prevent this by assuring that both pumps get equal run time. Figure 7-41 shows a typical alternating relay circuit used with a duplex pumping system where it is desirable to equalize pump run time. The operation of the circuit can be summarized as follows:

- In the off state, the float switch is open, the alternating relay is in the load A position, and both loads (M1 and M2) are off.
- When the float switch closes, it energizes the first load (M1) and remote PL1 to indicate that pump motor 1 is running. The circuit remains in this state as long as the float switch remains closed.
- When the float switch opens, the first load (M1) is turned off and the alternating relay toggles to the load B position.
- When the float switch closes again, it energizes the second load (M2) and remote PL2 to indicate that pump motor 2 is running.
- When the float switch opens, the second load (M2) is turned off, the alternating relay toggles back to the load A position, and the process can be repeated again.

DPDT **crossed-wired** alternating relays are used in applications where additional capacity may be required in addition to normal alternating operation. These relays have the ability to alternate the loads of a dual system during normal operation or operate both when demand is high. Figure 7-42 shows the cross-wired contact version

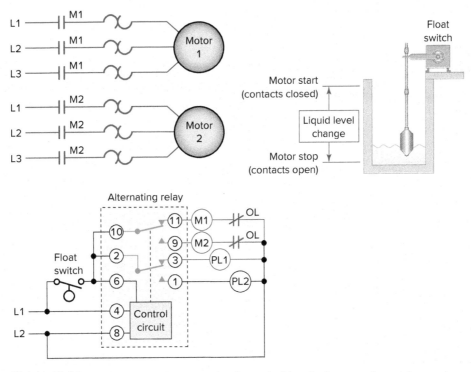

Figure 7-41 Typical alternating relay circuit used with a duplex pumping system.

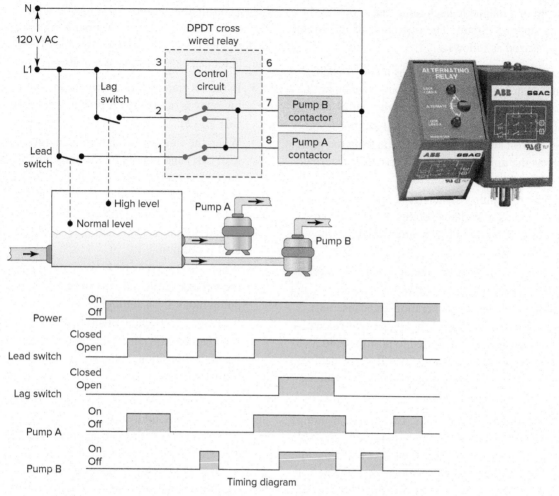

Figure 7-42 DPDT cross-wired contact version of a dual pumping application.
Photo courtesy of ABB, www.abb.com

of an alternating relay used in a dual pumping circuit. The operation of the circuit can be summarized as follows:

- The selector switch located on the relay allows selection of the alternation mode or either load for continuous operation.
- LEDs indicate the status of the output relay.
- With the alternation mode selected, if the level in the tank never reaches the high level, only the lead float switch cycles and normal alternating operation will occur.
- When both the lead and lag float switches close simultaneously, because of a heavy flow into the tank, both pumps A and B will be energized.
- This system saves on energy because only one pump is operating most of the time; yet the system has the capacity to handle twice the load.

Part 4 Review Questions

1. What two methods are used to hold the contacts of a latching relay in their last energized position?
2. Explain how a two-coil mechanical latching relay is latched and unlatched.
3. In what state are the contacts of a latching relay normally shown on diagrams?
4. Explain how a single-coil magnetic latching relay is latched and unlatched.
5. Assume that power is lost to a circuit that contains a latching relay. In what state will the contacts be when power is restored?
6. In what type of applications are alternating relays used?
7. What additional operating feature is available for use with cross-wired alternating relays?

PART 5 RELAY CONTROL LOGIC

Digital signals are the language of modern day computers. Digital signals comprise only **two states,** which can be expressed as on or off. A relay can be considered digital in nature because it is basically an on/off, two-state device. It is common practice to use relays to make logical control decisions in motor control circuits. The primary programming language for programmable logic controllers (PLCs) is based on relay control logic and ladder diagrams.

Control Circuit Inputs and Outputs

Most electrical control circuits can be divided into two separate sections consisting of an input and output. The **input** section provides the **signals** and includes such devices as manually operated switches and push buttons, automatically operated pressure, temperature, float, limit, and sensor switches, as well as relay contacts. In general, input signals initiate or stop the flow of current by closing or opening the control devices contacts.

The **output** section of the control circuit provides the action and includes such devices as contactors, motor starters, heater units, relay coils, indicator lights, and solenoids. Outputs are **load** devices that directly or indirectly carry out the actions of the input section. The action is considered direct when devices such as solenoids and pilot lights are energized as a direct result of the input logic. The action is considered indirect when the coils in relays, contactors, and starters are energized. This is because these coils operate contacts, which actually control the load.

Motor control circuits may have one or more inputs controlling one or more outputs. A combination of input devices that either manually or automatically sense a condition—and the corresponding change in condition performed by the output device—make up the core of motor control. Figure 7-43 illustrates typical inputs and outputs of a control ladder diagram. The control logic for the circuit can be summarized as follows:

- Relay coil CR is energized when the on/off switch is closed and acts to close the CR-1 contact and open the CR-2 contact.
- For the horn to energize and sound, both the CR-1 contact and the limit switch must be closed.
- The solenoid is energized and operates whenever the CR-2 contact or float switch is closed.
- When the temperature switch closes, the contactor coil energizes and acts to close the C1 contact. At the same time, the circuit is completed to the red pilot light, switching it on.

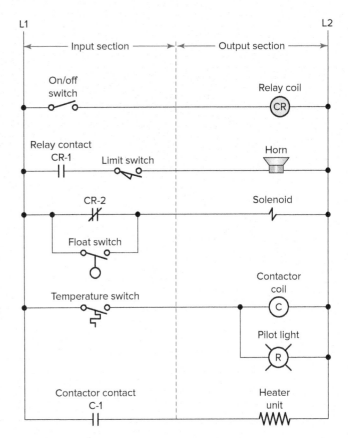

Figure 7-43 **Typical inputs and outputs of a control diagram.**

- The heater unit is energized and operates whenever contact C1 is closed.

AND Logic Function

Logic is the ability to make decisions when one or more different factors must be taken into consideration. Control logic functions describe how inputs interact with each other to control the outputs and include AND, OR, NOT, NAND, and NOR functions. In electronic circuits, these functions are implemented using digital circuits known as gates.

The *AND* logic function operates like a **series** circuit. AND logic is used when two or more inputs are connected in series and they all must be closed in order to energize the output load. Figure 7-44 shows a simple application of the AND logic function. Most AND logic circuits use normally open input devices connected in series. In this application, both the temperature switch and the float switch inputs must be closed to energize the solenoid output.

OR Logic Function

The *OR* logic function operates like a **parallel** circuit. OR logic is used when two or more inputs are connected in parallel and any one of the inputs can close to energize the

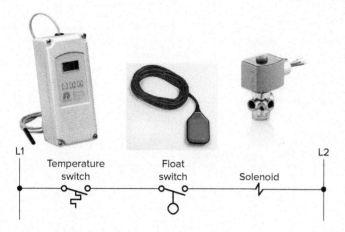

Figure 7-44 AND logic function.
Left photo courtesy of Robertshaw, www.robertshaw.com; center photo Madison Company—Float Switch; right photo courtesy of ASCO Valve Inc., www.ascovalve.com

output load. Figure 7-45 shows a simple application of the OR logic function. Most OR logic circuits use normally open input devices connected in parallel. In this circuit, any one of the two pushbutton inputs can close to energize the motor starter coil load.

Combination Logic Functions

Control circuits often require more than one type of logic function when more complex decisions need to be made. Figure 7-46 shows an example of an AND/OR **combination logic** circuit. In this control application, the output is a

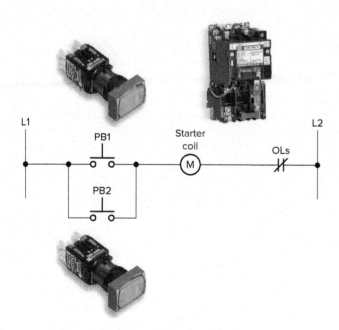

Figure 7-45 OR logic function.
This material and associated copyrights are proprietary to, and used with the permission of, Schneider Electric.

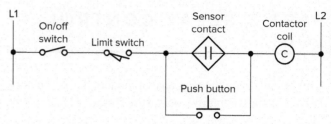

Figure 7-46 AND/OR combination logic.

contactor coil that is controlled by combined AND/OR logic functions. Both the on/off switch **and** limit switch in addition to the sensor contact **or** push button must be closed to energize the contactor coil.

NOT Logic Function

Unlike the AND logic and OR logic, the *NOT* logic function uses a single normally closed rather than a normally open input device. NOT logic energizes the load when the control signal is off. Figure 7-47 shows an example of the NOT logic function used to prevent accidental contact with live electrical connections. The normally closed safety switch operates by detecting the opening of guards such as doors and gates. Contacts of the normally closed safety switch are held open by the shut door. When the door is opened, the safety switch returns to its normally closed state and the trip solenoid of the circuit breaker is energized to remove all power from the circuit.

NAND Logic Function

NAND logic is a combination of AND logic and NOT logic in which two or more normally closed contacts are connected in parallel to control the load. Figure 7-48 shows an example of the NAND logic function used in the

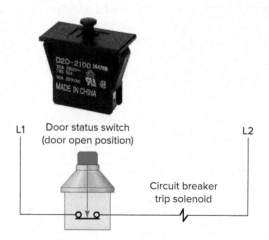

Figure 7-47 NOT logic function.
Photo courtesy of Omron Automation and Safety, www.Omron247.com

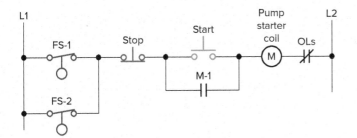

Figure 7-48 NAND logic function.

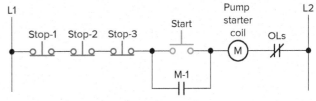

Figure 7-49 NOR logic function.

control circuit of a dual-tank liquid filling operation. The two tanks are interconnected and each is equipped with a float switch installed at the full level of the each tank. A three-wire control circuit is used in addition to the two parallel connected float switches that provide the NAND logic of the circuit. With either or both tanks below the full level, momentarily depressing the start push button energizes the motor starter coil, turning on the pump motor. Both float switches must open for the motor to shut off automatically. The stop push button will shut down the process at any time.

NOR Logic Function

NOR logic is a combination of OR logic and NOT logic in which two or more normally closed contacts are connected in series to control the load. Figure 7-49 shows an example of the NOR logic function used with three-wire control to energize a motor starter. In this circuit, the motor can be started from one location, but can be stopped from three locations. The three series-connected normally closed stop push buttons provide the NOR function of the circuit. Once the circuit is energized, if any one of the three stop push buttons is pressed, the starter coil M will de-energize.

Part 5 Review Questions

1. Why can a relay be considered to be digital in nature?
2. What type of controller is based on relay control logic?
3. Compare the functions of the input and output sections of an electrical control circuit.
4. Define the term *logic* as it applies to electrical control circuits.
5. What configuration of contacts is logically equivalent to the AND function?
6. What configuration of contacts is logically equivalent to the OR function?
7. What type of contact is logically associated with the NOT function?
8. What configuration of contacts is logically equivalent to the NAND function?
9. What configuration of contacts is logically equivalent to the NOR function?

Troubleshooting Scenarios

1. An electromechanical relay is suspected of being bad. How would you go about checking the relay coil in and out of the circuit? How would you go about checking the relay contacts in and out of the circuit?
2. A solid-state relay with a control circuit rated for 5 V DC has a positive sign marked on one of the control terminal connections. What does this mean as far as the operation of the relay is concerned?
3. Assume the switching semiconductor within a solid-state relay becomes faulted as a short-circuit. How would this affect the operation of the output circuit? How would the operation of the output circuit be affected if it became faulted as an open circuit?
4. The operation of an on-delay timer with a set of NOTC contacts is to be tested in-circuit by means of voltage measurements. Outline the procedures you would follow to determine if the circuit is operating properly.
5. Many alternating relays come equipped with a built-in switch that is used to manually force one motor of a dual motor system pump to run every time the circuit is actuated. Discuss in what troubleshooting instances this feature might be used.

Discussion Topics and Critical Thinking Questions

1. Why is the pickup voltage of an electromechanical relay normally higher than its dropout voltage?
2. During normal operation, solid-state relays generate more heat than equivalent electromagnetic types. Explain why the on-state resistance and off-state current leakage of the switching semiconductor contribute to this.
3. A latching relay is sometimes referred to as a memory relay. Why?
4. Design and draw a combination logic circuit that includes the following logic functions connected to control a solenoid output load:
   - Two push buttons connected to implement an AND logic function
   - Three limit switches connected to implement an OR logic function
   - A float switch connected to implement a NOT logic function
5. When troubleshooting a relay, if there is a diode connected across a relay coil, does this need to be tested? How is it tested?

CHAPTER EIGHT

Motor Control Circuits

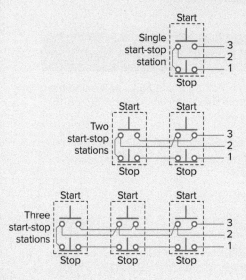
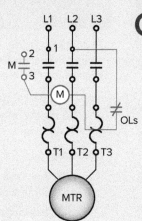

CHAPTER OBJECTIVES

This chapter will help you:

- Understand the recommended procedure for a basic motor installation based on NEC Article 430.
- List and describe the methods by which a motor can be started.
- Illustrate the operation of reversing and jogging motor control circuits.
- List and describe the methods of stopping a motor.
- Explain the operation of basic speed control circuits.

This chapter is intended to give students an insight into properly designed, coordinated, and installed motor control circuits. Topics covered include installation requirements of the NEC and motor starting, stopping, reversing, and speed control.

PART 1 NEC MOTOR INSTALLATION REQUIREMENTS

Understanding the rules detailed in the National Electrical Code (NEC) is critical to the proper installation of motor control circuits. **NEC Article 430** covers application and installation of motor circuits, including conductors, short-circuit and ground fault protection, starters, disconnects, and overload protection. Figure 8-1 illustrates the basic elements of a motor branch circuit that the NEC addresses. A motor branch circuit includes the final overcurrent device (disconnect switch and fuses or circuit breaker), the motor starter and associated control circuits, circuit conductors, and the motor.

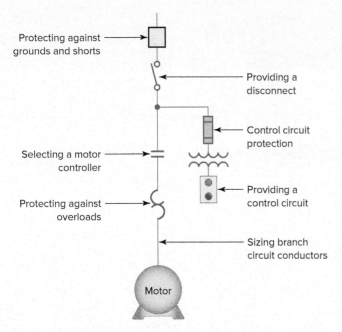

Figure 8-1 Basic elements of a motor branch circuit that the NEC addresses.

rating. Use of the term *full-load current* (FLC) *rating* indicates the table rating, while use of the term *full-load ampere* (FLA) *rating* indicates the actual nameplate rating. This makes it easier to clarify whether the table ampacity or the nameplate ampacity is being used.

EXAMPLE 8-1

Problem: Using your edition of the NEC, determine the minimum branch circuit conductor ampacity required for each of the following motors:

a. 2 hp, 230 V, single-phase motor

b. 30 hp, 230 V, three-phase motor with a nameplate FLA rating of 70 A

Solution:

a. NEC Table 430-248 shows the FLC as 12 A. Therefore, the conductor ampacity required is 12 × 125% = 15 A.

b. NEC Table 430.250 shows the FLC as 80 A. Therefore, the conductor ampacity required is 80 × 125% = 100 A.

Sizing Motor Branch Circuit Conductors

Installation requirements for motor branch circuit conductors are outlined in **NEC Article 430, Part II.** Generally, motor branch circuit conductors that supply a single motor used in a continuous-duty application must have an ampacity of not less than **125 percent** of the motor's full-load current (FLC) rating as determined by **Article 430.6**. This provision is based on the need to provide for a sustained running current that is greater than the rated full-load current and for protection of the conductors by the motor overload protective device set above the full-load current rating.

The full-load current rating shown on the motor nameplate *is not* permitted to be used to determine the ampacity of the conductors, the ampacity of switches, or motor branch circuit short-circuit and ground fault protection. The reasons for this are:

- The supply voltage normally varies from the voltage rating of the motor, and the current varies with the voltage applied.
- The actual full-load current rating for motors of the same horsepower may vary, and requiring the use of NEC tables ensures that if a motor must be replaced, this can be safely done without having to make changes to other component parts of the circuit.

Conductor ampacity must be determined by NEC Tables 430.247 through 430.250 and is based on the motor nameplate horsepower rating and voltage. Overload protection, however, is based on the marked motor nameplate

Feeder conductors supplying two or more motors must have an ampacity not less than 125 percent of the full-load current rating of the highest-rated motor plus the sum of the full-load current ratings of the other motors supplied. When two or more motors of equal rating are on a feeder, one of the motors will be considered the largest and calculated at 125 percent, and the others will be added at 100 percent.

Once the required ampacity of the conductor has been determined, **NEC Table 310.15(B)(16)** can be used to determine the American Wire Gauge (AWG) or thousand circular mil (kcmil) size of conductor required. NEC Table 310.15(B)(16) applies to situations where you have three or less current-carrying conductors in a single wireway. You must select from the column that shows the cable (identified by the insulating material letter designation) you intend to use, and choose between copper and aluminum. Bear in mind that all calculated conductor sizes based on ampacity are minimum, taking into consideration temperature rise only. The calculations do not take into account voltage dip during motor starting or voltage drop during motor running. Such considerations often require increasing the size of the branch circuit conductors.

Branch Circuit Motor Protection

Overcurrent protection for motors and motor circuits is a little different from that for nonmotor loads. The most common method for providing overcurrent protection for nonmotor loads is to use a circuit breaker that combines overcurrent protection with short-circuit and ground fault

EXAMPLE 8-2

Problem: Three 460 V, three-phase motors rated at 50, 30, and 10 hp share the same feeder (Figure 8-2). Using your edition of the NEC, determine the ampacity required to size the feeder conductors.

Solution:

50 hp motor—NEC Table 430.250 shows the FLC as 65 A.

30 hp motor—NEC Table 430.250 shows the FLC as 40 A.

10 hp motor—NEC Table 430.250 shows the FLC as 14 A.

Therefore, the required ampacity of the feeder conductors is $(1.25)(65) + 40 + 14 = 135.25$ A.

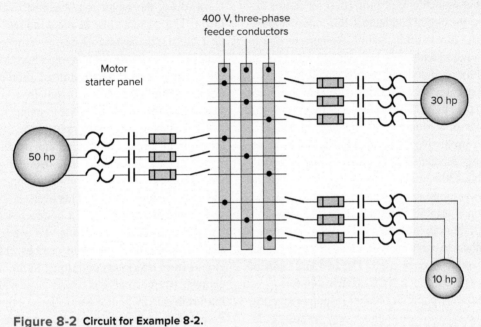

Figure 8-2 Circuit for Example 8-2.

protection. However, this isn't usually the best choice for motors because they draw a large amount of current at initial start-up, usually around **six times** the normal full-load current of the motor. With rare exceptions, the best method for providing overcurrent protection for motors is to separate the overload protection devices from the short-circuit and ground fault protection devices, as illustrated in Figure 8-3. Motor overload protection devices like heaters

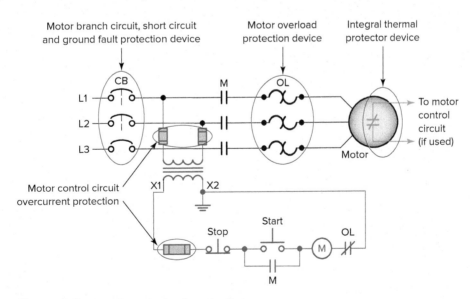

Figure 8-3 Motor branch circuit protection.

Part 1   NEC Motor Installation Requirements

and integral thermal devices protect the motor, the motor control equipment, and the branch circuit conductors from motor overload and the resultant excessive heating. They don't provide protection against short-circuit or ground fault currents. That's the job of the branch and feeder breakers. This arrangement makes motor calculations different from those used for other types of loads.

NEC Article 430, Part IV explains the requirements for branch circuit short-circuit and ground fault protection. The NEC requires that branch circuit protection for motor circuits must protect the circuit conductors, the control apparatus, as well as the motor itself against overcurrent due to short circuits or ground faults. The protection device (circuit breaker or fuse) for an individual branch circuit to a motor must be capable of carrying the starting current of the motor without opening the circuit. The Code places maximum values on the ratings or setting of these devices, as found in **Table 430.52**. A protective device that has a rating or setting not exceeding the value calculated according to the values given in Table 430.52 must be used. In cases where the values do not correspond to the standard sizes of fuses, to the ratings of nonadjustable circuit breakers, or to possible settings of adjustable circuit breakers, and the next lower value is not adequate to carry the motor load, the next higher size, rating, or setting may be used. The standard sizes of fuses and breakers are listed in **NEC Article 240.6**.

An instantaneous trip circuit breaker responds to a predetermined value of overload without any purposely delayed action. Most circuit breakers have an inverse time-delay characteristic. With an **inverse time** circuit breaker, the higher the overcurrent, the shorter the time required for the breaker to trip and open the circuit.

Non-time-delay fuses provide excellent short-circuit protection. When an overcurrent occurs, heat builds up rapidly in the fuse. Non-time-delay fuses usually hold about five times their rating for approximately ¼ second, after which the current-carrying element melts. **Time-delay** fuses provide overload and short-circuit protection. Time-delay fuses usually allow five times the rated current for up to 10 seconds to allow motors to start.

NEC Article 430, Part III deals with motor and branch circuit overload protection. A motor **overload** condition is caused by excessive load applied to the motor shaft. For example, when a saw is used, if the board is damp or the cut is too deep, the motor may become overloaded and slow down. The current flow in the windings will increase and heat the motor beyond its design temperature. A jammed pump or an extra-heavy load on a hoist will have the same effect on a motor. The overload protection also guards against failure of a motor to start from a locked rotor and loss of a phase on a three-phase system. Overload protection *is not* designed to break or may not be capable of breaking short-circuit current or ground fault current.

Motors are required to have overload protection, either within the motor itself or somewhere in close proximity to the line side of the motor. This overload protection is actually protecting the motor, the conductors, and much of the circuit ahead of the overloads. An overload in the circuit will trip the circuit overload devices, thus protecting the circuit from overload conditions. In the majority of applications, overcurrent protection is provided by the **overload relays** in the motor controller. All three-phase motors, except those protected by other approved means, such as integral-type detectors, must be provided with three overload units, one in each phase.

The **nameplate** full-load ampere (FLA) rating of the motor, rather than NEC tables, is used for sizing the overloads for the motor. Using this data, tables supplied by the starter manufacturer are consulted to find the correct overload relay thermal heater unit for the particular overload relay in use. A separate overload device that is responsive to motor current is required by **NEC Article 430.32**. This device shall be selected to trip or shall be rated at not more than 125 percent of the motor nameplate full-load current rating. When the thermal element selected in accordance with NEC 430.32 is not sufficient to start the motor or to carry the load, a higher-size thermal element can be used, provided the trip current of the overload relay does not exceed 140 percent of the motor nameplate full-load current rating.

Motor control circuits carry the current that controls the operation of the controller but do not carry the main power current to the motor. These circuits are permitted to be tapped from the motor branch circuit conductors or supplied from a separate source. **NEC 430.72** deals with overcurrent protection of motor control circuits tapped from the motor branch circuit, and **NEC 725.43** is used for those from other sources of power. Where a lower voltage

---

EXAMPLE 8-3

Problem: Determine the size of inverse time circuit breaker permitted to be used to provide motor branch circuit short-circuit and ground fault protection for a 10 hp, 208 V, three-phase squirrel-cage motor.

Solution:

NEC Table 430.250 shows the motor FLC as 30.8 A.

NEC Table 430.52 shows the maximum rating for an inverse time breaker as 250 percent of the FLC.

$$30.8 \times 2.5 = 77 \text{ A}$$

Because this is not a standard size, 80 A may be used if a 70 A inverse time circuit breaker is not adequate.

 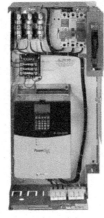

Magnetic starter   Manual snap switch   Adjustable-speed drive

Figure 8-4 Examples of motor controllers.
Photos courtesy of Rockwell Automation,
www.rockwellautomation.com

is desired, a control transformer may be installed in either of the two supply methods used.

Selecting a Motor Controller

A **motor controller** is any device that is used to directly start and stop an electric motor by closing and opening the main power current to the motor. The controller can be a switch, starter, or other similar type of control device. Examples of motor controllers are illustrated in Figure 8-4.

- A magnetic starter consisting of a contactor and overload relay is considered to be a motor controller.
- A properly rated snap switch that is permitted to turn a single-phase motor on and off is also considered to be a motor controller. Snap switches are permitted to serve as the motor controller as well as the disconnecting means.
- A solid-state starter or an AC or DC drive is also classified as a motor controller. With solid-state controllers, it is the power-circuit element, such as a triac or SCR, that meets the definition of controller.

The rating of the motor controller or starter is directly related to its NEMA size, with the electrical ratings of each being provided by the manufacturer's data sheets. A controller enclosure must be marked with the manufacturer's identification, the voltage rating, the current rating, or the horsepower rating. **NEC Article 430, Part VII** details the requirements for motor controllers. The following are some of the highlights of this section:

- The branch circuit and ground fault device can be used as the controller for stationary motors of 1/8 hp or less that are normally left running and cannot be damaged by overload or failure to start. A good example of this would be a clock motor.
- An attachment plug and receptacle can be used as a controller for a portable motor of 1/3 hp or less.
- A controller must be capable of starting and stopping the motor it controls as well as being able to interrupt the locked-rotor current of the motor.
- Unless an inverse-time circuit breaker or molded case switch is used, controllers must have horsepower rating at the applied voltage not lower than the horsepower rating of the motor.
- For stationary motors rated at 2 hp or less and 300 V or less, the controller can be either of the following:
  1. A general-use switch having an ampere rating not less than twice the full-load current rating of the motor.
  2. On AC circuits, a general-use snap switch suitable only for use on AC (not general use AC/DC snap switches) where the full-load current rating is not more than 80 percent of the amperage rating of the switch.
- A controller that does not also serve as a disconnecting means must open only as many motor circuit conductors as may be necessary to stop the motor, that is, one conductor for DC or single-phase motor circuits and two conductors for a three-phase motor circuit.
- Individual controllers must be provided for each motor unless the motor is under 1000 V, and there is a single machine with several motors, or a single overcurrent device, or a group of motors located in a single room within sight of the controller.

Disconnecting Means for Motor and Controller

The ability to safely work on a motor, a motor controller, or any motor-driven machinery starts with being able to turn the power off to the motor and its related equipment. **NEC Article 430, Part IX** covers the requirements for the motor disconnecting means. The Code requires that a means (a motor circuit switch rated in horsepower or a circuit breaker) must be provided in each motor circuit to disconnect both the motor and its controller from all ungrounded supply conductors. All disconnecting switches must plainly indicate whether they are open (off) or closed (on) and no pole is allowed to operate independently. (Figure 8-5).

The disconnecting means other than the branch circuit short-circuit and ground fault protective device is used as a safety switch to disconnect the motor circuit. It should be in sight of the motor and can be unfused. If a person is working on the motor, the disconnect will be where he or she can see it; that protects the person from a motor accidentally starting. If the branch circuit short-circuit and

Figure 8-5 Disconnecting switch.
This material and associated copyrights are proprietary to, and used with the permission of, Schneider Electric.

40 hp DC or 100 hp AC, a general-use or isolating switch can be used but should be plainly marked "**DO NOT OPERATE UNDER LOAD.**" An isolating switch is intended to isolate an electric circuit from its source of power; it has no interrupting rating and is intended to be operated only after the circuit has been opened by some other means.

EXAMPLE 8-4

Problem: Determine the current rating of the motor disconnect switch required for a 460 V, three-phase, 125 hp motor.

Solution:
NEC Table 430.250 shows the motor FLC as 156 A.
NEC 430.110 requires the motor disconnecting means to have an ampere rating of at least 115 percent of the FLC rating of the motor.

$$156 \text{ A} \times 1.15 = 179 \text{ A}$$

Therefore, a 200 A disconnect switch is required.

ground fault protective device is used as a disconnecting means and is not within sight of the motor, it must be capable of being locked in the open position. The NEC defines **"within sight"** as being visible and not more than 50 ft (15 m) distant from the other as illustrated in Figure 8-6.

For motor circuits rated at 1000 V or less, NEC 430.110 requires the disconnecting means must be at least 115 percent of the full-load rating (FLC) of the motor. The disconnecting means can be branch circuit fuses or circuit breakers. Motor disconnect switches are rated in volts, amperes, and horsepower. If rated in horsepower, the disconnect switch must have a horsepower rating equal to, or greater than, the horsepower rating of the motor at the applicable voltage. For stationary motors rated more than

Providing a Control Circuit

A motor control circuit carries electrical signals directing the action of the controller but does not carry the main power circuit. The control circuit commonly has as its load device the operating coil of a magnetic motor starter, a magnetic contactor, or a relay. **NEC Article 430, Part VI** covers the requirements for motor control circuits. Control circuits associated with motor controls can be extremely complex and vary greatly with application. The elements of a control circuit include all the equipment and devices concerned with the function of the circuit: conductors, raceway, contactor operating coil, source of energy supply to the circuit, overcurrent protection devices, and all switching devices that govern energization of the operating coil.

Motor control circuits can be the same voltage as the motor up to 600 V or can be reduced by means of a control transformer. Often a control transformer is used, especially when the control circuit extends beyond the controller. For example, a 460 V motor with an external 120 V control circuit is much easier and safer to deal with.

Where one side of the motor control circuit is grounded, the design of the control circuit must prevent the motor from being started by a ground fault in the control circuit wiring. This rule must be observed for any control circuit that has one leg grounded. If one side of the start button is in the ground leg of the circuit, as shown in Figure 8-7a, a ground fault on the coil side of the start button can short-circuit the start circuit and start the motor. By switching

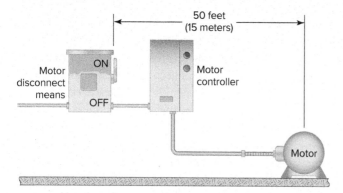

Figure 8-6 The disconnecting means must be located within sight from the controller, motor, and the driven machinery location.

216 Chapter 8 Motor Control Circuits

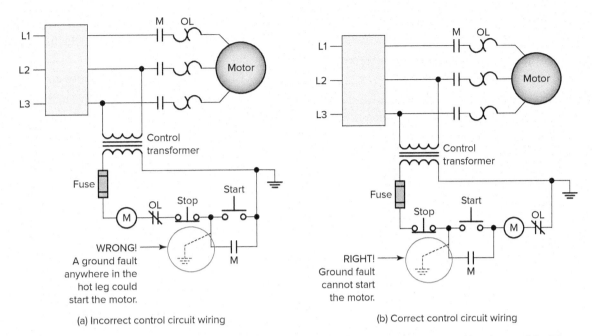

Figure 8-7 The design of the control circuit must prevent the motor from being started by a ground fault in the control circuit wiring.

the hot leg, as shown in Figure 8-7b, the starting of the motor by an accidental ground fault can be effectively eliminated. Another requirement for control circuits is that a ground fault will not bypass manual shutdown devices or automatic safety shutdown devices.

NEC Article 430.75 requires that motor control circuits be arranged so that they will be disconnected from all sources of supply when the disconnecting means is in the open position.

- Where the motor control circuit is supplied from the motor branch circuit, the controller disconnecting means may serve as the motor control circuit disconnecting means.

- Where the motor control circuit is supplied from a source other than the motor branch circuit, the motor control circuit disconnecting means must be located immediately adjacent to the controller disconnecting means.

- Where a transformer is used to obtain a reduced voltage for the motor control circuit and the transformer is located within the motor controller enclosure, the transformer must be connected to the load side of the motor control circuit disconnecting means. The control transformer must be protected in accordance with NEC Article 430.72.

Part 1 Review Questions

1. List seven basic elements of a motor branch circuit that Article 430 of the NEC addresses.
2. Give two reasons why motor branch circuit conductors are required to have an ampacity not less than 125 percent of the motor FLC.
3. Compare the terms of reference used in defining the terms FLC and FLA.
4. Determine the FLC and the ampacity required to size a 15 hp, 575 V, three-phase squirrel-cage motor.
5. Two 25 hp, 460 V, three-phase motors share the same feeder. What is the ampacity needed to size the feeder conductors?
6. Two 30 hp motors and a 40 hp motor, all 460 V, three-phase, continuous-duty squirrel-cage motors, are on a single feeder.
   a. What is the ampacity needed to size the feeder conductors?
   b. What size THWN copper conductors are required?

7. In what way are motor load characteristics different from those of general lighting and other loads?
8. Compare the type of branch overcurrent protection used for nonmotor and motor loads.
9. A 460 V, three-phase, 30 hp motor is to be short-circuit and ground fault–protected by non-time-delay fuses.
   a. What is the full-load current of this motor according to Table 430.250?
   b. What is the maximum fuse rating required according to Table 430.52?
   c. What standard fuse size, according to Article 240.6, would be selected?
10. Compare the operation of instant trip and inverse time circuit breakers.
11. Compare the operation of non-time-delay and time-delay fuses.
12. List three things motor overload protection guards against.
13. How many overload units are required for a three-phase motor?
14. Explain the procedure followed in selecting the size of the thermal heating elements for a given motor and starter.
15. List three common types of motor controllers.
16. How is a controller sized relative to the motor horsepower rating?
17. An attachment plug and receptacle can be used as a controller for what type of motor?
18. What is the basic rule with regard to location of the motor disconnecting means?
19. What provisions must be made if the branch circuit short-circuit and ground fault protective device is used as the disconnecting means and is not within sight of the motor?
20. Determine the current rating of the motor disconnect switch required for a 460 V, three-phase, 40 hp motor.
21. List three devices that commonly serve as the load for a motor control circuit.
22. What safety issues must be addressed in the design of a motor control circuit when one side of the control circuit is grounded?

PART 2 MOTOR STARTING

Every motor, when turning, acts somewhat like a generator. This generator action produces a voltage opposite, or opposed to, the applied voltage that reduces the amount of current supplied to the motor. The voltage generated in a motor is called **counter EMF (CEMF)** and results from the rotor cutting through magnetic lines of force.

When the motor is being started and before it has begun to turn, however, there is no CEMF to limit the current, so initially there is a high starting, or locked-rotor, current. The term **locked-rotor current** is based on the fact that its value is determined by locking the motor shaft so that it cannot turn, then applying rated voltage to the motor, and measuring the current. Although the starting current may be up to six times the normal running current, it normally lasts for only a fraction of a second (Figure 8-8).

The main factor in determining the amount of counter-generated voltage and current in the motor is its speed.

- All motors tend to draw much more current during the starting period (starting current) than when rotating at operating speed (running current).
- If the load placed on a motor reduces the speed, less generated counter emf will be developed and more applied current will flow.
- The greater the load on the motor, the slower the motor will rotate and the more applied current will flow through its windings.
- If the motor is jammed or prevented from rotating in any way, a locked-rotor condition is created and the applied current becomes very high. This high current will cause the motor to burn out quickly.

Full-Voltage Starting of AC Induction Motors

A full-voltage, or across-the-line, starter is designed to apply **full line voltage** to the motor upon starting. If the high starting current does not affect the power supply system and the machinery will stand the high starting torque, then full-voltage starting may be acceptable. Full-voltage starters may be either manual or magnetic. Manual motor starters are hand-operated and consist of an on/off switch with one set of contacts for each phase and motor overload protection. Since no electrical closing coil is used, the starter's contacts remain closed during a power interruption. When power is restored, the motor immediately restarts.

Manual starters for fractional-horsepower, single-phase motors are found in a variety of residential, commercial, and industrial applications. Figure 8-9 shows a single-pole fractional horsepower manual motor starter

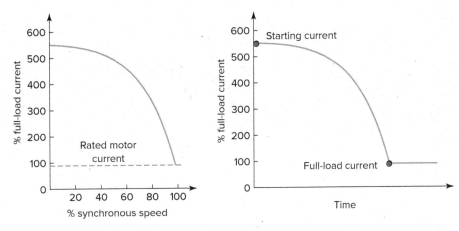

Figure 8-8 Starting current is reduced as the motor accelerates.

consisting of a manually operated on/off snap-action switch with overload protection.

- When the switch is moved to the on or start position, the motor is connected directly across the line and in series with the starter contact and the thermal overload (OL) protection device.
- As more current flows through the circuit, the temperature of the thermal overload rises, and at a predetermined temperature point, the device actuates to open the contact.
- When an overload is sensed, the starter handle automatically moves to the center position to signify that the contacts have opened because of overload and the motor is no longer operating.
- The starter contacts cannot be reclosed until the overload relay is reset manually. The starter is reset by moving the handle to the full off position after allowing about two minutes for the heater to cool.

Manual motor starters are available in single-pole, double-pole, and three-pole designs. Figure 8-10 shows a double-pole manual motor starter with a single overload heater to protect the motor windings. Line-rated control devices such as thermostats, float switches, and relays are used to connect and disconnect the motor when automatic operation is desired.

The three-pole manual starter shown in Figure 8-11 provides three overload heaters to protect the motor windings.

- This starter is operated by pushing a button on the starter enclosure cover that mechanically operates the starter.
- When an overload relay trips, the starter mechanism unlatches, opening the contacts to stop the motor.
- The contacts cannot be reclosed until the starter mechanism has been reset by pressing the stop button; first, however, the thermal unit needs time to cool.

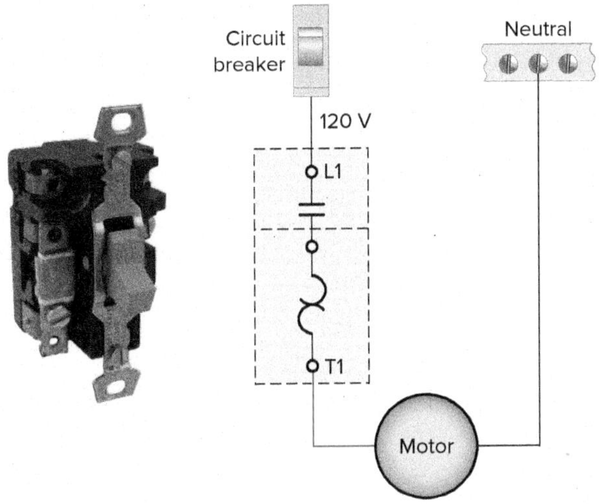

Figure 8-9 Single-pole fractional horsepower manual starter.
This material and associated copyrights are proprietary to, and used with the permission of, Schneider Electric.

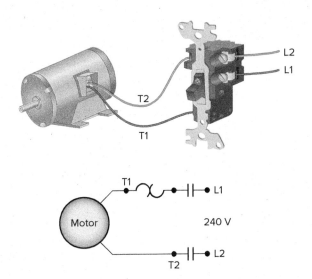

Figure 8-10 Double-pole manual motor starter.

Part 2 Motor Starting 219

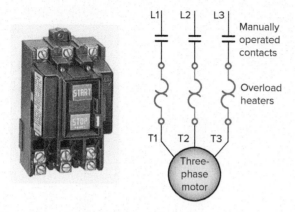

Figure 8-11 Three-pole manual starter.
Photo courtesy of Rockwell Automation,
www.rockwellautomation.com

- These starters are designed for infrequent starting of AC motors at voltages ranging from 120 to 600 V.

The power-circuit contacts of manual motor starters are unaffected by the loss of voltage, so consequently will remain in the closed position when the supply voltage fails. When the motor is running and the supply voltage fails, the motor will stop and restart automatically when the supply voltage is restored. This places these starters in the classification of **low-voltage release**. Also, manual starters must be mounted near the motor that is being controlled. Remote control operation is not possible as it would be with a magnetic starter.

Unlike the manual starter in which the power contacts are closed manually, the magnetic motor starter contacts are closed by energizing a holding coil. This enables the use of **automatic** and **remote** control of the motor. With magnetic control, pushbutton stations are mounted nearby, but automatic control pilot devices can be mounted almost anywhere on the machine.

Figure 8-12 shows a typical three-phase magnetic across-the-line AC starter diagram. The circuit's operation can be summarized as follows:

- The control transformer is powered by two of the three phases. This transformer lowers the voltage to a more common value useful when adding lights, timers, or remote switches not rated for the higher voltages.

- When the start button is pressed, coil M energizes to close all M contacts. The M contacts in series with the motor close to complete the current path to the motor. These contacts are part of the power circuit and must be designed to handle the full-load current of the motor. Memory contact M (connected across the start button) also closes to seal in the coil circuit when the start button is released. This contact is part of the control circuit; as such, it is required to handle the small amount of current needed to energize the coil.

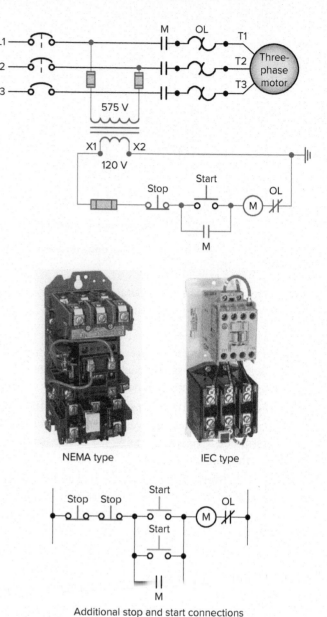

Figure 8-12 Typical magnetic across-the-line starter.
Photos courtesy of Rockwell Automation,
www.rockwellautomation.com

- The starter has three overload heaters, one in each phase. The normally closed (NC) relay contact OL opens automatically when an overload current is sensed on any phase to de-energize the M coil and stop the motor.

- The motor can be started or stopped from a number of locations by connecting additional start buttons in parallel and additional stop buttons in series.

- These starters are available in both IEC and NEMA ratings.

Different manufacturers use different methods of showing the control circuit wiring. Figure 8-13 shows a typical

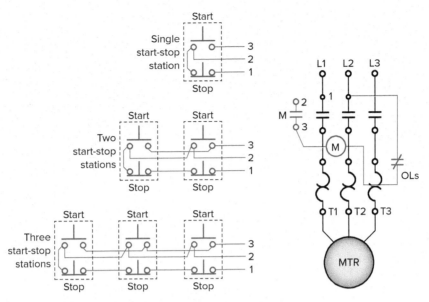

Figure 8-13 Connection diagram for motor pushbutton stations.

connection diagram for one, two, and three remote motor start-stop stations. The control wiring from the remote start/stop station runs to the connection points 1, 2, and 3 of the starter. Instead of showing the control wires running to the actual connection points, arrowed lines are used to represent the connections that must be made.

The circuit of Figure 8-14 is used to start two motors at full line voltage. In order to reduce the amount of starting current, the circuit has been designed so that there will be a short time delay period between the starting of motor 1 and motor 2. Its operation can be summarized as follows:

- The first motor is started by pressing the start button connected in a three-wire control configuration to motor starter M1.

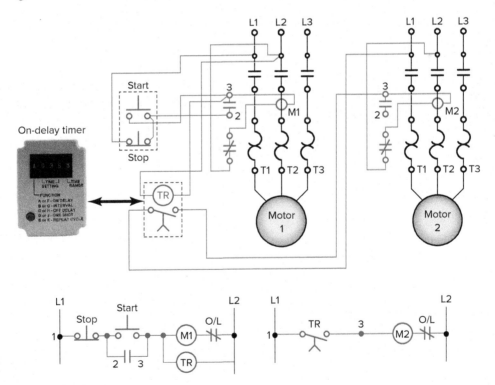

Figure 8-14 Timed starting of two motors.
This material and associated copyrights are proprietary to, and used with the permission of, Schneider Electric.

Part 2 Motor Starting   221

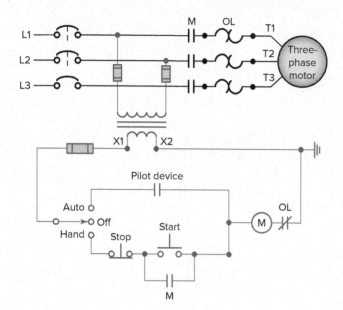

Figure 8-15 Hand-off-auto (HOA) motor control circuit.

- Power is applied to both motor 1 and on-delay timer coil TR.
- After the preset time, the normally open timer contacts on TR close to energize starter coil M2 and the second motor starts.
- Both motors can be stopped by pressing the stop button.

A common control circuit used in motor control center buckets is the **hand-off-auto (HOA)** circuit shown in Figure 8-15. The function of an HOA is to control whether a motor is in hand (manual) mode, auto mode, or off. Stop/start push buttons are selected to operate the motor in a three-wire hand mode. Auto mode allows a separate pilot device (such as a thermostat or PLC output) to operate the motor in a two-wire automatic mode.

At times a control process may have to be interlocked so that motors must be turned on in order. The control circuit of Figure 8-16 shows three motors interlocked so that they must be turned on manually in the order 1-2-3. Its operation can be summarized as follows:

- The auxiliary NO contact M1 from the first starter is connected in series with the stop button of the second starter.
- The auxiliary NO contact M2 from the second starter is connected in series with the stop button of the third starter.
- As a result, motor 2 or 3 will not be able to be turned on without the previous motor starter being energized.
- If motor 1 is turned off, then all three motor starters will be de-energized.

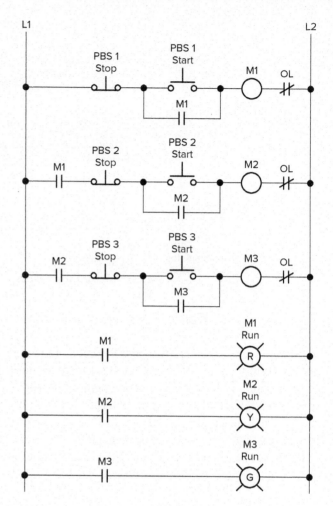

Figure 8-16 Three motor sequential interlocking.

Some process requires that the motors be turned off in sequence instead of turning on in sequence. The circuit of Figure 8-17 is an example of this type of application applied to a two-stage motor conveyor system. The operation of the control circuit can be summarized as follows:

- When the start button is momentarily closed, starter coil M1, starter coil M2, and timer TR are energized.
- Both motors start at the same time.
- M1 normally closed contact opens to open the circuit to the horn.
- TR normally open, timed open contact closes to keep starter coil M2 energized.
- When the stop button is activated, motor M1 stops immediately but motor M2 continues to operate until the time delay period has elapsed.
- The siren comes on and remains on at all times that motor M2 is running and M1 is not running.

Emergency stop (E-stop) switches are used to ensure the safety of persons and machinery. The circuit of

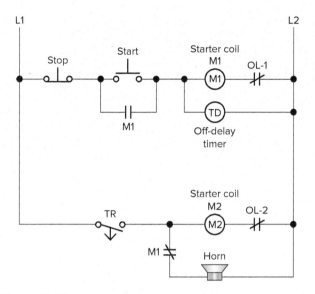

Figure 8-17 Motors turned off in sequence.

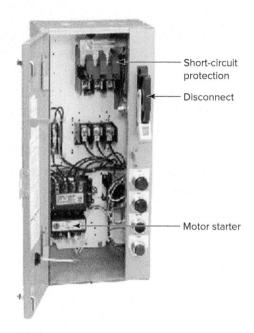

Figure 8-19 Combination starter.
Photo courtesy of Siemens, www.siemens.com

Figure 8-18 shows a two-starter motor control circuit that includes an emergency stop push button. The emergency stop function is initiated by a single human actuation of the E-stop button. When it is activated, both motor starters drop out or de-energize regardless of their original state of operation. Until the E-stop is reset, both starters will remain de-energized. To reset the E-stop, the actuator must be manually unlatched with a twist or a key release. The E-stop itself does not restart the motors; it only permits restarting through normal procedures appropriate for the process involved.

The NEC requires all motors to have a disconnecting means designed to disengage power to the motor or motor starter. A **combination starter,** such as shown in Figure 8-19,

consists of a safety switch and a magnetic motor starter placed in a common enclosure. The cover of the enclosure is interlocked with the external operating handle of the disconnect means. The door cannot be opened while the disconnect means is closed. When the disconnect means is open, all parts of the starter are accessible; however, the hazard is reduced because the readily accessible parts of the starter are not connected to the power line. Pilot devices such as push button and indicator lights may also be mounted on the panel. Combination starters offer space and cost savings over the use of separate components.

Reduced-Voltage Starting of Induction Motors

There are two primary reasons for using a reduced voltage in starting a motor:

- It limits line disturbances.
- It reduces excessive torque to the driven equipment.

When a motor is started at full voltage, the current drawn from the power line is typically 600 percent of normal full-load current. The large starting inrush current of a big motor could cause line voltage dips and brownout. In addition to high starting currents, the motor also produces starting torques that are higher than full-load torque. In many applications, this starting torque can cause mechanical damage such as belt, chain, or coupling breakage. When a reduced voltage is applied to a motor at rest, both the current drawn by the motor and the torque produced by the motor are reduced. Table 8-1 shows the typical relationship of voltage, current, and torque for a NEMA design B motor.

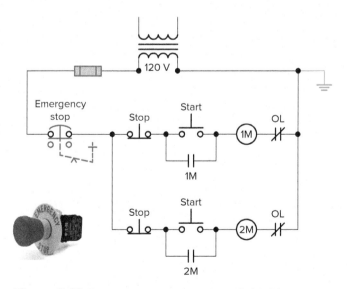

Figure 8-18 Emergency stop motor control circuit.
Photo courtesy of Rockwell Automation,
www.rockwellautomation.com

Part 2 Motor Starting

Table 8-1 Typical Voltage, Current, and Torque Characteristics for NEMA Design B Motors

Starting method	% voltage at motor terminals	Motor starting current as a percent of:		Line current as a percent of:		Motor starting torque as a percent of:	
		Locked-rotor current	Full-load current	Locked-rotor current	Full-load current	Locked-rotor torque	Full-load torque
Full voltage	100	100	600	100	600	100	180
Autotransformer 80% tap 65% tap 50% tap	80 65 50	80 65 50	480 390 300	64 42 25	307 164 75	64 42 25	115 76 45
Part-winding	100	65	390	65	390	50	90
Star delta	100	33	198	33	198	33	60
Solid-state	0–100	0–100	0–600	0–100	0–600	0–100	0–180

Electric utility current restrictions, as well as in-plant bus capacity, may require motors above a certain horsepower to be started with reduced voltage. High-inertia loads may require control of the acceleration of the motor and load. If the driven load or the power distribution system cannot accept a full-voltage start, some type of reduced-voltage or soft-starting scheme must be used. Typical reduced-voltage starters include primary-resistance starters, autotransformers, wye-delta starters, part-winding starters, and solid-state starters. These devices can be used only where low starting torque is acceptable.

Primary-Resistance Starting Reduced voltage is obtained in the **primary-resistance starter** by means of resistances that are connected in series with each motor stator lead during the starting period. The voltage drop in the resistors produces a reduced voltage at the motor terminals. At a definite time after the motor is connected to the line through the resistors, timer contacts close; this short-circuits the starting resistors and applies full voltage to the motor. Typical applications include conveyors, belt-driven equipment, and gear-driven equipment.

Figure 8-20 shows a typical primary-resistance reduced-voltage starter. Its operation can be summarized as follows:

- Pressing the start push button energizes motor starter coil M and timer coil TR. The motor is initially started through a resistor in each of the three incoming lines. Part of the line voltage is dropped through the resistors with the motor receiving about 75 to 80 percent of the full line voltage.
- At a preset time, the timer on-delay contacts close to energize contactor R coil. This causes contacts R to close, shorting out the resistors and applying full voltage to the motor. The resistors' value is chosen to provide adequate starting torque while minimizing starting current.
- Improved starting characteristics with some loads can be achieved by the use of several stages of resistance shorting. This type of reduced-voltage starting is limited by the amount of heat the resistors can dissipate.

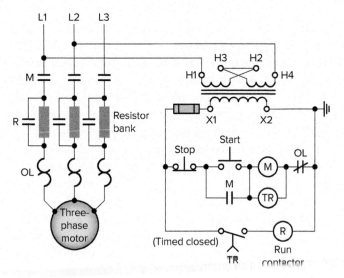

Figure 8-20 Primary resistance starter.

Autotransformer Starting Instead of resistors, autotransformer starting uses a step-down autotransformer (single-winding transformer) to reduce the line voltage. **Autotransformer starters** offer the greatest reduction of line current of any reduced-voltage starting method. Multiple taps on the transformer permit the voltage, current, and torque to be adjusted to satisfy many different starting conditions. In closed transition starting, the motor is never disconnected from the line source during acceleration. Typical applications include crushers, fans, conveyors, and mixers.

Figure 8-21 shows a typical closed transition autotransformer starting circuit. Its operation can be summarized as follows:

- Closing the start button energizes on-delay timer coil TR.
- Memory control contact TR1 closes to seal in and maintain timer coil TR.
- Contact TR2 closes to energize contactor coil C2.
- Normally open C2 auxiliary contact closes to energize contactor coil C3.
- Main power contacts of C2 and C3 close, and the motor is connected through the transformer's taps to the power line.
- Normally closed C2 auxiliary contacts are opened at this point, providing an electrical interlock that prevents C1 and C2 from both being energized at the same time. A mechanical interlock is also provided between these two contactors as this circuit condition would overload the transformer. In addition, normally open control contact C3 closes to seal-in and maintain contactor coil C3.
- After a preset time, the on-delay timer times out.
- Normally closed timed TR4 contacts open to de-energize contactor coil C2 and return all C2 contacts to their de-energized state.
- Normally open timed TR3 contact closes to energize contactor coil C1.
- Normally closed C1 auxiliary contact opens to de-energize contactor coil C3.
- The net result is the de-energizing of contactors C2 and C3 and the energizing of contactor C1, resulting in the connection of the motor to full line voltage.
- During the transition from starting to full line voltage, the motor is never disconnected from the circuit, providing closed circuit transition.

Wye-Delta Starting Wye-delta starting (also referred to as star-delta starting) involves connecting the motor windings first in wye during the starting period and then in delta after the motor has begun to accelerate (Figure 8-22). **Wye-delta starters** can be used with three-phase AC motors where all six leads of the stator windings are available (on some motors only three leads are accessible). Connected in a wye configuration, the motor starts with a significantly lower inrush current than if the motor windings had been connected in a delta configuration. Typical

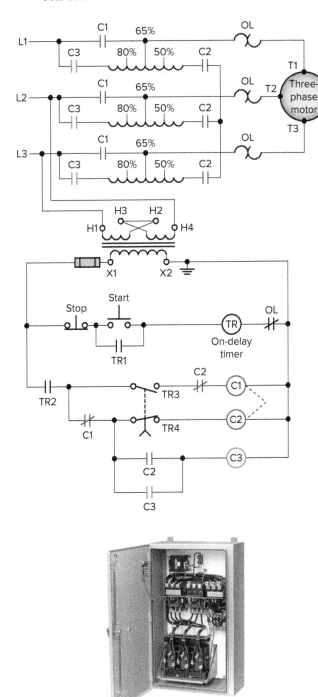

Figure 8-21 Autotransformer starter.
Photo courtesy of Rockwell Automation, www.rockwellautomation.com

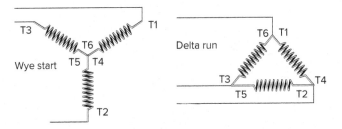

Figure 8-22 Wye and delta motor winding connections.

Part 2 Motor Starting 225

applications include central air conditioning equipment, compressors, and conveyors.

Figure 8-23 shows a typical wye-delta starting circuit. The transition from wye to delta is made using three contactors and a timer. The two contactors that are closed during run are often referred to as the main contactor (M1) and the delta contactor (M2). The third contactor (S) is the wye contactor and that carries wye current only while the motor is connected in wye. Operation of the circuit can be summarized as follows:

- When the start push button is pressed, contactor S coil is energized.
- The S main power contacts close to connect the motor windings in a wye (or star) configuration.
- The normally open S auxiliary contact closes to energize timer coil TR and contactor coil M1.
- The M1 main power contacts close to apply voltage to the wye-connected motor windings.
- Normally open auxiliary contacts S and M1 close to seal in and maintain timer coil TR.
- After the time delay period has elapsed, the TR contacts change state to de-energize contactor coil S and energize contactor coil M2.
- S main power contacts, which hold the motor windings in a wye arrangement, open.
- The M2 contacts close and cause the motor windings to be connected in a delta configuration. The motor then continues to run with the motor connected in a delta arrangement.
- In most wye-delta starters, contactors S and M2 are electrically and mechanically interlocked. If both contactors were to be energized at the same time, the result would be a line-to-line short.
- With this type of "open transition" starter, there is a very short period of time where no voltage is applied to the motor during transition from wye to delta connections. This condition can cause current surges or disturbances to be fed back into the main power source. The magnitude of the surges is proportional to the phase difference between the voltage generated by the running motor and the power source. These transients can in some instances affect other equipment that is sensitive to current surges.

Part-Winding Starting Part-winding reduced-voltage starters are used on squirrel-cage motors wound for dual-voltage operation, such as a 230/460 V motor. Power is applied to part of the motor windings on start-up and then is connected to the remaining coils for normal speed. These motors have two sets of windings connected in parallel for low-voltage operation and connected in series for high-voltage operation. When used on the **lower voltage,** they can be started by first energizing only one winding, limiting starting current and torque to approximately one-half of the full-voltage values. The second winding is then

Figure 8-23 Wye-delta starter.
Photo courtesy of Rockwell Automation,
www.rockwellautomation.com

connected in parallel, once the motor nears operating speed. Since one set of windings has higher impedance (AC resistance) than the two connected in parallel, less inrush current flows to the motor on start-up. By strict definition, part-winding starting is not true reduced-voltage starting since full voltage is applied to the motor at all times through acceleration and up to normal speed. The motor must be operated at the lower voltage, as the higher voltage would quickly damage the motor.

Part-winding starters are the least expensive type of reduced-voltage starters and use a simplified control circuit. However, they require a special motor design and do not have adjustments for current or torque. This starting method may not be suited for heavy-load applications because of the reduction of starting torque. Typical applications include low-inertia fans and blowers, low-inertia pumps, refrigeration, and compressors.

Figure 8-24 shows a typical **part-winding starting circuit.** Operation of the circuit can be summarized as follows:

- In most cases, the starter operates a 230/460 V dual-voltage wye-connected motor operating at 230 V.
- When the start push button is pressed, motor starter coil M1 and on-delay timer coil TR1 are energized.
- Auxiliary memory contact 1M-1 closes to seal in and maintain M1 and TR1 coils.
- The three M1 main contacts close, starting the motor at reduced current and torque through one-half the wye windings.
- After a preset time delay, timed contact 1-TR1 closes and energizes starter coil M2.
- The three M2 main contacts close, applying voltage to the second set of wye windings.
- Both windings of the motor are now connected in parallel to the supply voltage for full current and torque.
- Once the motor is in normal operation, the motor full-load current (FLC) is divided between the two sets of windings and starters. The overload device must be sized to the winding it serves.
- It is of utmost importance to connect the motor terminals (T1, T2, T3, T7, T8, and T9) to the proper terminals on the motor starter. The motor winding, T1-T2-T3, must be treated as a three-phase motor that when connected will have a definite direction of rotation. When motor winding T7-T8-T9 is connected, it must produce the same rotation. If by chance an error has been made, and T8 and T9 were interchanged, the second winding will attempt to change the rotation of the motor. Extremely high current will then flow, damaging the equipment.

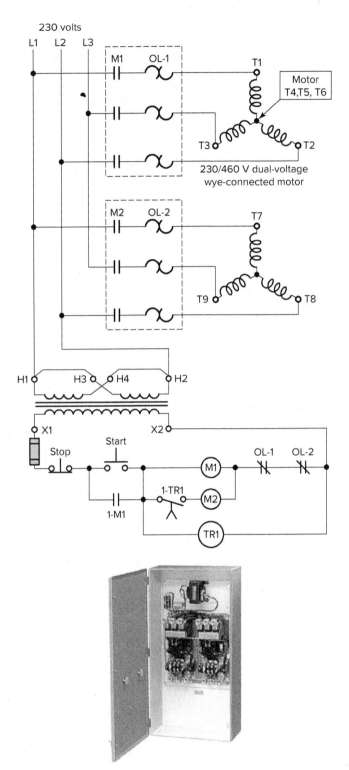

Figure 8-24 Typical part-winding starting circuit.
Photo courtesy of Rockwell Automation,
www.rockwellautomation.com

Open and Closed Transition from Start to Run Electromechanical reduced-voltage starters must make a transition from reduced voltage to full voltage at some point in the starting cycle. At this point, there is normally a line current surge. The amount of surge depends on the

Part 2 Motor Starting 227

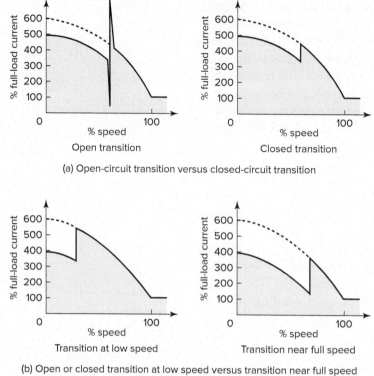

Figure 8-25 Transition from reduced voltage to full voltage.

type of transition used and the speed of the motor at the transition point.

Figure 8-25 illustrates typical transition current curves for reduced-voltage starters. There are two methods of transition from reduced voltage to full voltage, namely open-circuit transition and closed-circuit transition. **Open transition** means that the motor is actually disconnected from the line for a brief period of time when the transition takes place. With closed transition, the motor remains connected to the line during transition. Open transition will produce a higher surge of current because the motor is momentarily disconnected from the line. **Closed transition** is preferred over open transition because it causes less electrical disturbance. The switching, however, is more expensive and complex.

Soft Starting Electronic solid-state soft starters limit motor starting current and torque by **ramping** (gradually increasing) the voltage applied to the motor during the selected starting time. They are commonly used in operations requiring smooth starting and stopping of motors and driven machinery. Figure 8-26 illustrates typical transition voltage and current curves for soft starters. The time to full voltage can be adjustable, usually from 2 to 30 seconds. As a result there is no large current surge when the controller is set up and correctly matched to the load. Current limiting is used when it is necessary to limit the maximum starting current and is usually adjustable from 200 to 400 percent of full-load amperes.

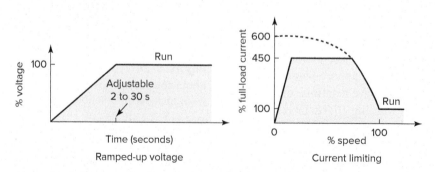

Figure 8-26 Soft start ramped-up voltage and current limiting.

228   Chapter 8   Motor Control Circuits

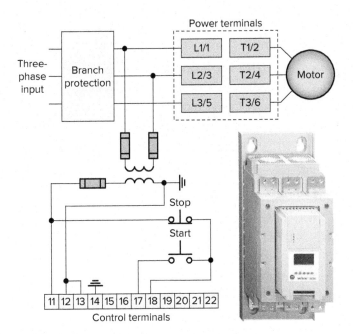

Figure 8-27 Typical soft start starter.
Photo courtesy of Rockwell Automation,
www.rockwellautomation.com

Figure 8-27 shows the wiring for a typical soft start starter. The different standard modes of operation for this controller are:

Soft start This method covers the most general applications. The motor is given an initial torque setting, which is user-adjustable. From the initial torque level, the output voltage to the motor is steplessly increased during the acceleration ramp time, which is user-adjustable.

Selectable kick start The kick start feature provides a boost at start-up to break away loads that may require a pulse of high torque to get started. It is intended to provide a current pulse, for a selected period of time.

Current limit start This method provides current limit start and is used when it is necessary to limit the maximum starting current. The starting current is user-adjustable. The current limit starting time is user-adjustable.

Dual-ramp start This starting method is useful on applications with varying loads, starting torque, and start time requirements. Dual-ramp start offers the user the ability to select between two separate start profiles with separately adjustable ramp times and initial torque settings.

Full-voltage start This method is used in applications requiring across-the-line starting. The controller performs like a solid-state contactor. Full inrush current and locked-rotor torque are realized. This controller may be programmed to provide full-voltage start in which the output voltage to the motor reaches full voltage in ¼ second.

Linear speed acceleration With this type of acceleration mode, a closed-loop feedback system maintains the motor acceleration at a constant rate. The required feedback signal is provided by a DC tachometer coupled to the motor.

Preset slow speed This method can be used on applications that require a slow speed for positioning material. The preset slow speed can be set for either low, 7 percent of base speed, or high, 15 percent of base speed.

Soft Stop The soft stop option can be used in applications requiring an extended stop time. The voltage ramp-down time is adjustable from 0 to 120 seconds. The load will stop when the voltage drops to a point where the load torque is greater than the motor torque.

The soft starter is a reduced-voltage starter that restricts starting current by applying the voltage in a ramp. It can only be applied to constant-speed motor applications. A **variable-frequency drive (VFD)** can change output **voltage** and **frequency** at the same time to vary the speed of the motor. The variable-frequency drive has all the features of a soft starter but also allows the speed to be varied, while offering more flexibility and features.

DC Motor Starting

As was the case with AC motors, fractional horsepower manual starters or magnetic contactors and starters can be used for across-the-line starting of smaller DC motors. One major difference between AC and DC starters is the electrical and mechanical requirements necessary for suppressing the arcs created in opening and closing contacts under load. To combat prolonged arcing in DC circuits, the contactor switching mechanism is constructed so that the contacts will separate rapidly and with enough air gap to extinguish the arc as soon as possible on opening. Figure 8-28

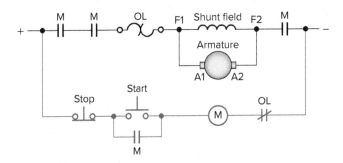

Figure 8-28 Across-the-line DC motor starter.

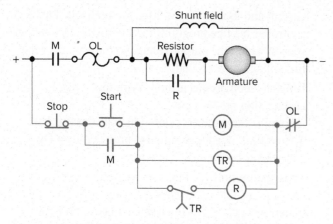

Figure 8-29 Definite-time reduced-voltage DC starter.

shows the schematic diagram for a typical DC across-the-line starter that uses three-wire control. To help extinguish the arc, the starter is equipped with three power contacts connected in series.

At the moment a DC motor is started, the armature is stationary and there is no counter EMF being generated. The only component to limit starting current is the armature resistance, which in most DC motors is a very low value. Common types of reduced-voltage DC starters include definite-time acceleration, current acceleration, counter-EMF acceleration, and variable-voltage acceleration. Figure 8-29 illustrates a two-stage **DC definite-time resistance starter**. When the power contact M closes, full line voltage is applied to the shunt field while the resistor is connected in series with the armature. After a preset time period, the contactor R closes, bypassing the resistor, allowing the motor to operate at base speed. This gives the motor smooth torque without creating a large surge of current. Operation of the circuit can be summarized as follows:

- Pressing the start button energizes the M and TR coils.
- Auxiliary memory contact M closes to seal in and maintain M and TR coils.
- The M main contact closes, starting the motor at reduced current and torque with the resistor connected in series with the armature.
- After a preset time delay, timed contact TR closes to energize contactor R coil.
- Contact R closes, bypassing the resistor and allowing full line voltage to be applied to the armature.
- The starting method is closed transition.
- Starting resistance can be shorted out in one or more steps, depending on motor size and the smoothness of acceleration desired.
- The shunt field has full line voltage applied to it any time the motor is on.

Figure 8-30 illustrates **variable-voltage acceleration** of a DC shunt motor using a silicon-controlled rectifier (SCR) armature voltage controller. The SCR provides a useful method of converting AC voltage to variable DC voltage. An SCR is a semiconductor device that has three elements: anode, cathode, and gate. By applying a signal to the gate element at precisely the right time, you can control how much current the SCR will either pass or block during a cycle. This is known as phase control. The shorter the on time, the lower the DC voltage applied to the armature. The shunt field is fed from a separate DC source and has full voltage applied any time the motor is on.

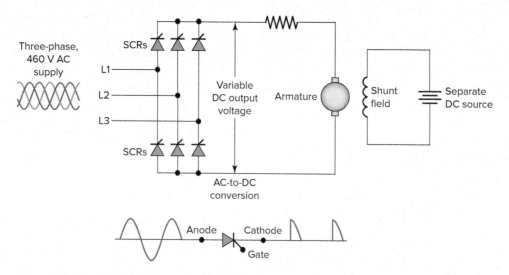

Figure 8-30 Variable-voltage acceleration of a DC shunt motor.

Part 2 Review Questions

1. Why is there initially a high inrush of current while a motor is being started?
2. How do the motor starting current value and the normal full-load current compare?
3. What will create a locked-rotor condition in a motor?
4. How is a full-voltage starter designed to start a motor?
5. Compare the way main contacts of a manual and magnetic starter are operated.
6. Explain the term *low-voltage release* as it applies to manual starters.
7. A magnetic across-the-line starter is operated by a single start/stop pushbutton station. If a second start/stop pushbutton station is added, how would the additional push buttons be connected relative to the existing ones?
8. What exactly is a combination motor starter?
9. State two reasons for the use of reduced-voltage starting.
10. Outline the operation of a primary-resistance induction motor starter.
11. Outline the operation of an autotransformer induction motor starter.
12. Outline the operation of a wye-delta induction motor starter.
13. Outline the operation of a part-winding induction motor starter.
14. Which type of reduced-voltage transition results in the least amount of electrical disturbance?
15. Explain the term *ramping* as it applies to soft starting of a motor.
16. Outline the operation of a definite-time reduced-voltage DC starter.
17. Explain the term *phase control* as it applies to an SCR armature voltage controller.
18. What control technique is used to turn motors on or off in sequence?

PART 3 MOTOR REVERSING AND JOGGING

Reversing of AC Induction Motors

Reversing Three-Phase Induction Motor Starter
Certain applications require a motor to operate in either direction. **Interchanging any two leads** to a three-phase induction motor will cause it to run in the reverse direction. The industry standard is to interchange phase A (line 1) and phase C (line 3), while phase B (line 2) remains the same. Reversing starters are used to automatically accomplish this phase reversal.

The power circuit of a magnetic full-voltage three-phase reversing motor starter is shown in Figure 8-31.

- This starter is constructed using two 3-pole contactors with a single overload relay assembly.
- The contactor on the left is usually designated as the forward contactor, and the right contactor is usually designated as the reverse contactor.
- The power circuit of the two contactors is interconnected using bus bars or jumper wires.
- Power contacts (F) of the forward contactor, when closed, connect L1, L2, and L3 to motor terminals T1, T2, and T3, respectively.
- Power contacts (R) of the reverse contactor, when closed, connect L1 to motor terminal T3 and connect L3 to motor terminal T1, causing the motor to run in the opposite direction.
- Whether operating through either the forward or reverse contactor, the power connections are run through the same set of overload relays.
- Only one overload relay assembly is required because the motor windings must be protected for the same current level regardless of the direction of rotation.

When the motor is reversed, it is vital that both contactors **not** be energized at the same time. Activating both contactors would cause a **short circuit,** since two of the line conductors are reversed on one contactor. Both mechanical and electrical interlocks are used to prevent the forward and reverse contactors from being activated at the same time. **Mechanical interlocking** is normally factory-installed and uses a system of levers to prevent both contactors from

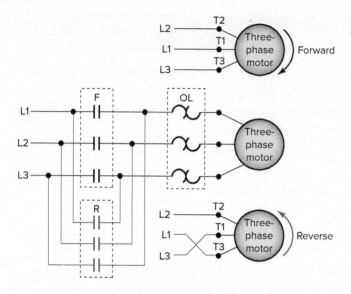

way as to physically block the movement of the reverse contactor. Even if the reverse contactor coil should become energized, the contacts will not close because the mechanical interlock is physically blocking the reverse contactor. The forward contactor coil must be de-energized before the reverse contactor can operate. The same scenario applies if the reverse contactor is energized. Mechanical interlocks have been known to fail, and for this reason additional electrical interlocking is used for added protection.

Most reversing starters utilize auxiliary contacts operated by the forward and reverse coils to provide **electrical interlocking.** When the coil is energized, the frame of the contactor moves and activates the auxiliary contacts mounted on the contactor. The auxiliary contacts are connected to the motor control circuit, and the state of the contacts (normally open or closed) is associated with the state of the coil of the contactor.

The control circuit of Figure 8-33 illustrates how auxiliary contact interlocking works and can be summarized as follows:

- The normally closed contact controlled by the forward coil is connected in series with the reverse coil.

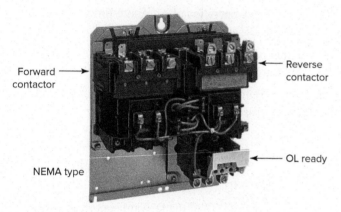

Figure 8-31 Magnetic full-voltage three-phase reversing motor starter.

Photo courtesy of Rockwell Automation, www.rockwellautomation.com

being engaged at the same time. The broken line, as illustrated in Figure 8-32, indicates that the coils F and R cannot close contacts simultaneously because of the mechanical interlocking action of the device. For example, energizing the coil of the forward contactor moves a lever in such a

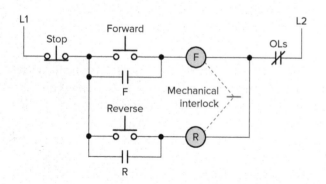

Figure 8-32 Mechanical interlocking of forward and reverse contactors.

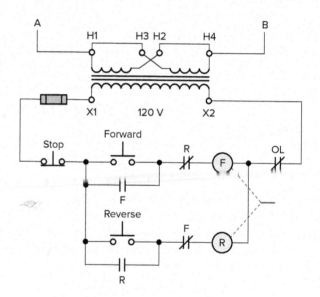

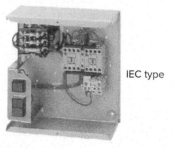

Figure 8-33 Magnetic reversing starter with electrical interlock in the motor starter.

Photo courtesy of Rockwell Automation, www.rockwellautomation.com

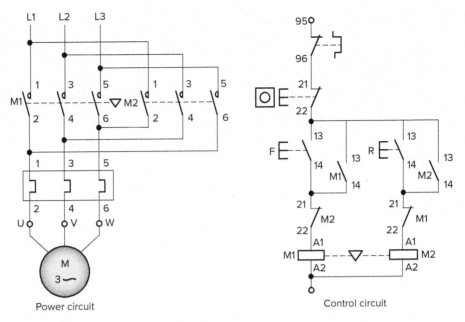

Figure 8-34 Reversing starter circuit implemented using IEC symbols.

- The normally closed contact controlled by the reverse coil is connected in series with the forward coil.
- When the forward coil is energized, the normally closed contact in series with the reverse coil is opened to prevent the reverse coil from being energized.
- When the reverse coil is energized, the normally closed contact in series with the forward coil opens to prevent the forward coil from being energized.
- To reverse the motor with this control circuit, the operator must press the stop button to de-energize the respective coil, reclosing the respective normally closed contact.
- Reversing starters are usually factory-wired for electrical interlocking.
- Starter mechanical and electrical interlocking offers sufficient protection for most reversing motor control circuits.
- Figure 8-34 shows the reversing starter circuit implemented using IEC symbols.

Electrical **pushbutton interlocking** utilizes break-make, normally closed and normally open switch contacts on the forward and reverse buttons. The control circuit of Figure 8-35 illustrates how pushbutton interlocking works and can be summarized as follows:

- Interlocking is achieved by connecting the normally closed contact of the reverse button in series with the normally open contact of the forward push button.
- The normally closed contact of the reverse push button acts like another stop push button in the forward circuit.
- The normally open contact on the reverse push button is used as the start button for the reverse circuit.
- When the reverse button is pressed, its normally closed contact opens the circuit to the forward coil and then its normally open contact completes the circuit to the reverse coil.
- When the forward button is pressed, its normally closed contact opens the circuit to the reverse coil and then its normally open contact completes the circuit to the forward coil.
- The motor reverses direction immediately without the stop button being pressed. Take care when

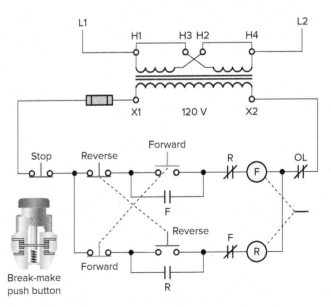

Figure 8-35 Pushbutton interlocking.

Part 3 Motor Reversing and Jogging

reversing large motors, as the sudden jar of reversal can damage the equipment the motor is driving. High inrush currents can cause damage to both the motor and the controller if the motor is reversed without allowing enough time for the speed of the motor to decrease.

- Pushbutton interlocking should be used in conjunction with both mechanical and auxiliary electrical interlocking and is intended to supplement these methods, not replace them.

Limit switches are used in motor control applications as a safety device, for counting or sorting products, to initiate another operating sequence, and for reversing machine travel. The contacts of a momentary limit switch change state when a predetermined force or torque is applied to the actuator. A spring returns its contacts to their original position when the operating force is removed.

Limit switches can be used to limit the **travel** of electrically operated doors, conveyors, hoists, machine tool worktables, and similar devices. The control circuit of Figure 8-36 illustrates how limit switches can be incorporated into a reversing starter circuit to limit travel. The operation of the circuit can be summarized as follows:

- Pressing the forward push button energizes coil F.
- Auxiliary memory contact F closes to seal in and maintain F coil.

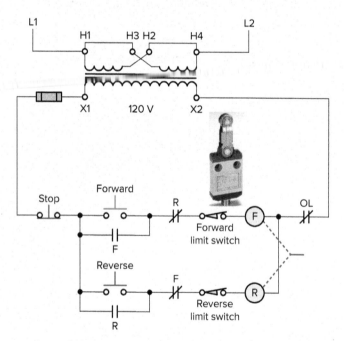

Figure 8-36 Limit switches incorporated into a reversing starter circuit to limit travel.
Photo courtesy of Omron Automation and Safety, www.Omron247.com

- Auxiliary interlock contact F opens to isolate the reversing circuit.
- Power contacts F close and the motor runs in the forward direction.
- If either the stop button or forward limit switch is actuated, the holding circuit to coil F opens, thus de-energizing the coil and returning all F contacts to their normal de-energized state.
- Pressing the reverse push button energizes coil R.
- Auxiliary memory contact R closes to seal in and maintain R coil.
- Auxiliary interlock contact R opens to isolate the forward circuit.
- Power contacts R close and the motor runs in the reverse direction.
- If either the stop button or reverse limit switch is actuated, the holding circuit to coil R opens, thus de-energizing the coil and returning all R contacts to their normal de-energized state.
- The location of the limit switches in the circuit allows the one direction of travel to be stopped if the motor is driving a device that has limits to its travel. The opposite direction is not affected by one travel limit being opened. As soon as the motor is reversed and the actuator is no longer holding the limit switch open, it will return to its normally closed position.

Reversing of Single-Phase Motors

Figure 8-37 illustrates how a single-phase capacitor-start motor is wired to operate in the forward and reverse directions using a reversing starter. The direction of rotation is changed by interchanging the start winding leads, while those of the run winding remain the same. Unlike the three-phase motor, the single-phase capacitor-start motor must be allowed to slow down before any attempt to reverse the direction of rotation. The centrifugal switch in the start winding circuit opens at approximately 75 percent of the motor speed and must be allowed to reclose before the motor will reverse.

Certain machine tool operations require a repeated forward and reverse action in their operation. Figure 8-38 illustrates a **reciprocating machine process** that uses two limit switches to provide automatic control of the motor. Each limit switch (LS1 and LS2) has two sets of contacts, one normally open and the other normally closed. The operation of the circuit can be summarized as follows:

- The start and stop push buttons are used to initiate and terminate the automatic control of the motor by limit switches.

234 Chapter 8 Motor Control Circuits

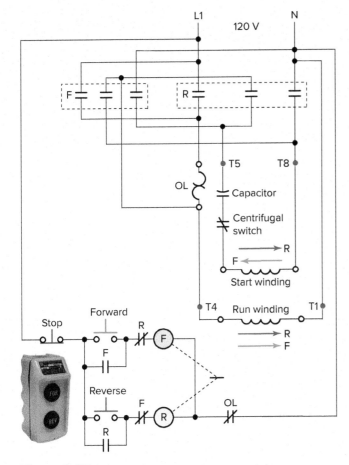

Figure 8-37 Reversing a single-phase motor.
This material and associated copyrights are proprietary to, and used with the permission of, Schneider Electric.

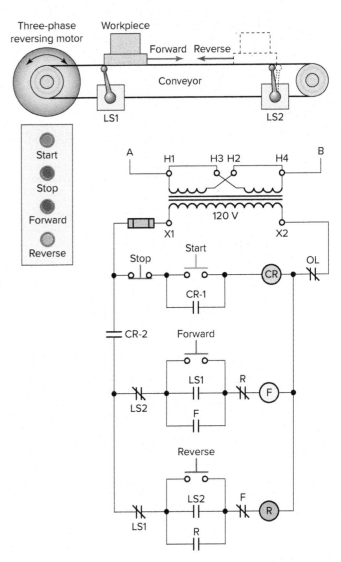

Figure 8-38 Reciprocating machine process.

- Contact CR1 is used to maintain the circuit to the control relay during the running operation of the circuit.
- Contact CR2 is used to make and break the line circuit to the forward and reverse control circuit.
- Using the control relay and its start and stop buttons also provides low-voltage protection—that is, the motor will stop when there is a supply voltage failure and the motor will not restart automatically when the supply voltage is restored.
- The normally closed contact of limit switch LS2 acts as the stop for the forward controller, and the normally open contact of limit switch LS1 acts as the start contact for the forward controller. The auxiliary contact of the forward starter is connected in parallel with the normally open contact of limit switch LS1 to maintain the circuit during the running of the motor in the forward direction.
- The normally closed contact of limit switch LS1 is wired as a stop contact for the reverse starter, and the normally open contact of limit switch LS2 is wired as a start contact for the reverse starter. The auxiliary contact on the reverse starter is wired in parallel with the normally open contacts of limit switch LS2 to maintain the circuit while the motor is running in reverse.
- Electrical interlocking is accomplished by the addition of a normally closed contact in series with each starter and operated by the starter for the opposite direction of rotation of the motor.
- Reversal of the direction of rotation of the motor is provided by the action of the limit switches. When limit switch LS1 is moved from its normal position, the normally open contact closes energizing coil F and the normally closed contact opens and drops out coil R. The reverse action is performed by limit switch LS2, and thus reversing in either direction is provided.

Part 3 Motor Reversing and Jogging

- The forward and reverse push buttons provide a means of starting the motor in either forward or reverse so that the limit switches can take over automatic control.

Reversing of DC Motors

The reversal of a DC motor can be accomplished in two ways:

- Reversing the direction of the **armature current** and leaving the field current the same.
- Reversing the direction of the **field current** and leaving the armature current the same.

Most DC motors are reversed by switching the direction of current flow through the armature. The switching action generally takes place in the armature because the armature has a much lower inductance than the field. The lower inductance causes less arcing of the switching contacts when the motor reverses its direction.

Figure 8-39 shows the power circuit for DC motor reversing using electromechanical and electronic control. For electromechanical operation, the forward contactor causes current to flow through the armature in one direction, and the reverse contactor causes current to flow through the armature in the opposite direction. For solid-state electronic control, two sets of SCRs are provided. One set is used for current flow in one direction through the armature, and the second set is used for current flow in the opposite direction.

Jogging

Jogging (sometimes called inching) is the **momentary operation** of a motor for the purpose of accomplishing small movements of the driven machine. It involves an operation in which the motor runs when the push button is pressed and will stop when the push button is released. Jogging is used for frequent starting and stopping of a motor for short periods of time.

The pushbutton jog circuit shown in Figure 8-40 uses a standard start/stop control circuit with a double-contact jog push button: one normally closed contact and one normally open contact. The operation of the circuit can be summarized as follows:

- Pressing the start push button energizes starter coil M, causing the M main contacts to close to start the motor and the M auxiliary contact to close to maintain the M coil circuit.
- With the M coil de-energized and the jog push button then pressed, a circuit is completed for the M coil around the M auxiliary maintaining contact.
- The M main contacts close to start the motor, but the maintaining circuit is incomplete as the normally closed jog contact is open.
- As a result, starter coil M will not seal in; instead, it can stay energized only as long as the jog button is fully depressed.
- On quick release of the jog push button, should its normally closed contacts reclose before the starter maintaining contact M opens, the motor would continue to run. In certain applications, this could be hazardous to workers and machinery.

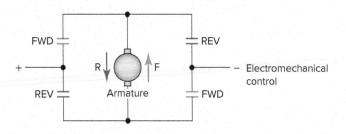

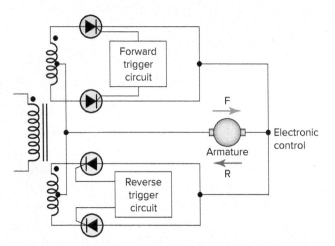

Figure 8-39 DC motor reversing power circuits.

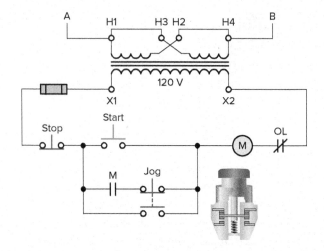

Figure 8-40 Pushbutton job circuit.

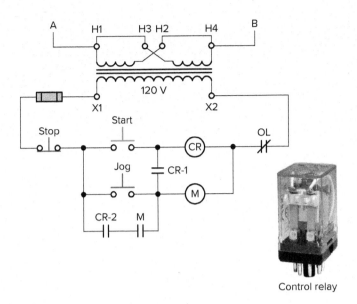

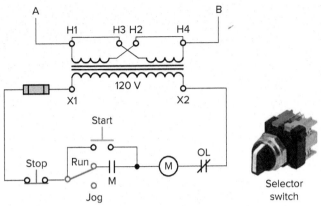

Figure 8-42 Start/stop/selector jog control circuit.
Photo courtesy of Rockwell Automation, www.rockwellautomation.com

Figure 8-41 Jog circuit with control relay.
Photo courtesy of IDEC Corporation, www.IDEC.com/usa, RR Relay

The **control relay jogging** circuit shown in Figure 8-41 is much safer than the previous circuit. A single-contact jog push button is used; in addition, the circuit incorporates a jog control relay (CR). The operation of the circuit can be summarized as follows:

- Pressing the start push button completes a circuit for the CR coil, closing the CR-1 and CR-2 contacts.
- The CR-1 contact completes the circuit for the M coil, starting the motor.
- The M maintaining contact closes; this maintains the circuit for the M coil.
- Pressing the jog button energizes the M coil only, starting the motor. Both CR contacts remain open, and the CR coil is de-energized. The M coil will not remain energized when the jog push button is released.

Figure 8-42 shows the use of a **selector switch** in the control circuit to obtain jogging. The start button doubles as a jog button. The operation of the circuit can be summarized as follows:

- When the selector switch is placed in the run position, the maintaining circuit is not broken. If the start button is pressed, the M coil circuit is completed and maintained.
- Turning the selector switch to the jog position opens the maintaining circuit. Pressing the start button completes the circuit for the M coil, but the maintaining circuit is open. When the start button is released, the M coil is de-energized.

A circuit for implementing the jogging function for a three-phase motor reversing starter, using a jog/run selector switch, is shown in Figure 8-43. The operation of the circuit can be summarized as follows:

- When the jog/run switch is in the **run** position, the circuit operates as a reversing starter with electrical interlocking of the forward and reverse starter coils in addition to the forward and reverse push buttons.
- When the jog/run switch is in the **jog** position, the jogging function is initiated through the selector switch by opening the seal-in circuit to both the forward and reverse starter coils.

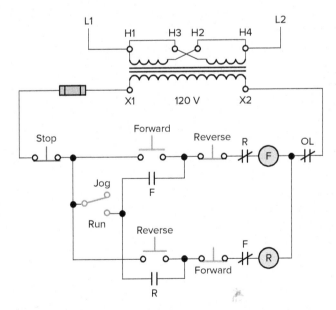

Figure 8-43 Jogging of a three-phase reversing starter using a jog/run selector switch.

Part 3 Motor Reversing and Jogging

Part 3 Review Questions

1. How can the direction of rotation of a three-phase induction motor be reversed?
2. An electromagnetic reversing motor starter is made up of what components?
3. What would occur if both contactors in a reversing motor starter were to become energized at the same time?
4. Explain the operation of the mechanical interlock in a reversing magnetic motor starter.
5. Explain how to provide electrical interlocking using auxiliary contacts.
6. What types of forward and reverse push buttons are utilized for pushbutton interlocking?
7. How is reversing a single-phase capacitor start motor accomplished using a magnetic motor starter?
8. Why are most DC motors reversed by switching the direction of current flow through the armature rather than the field?
9. What is a jog control used for?
10. A jog/run selector switch is to be used to implement the jogging function for a three-phase reversing motor starter. Explain how this is accomplished.

PART 4 MOTOR STOPPING

The most common method of stopping a motor is to remove the supply voltage and allow the motor and load to coast to a stop. In some applications, however, the motor must be stopped more quickly or held in position by some sort of braking device. Electric braking uses the windings of the motor to produce a retarding torque. The kinetic energy of the rotor and the load is dissipated as heat in the rotor bars of the motor.

Plugging and Antiplugging

Plugging is a name given to a method of quickly stopping a motor. To plug a motor, you reverse the motor while it is running in one direction, but then de-energize it before it can turn in the opposite direction. This acts as a retarding force for rapid stop. Plugging produces more heat than most normal-duty applications. NEMA specifications call for starters used for such applications to be derated. That is, the next size larger reversing starter must be selected when it is used for plugging to stop or reverse at a rate of more than five times per minute.

The control scheme for plugging includes a reversing magnetic starter and a rotary-motion switch called a **plugging switch**. Since this switch can be used for applications other than plugging, it is also known as a **zero-speed switch**. A typical plugging switch is shown in Figure 8-44. The switch is coupled to a moving shaft on the machinery whose motor is to be plugged. It consists of a set of contacts that are activated by centrifugal force resulting from rotation of the shaft mechanism. The contacts open or close depending on the intended application. Each zero-speed switch has a rated operating speed

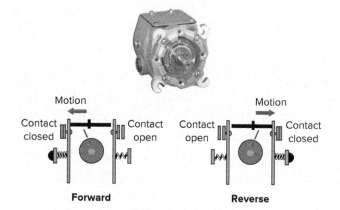

Figure 8-44 Plugging switch.
Photo courtesy of Rockwell Automation, www.rockwellautomation.com

range within which the contacts will be switched, for example, 50 to 200 rpm.

Figure 8-45 shows the common symbol used to represent a zero-speed switch. The curved arrow represents the rotation of the motor to either close or open the contacts, depending on their normal state. The letters F and R stand for forward and reverse, and indicate the direction of rotation. The speed at which the contacts open or close are adjusted with a screw through the switch housing. Since the slightest rotation can activate the switch contacts, some

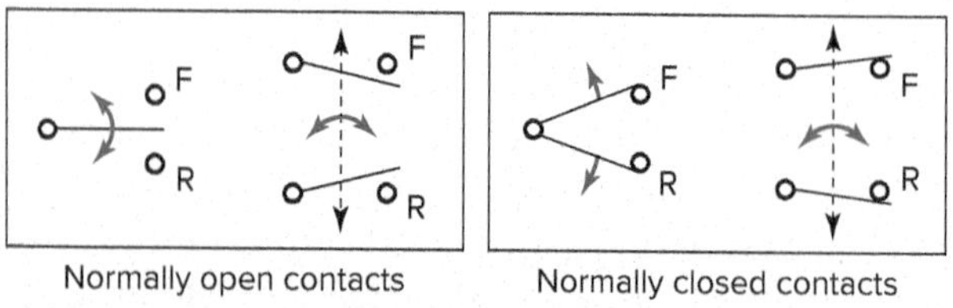

Figure 8-45 Zero-speed switch symbol.

plugging switches are equipped with a special solenoid coil called a **lockout relay** or **safety-latch** relay. This relay incorporates a mechanism to hold the plugging contacts open until the motor-control circuit itself has been energized. This is a safeguard in case someone rotates the shaft. The leads of this relay coil are usually connected to the T1 and T2 terminals of the motor.

The control schematic of Figure 8-46 shows a typical control circuit for forward-direction plugging. The operation of the circuit can be summarized as follows:

- Pressing the start button closes and seals in the forward contactor. As a result, the motor rotates in the forward direction.
- The normally closed auxiliary contact F opens the circuit to the reverse contactor coil.
- The forward contact on the speed switch closes.
- Pressing the stop button de-energizes the forward contactor.
- The reverse contactor is energized, and the motor is plugged.
- The motor speed decreases to the setting of the speed switch, at which point its forward contact opens and de-energizes the reverse contactor.
- This contactor is used only to stop the motor by using the plugging operation; it is not used to run the motor in reverse.

The sudden reversing torque applied when a large motor is reversed (without slowing the motor speed) could

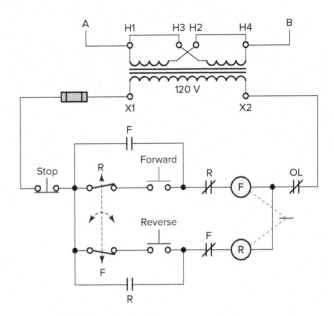

Figure 8-47 Antiplugging protection circuit.

damage the driven machinery, and the extremely high current could affect the distribution system. **Antiplugging** protection is obtained when a device prevents the application of a counter torque until the motor speed is reduced to an acceptable value.

The antiplugging circuit of Figure 8-47 may be used to prevent reversing the motor before the motor has slowed to near zero speed. In this application, the motor can be reversed but not plugged. The operation of the circuit can be summarized as follows:

- Pressing the forward button completes the circuit for the F coil, closing the F power contacts and causing the motor to run in the forward rotation.
- The F zero-speed switch contact opens because of the forward rotation of the motor.
- Pressing the stop button de-energizes the F coil, which opens the F power contacts, causing the motor to slow down.
- Pressing the reverse button will not complete a circuit for the R coil until the F zero-speed switch contact recloses.
- As a result, when the rotating equipment reaches near zero speed, the reverse circuit may be energized, and the motor will run in the reverse rotation.

Figure 8-48 illustrates how timing relays can be used to execute antiplugging control of a motor. The operation of the circuit can be summarized as follows:

- An off-delay timer is connected in parallel with the forward and reverse contactor coils of the reversing starter.

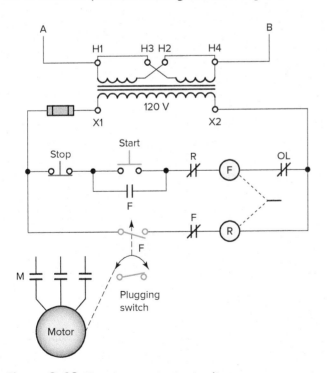

Figure 8-46 Plugging a motor to stop it.

Part 4 Motor Stopping 239

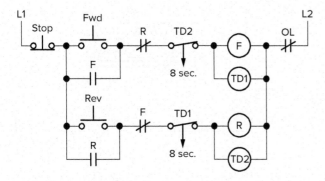

Figure 8-48 Antiplugging executed using time-delay relays.

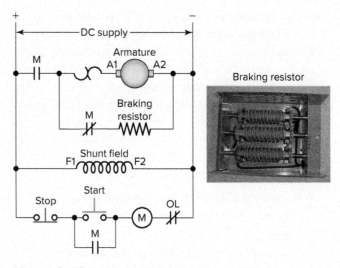

Figure 8-49 Dynamic braking applied to a DC motor.
Photo courtesy of Transfab TMS, www.transfabtms.com

- The normally closed, timed closed, contact from the forward timer coil (TD1) is interlocked with the reversing circuit.
- The normally closed, timed closed, contact from the reverse timer coil TD2 is interlocked with the forward circuit.
- When the motor is started in the forward direction, the timer coil TD1 is energized at the same time as the forward coil.
- The relay contact TD1 opens immediately in the reversing coil circuit, as does the auxiliary F-interlock.
- When the stop button is activated, both the F coil and the TD1 coil are de-energized.
- The F interlock returns to the normally closed state. However, the TD1 interlock remains open for an additional 8 seconds before returning to its normally closed position.
- As a result, the reversing coil (R) cannot be activated until the drive motor has been allowed 8 seconds to come to a stop.
- The timing relay TD2 provides the identical action in the reversing coil operation.

Dynamic Braking

Dynamic braking is achieved by reconnecting a running motor to act as a **generator** immediately after it is turned off, rapidly stopping the motor. The generator action converts the mechanical energy of rotation to electrical energy that can be dissipated as heat in a resistor. Dynamic braking of a DC motor may be needed because DC motors are often used for lifting and moving heavy loads that may be difficult to stop.

The circuit shown in Figure 8-49 illustrates how dynamic braking is applied to a DC motor. The operation of the circuit can be summarized as follows:

- Assume the motor is operating and the stop button is pressed.
- Starter coil M de-energizes to open the normally open M power contact to the motor armature.
- At the same time, the normally closed M power contact closes to complete a braking circuit around the armature through the braking resistor, which acts like a load.
- The shunt field winding of the DC motor remains connected to the power supply.
- The armature generates a counter-EMF voltage. This counter EMF causes current to flow through the resistor and armature. The smaller the ohmic value of the braking resistor, the greater the rate at which energy is dissipated and the faster the motor comes to rest.

DC Injection Braking

DC injection braking is a method of braking in which direct current is applied to the stationary windings of an AC motor after the applied AC voltage is removed. The **injected DC voltage** creates a magnetic field in the motor stator winding that does not change in polarity. In turn, this constant magnetic field in the stator creates a magnetic field in the rotor. Because the magnetic field of the stator is not changing in polarity, it will attempt to stop the rotor when the magnetic fields are aligned (N to S and S to N).

The circuit of Figure 8-50 is one example of how DC injection braking can be applied to a three-phase AC induction motor. The operation of the circuit can be summarized as follows:

- The DC injection voltage is obtained from the full-wave bridge rectifier circuit, which changes the line voltage from AC to DC.

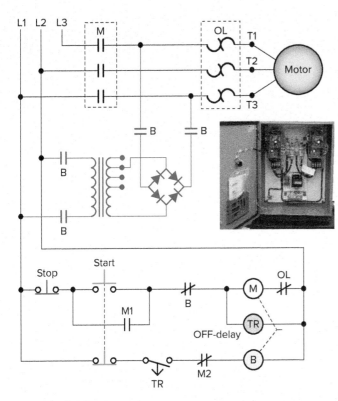

Figure 8-50 DC injection braking applied to an AC induction motor.

Photo courtesy of System Directions Ltd., 42 Fawcett Road, Coquitlam, B.C. V3K 6X9, Canada. Fax: 604-526-4618.

- Pressing the start button energizes starter coil M and off-delay timer coil TR.
- Normally open M1 auxiliary contact closes to maintain current to the starter coil and normally closed M2 auxiliary contact opens to open the current path to braking coil B.
- Normally open off-delay timer contact TR remains closed at all times while the motor is operating.
- When the stop button is pressed, starter coil M and off-delay timer coil TR are de-energized.
- Braking coil B becomes energized through the closed TR contact.
- All B contacts close to apply DC braking power to two phases of the motor stator winding.
- Coil B is de-energized after the timer contact time out. The timing contact is adjusted to remain closed until the motor comes to a complete stop.
- A transformer with tapped windings is used in this circuit to adjust the amount of braking torque applied to the motor.
- The motor starter (M) and braking contactor (B) are mechanically and electrically interlocked so that the AC and DC supplies are not connected to the motor at the same time.

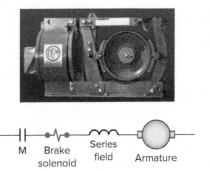

Figure 8-51 Electromechanical drum and shoe-type friction brake used on DC series motor drives.

Photo courtesy of EC&M, The Electric Controller and Manufacturing Company, www.ecandm.net

Electromechanical Friction Brakes

Unlike plugging or dynamic braking, electromechanical friction brakes can **hold** the motor shaft stationary after the motor has stopped. Figure 8-51 shows an electromechanical drum and shoe-type friction brake used on DC series motor drives. The brake drum is attached to the motor shaft, and the brake shoes are used to hold the drum in place. The brake is set with a spring and released by a solenoid. When the motor is running, the solenoid is energized to overcome the tension of the spring, thus keeping the brake shoes clear of the drum. When the motor is turned off, the solenoid is de-energized and the brake shoes are applied to the drum through the spring tension. The brake operating coil is connected in series with the motor armature and release and sets in response to motor current. This type of braking is fail-safe in that the brake is applied in case of an electrical failure.

AC motor brakes are commonly used as **parking brakes** to hold a load in place or as stopping brakes to decelerate a load. Applications include material-handling, food-processing, and baggage-handling equipment. These motors are directly coupled to an AC electromagnetic brake, as shown in Figure 8-52. When the power source is turned off, the motor stops instantaneously and holds the load. Most come equipped with an external manual release device, which allows the driven load to be moved without energizing the motor.

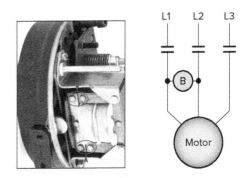

Figure 8-52 AC electromagnetic brake.

Photo courtesy of Warner Electric, www.warnerelectric.com.

Part 4 Motor Stopping

Part 4 Review Questions

1. How is plugging used to stop a motor?
2. Explain how a zero-speed switch functions in anti-plugging.
3. What type of starter is required for forward plugging of a motor?
4. How is dynamic braking used to stop a motor?
5. How is DC injection braking used to stop a motor?
6. How is an electromechanical friction brake set and released?
7. What problem can be created by excessive motor plugging operations?

PART 5 MOTOR SPEED

Multispeed Motors

The speed of an induction motor depends on the number of poles built into the motor and the frequency of the electrical power supply. A single-speed motor has one rated speed at which it runs when supplied with the nameplate voltage and frequency. A multispeed motor will run at more than one speed, depending on how you connect it to the supply. **Multispeed motors** typically have two speeds to choose from, but may have more.

The different speeds of a multispeed motor are selected by connecting the external motor winding stator leads to a multispeed starter. One starter is required for each speed of the motor, and each starter must be **interlocked** to prevent more than one starter from being on at the same time. Multispeed motors are available in two basic versions: consequent pole and separate winding. A separate winding motor has a winding for each speed while a consequent pole motor has a winding for every two speeds.

The starter for a consequent-pole two-speed motor requires a three-pole unit and a five-pole unit. Starters for separate-winding two-speed motors consist of two standard three-pole starter units that are electrically and mechanically interlocked and mounted in a single enclosure. Figure 8-53 shows a factory-wired two-speed separate winding IEC starter. A variety of factory and field-installed wiring configurations are used. The makeup of the starter is as follows:

- A high-speed and a low-speed starter, mechanically and electrically interlocked with each other.
- Two sets of overload relays, one for the high-speed circuit and one for the low-speed circuit to ensure adequate protection on each speed range.

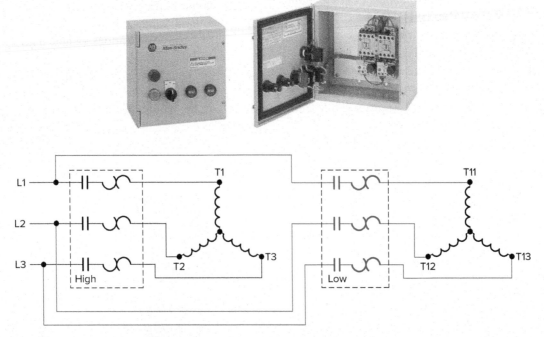

Figure 8-53 Two-speed separate winding across-the-line motor starter.
Photos courtesy of Rockwell Automation, www.rockwellautomation.com

- A hinged control panel containing high-speed push button, low-speed push button, off/high/low selector switch, high-speed OL reset, and low-speed OL reset.

In most instances, the three-phase, two-speed, six-lead squirrel-cage induction motor is a common application of a multispeed motor. A typical example would be a four-pole machine (with synchronous speed of 1,800 rpm) connected to operate at 1,800 rpm (high) and 900 rpm (low). It is important to carefully connect the motor leads to the starter as shown on the motor nameplate or on the connection diagram. Be sure to test each speed connection separately for direction of rotation before connecting the mechanical load.

The NEC requires that you protect each winding or connection against overloads and shorts. To meet this requirement you must:

- Use separate overloads for each winding.
- Size the branch circuit conductors that supply each winding for the full-load current (FLC) of the highest FLC winding.
- Ensure the controller horsepower rating isn't less than that required for the winding with the highest horsepower rating.

Wound-Rotor Motors

Construction of wound-rotor motors is different from that of squirrel-cage motors, basically in the design of the rotor. The **wound rotor** is constructed with windings that are brought out of the motor through slip rings on the motor shaft. These windings are connected to a controller, which places variable resistors in series with the windings. By changing the amount of external resistance connected to the rotor circuit, the motor speed can be varied (the lower the resistance, the higher the speed). Wound-rotor motors are most common in the range of 300 hp and above in applications where using a squirrel-cage motor may result in a starting current that's too high for the capacity of the power system.

Figure 8-54 shows the power circuit for a wound-rotor magnetic motor controller. It consists of a magnetic starter (M), which connects the primary circuit to the line, and

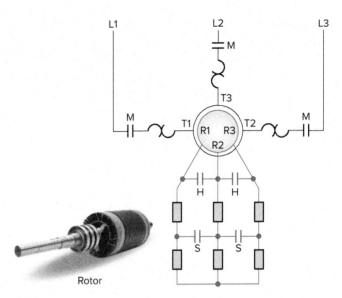

Figure 8-54 Wound-rotor magnetic motor controller.
©General Electric

two secondary accelerating contactors (S and H), which control the speed. The operation of the circuit can be summarized as follows:

- When operating at low speed, contactors S and H are both open, and full resistance is inserted in the rotor's secondary circuit.
- When contactor S closes, it shunts out part of the total resistance in the rotor circuit; as a result, the speed increases.
- When contactor H closes, all resistance in the secondary circuit of the motor is bypassed; thus, the motor runs at maximum speed.

One disadvantage of using resistance to control the speed of a wound-rotor induction motor is that a lot of heat is dissipated in the resistors; the efficiency, therefore, is low. Also, speed regulation is poor; for a given amount of resistance, the speed varies considerably if the mechanical load varies. Modern wound-rotor controllers use solid-state devices to obtain stepless control. These may incorporate thyristors (semiconductors) that serve in the place of magnetic contactors.

Part 5 Review Questions

1. How are the different speeds of a multispeed motor determined?
2. Compare the number of poles required for the starters of a consequent-pole and a separate-winding three-phase two-speed motor.
3. According to the NEC, what current rating must be used when calculating the size of the branch circuit conductors for a multispeed motor installation?
4. In what way is the construction of wound-rotor motors different from that of squirrel-cage motors?
5. Explain the relationship between the speed and resistance of the external resistors of a wound-rotor induction motor.

Troubleshooting Scenarios

1. What problems may be encountered when fuses or circuit breakers are sized too small for a specific application?
2. What might be the consequence if a DC starter is replaced by an AC starter with similar main contact voltage and current ratings?
3. In what way can excessive jogging have a negative effect on the operation of the starter and motor?
4. Why is it important to test each multispeed motor connection separately for direction of rotation before connecting the mechanical load?
5. The forward and reverse auxiliary electrical interlock contacts of a reversing motor starter are to be checked using an ohmmeter. Outline the procedure to be followed in carrying out this test.

Discussion Topics and Critical Thinking Questions

1. Determine each of the following for a 10 hp, three-phase, 208 V motor with a service factor of 1.15.
   a. Motor full-load current (FLC)
   b. Size of THWN copper branch circuit conductors required
   c. Fuse size (dual-element) to be used as motor branch circuit short-circuit and ground fault protection
   d. Current rating required for the motor disconnect switch
   e. Current rating for the overload relay located in the motor controller
2. Explain why standard fuses and circuit breakers cannot be used to protect against overloads.
3. Why should stop push buttons be of the normally closed type?
4. Why must the AC and DC supplies of a DC injection braking circuit not be connected to the motor at the same time?
5. The following is a typical plan of attack for solving a motor control failure problem. Outline what checks you would make for each of the following steps used to resolve the problem:
   a. Initial inspections
   b. Fuse checks
   c. Line voltage checks
   d. Motor junction box checks

CHAPTER NINE

Motor Control Electronics

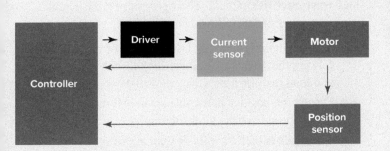

CHAPTER OBJECTIVES

This chapter will help you:

- Understand the operation and applications of different types of diodes.
- Understand the operation and applications of different types of transistors.
- Understand the operation and applications of different types of thyristors.
- Understand the operation and circuit function of different types of integrated circuits.
- Understand basic digital logic fundamentals and functions.
- Demonstrate the application of inverters.
- Explain the difference between sourcing and sinking sensors connections.

Electronic systems and controls have gained wide acceptance in the motor control industry; consequently it has become essential to be familiar with power electronic devices. This chapter presents a broad overview of diodes, transistors, thyristors, and integrated circuits (ICs) along with their applications in motor control circuits.

PART 1 SEMICONDUCTOR DIODES

Diode Operation

The **PN-junction diode,** shown in Figure 9-1, is the most basic of semiconductor devices. This diode is formed by a doping process, which creates P-type and N-type semiconductor materials on the same component. An N-type semiconductor material has electrons (represented as negative charges) as the current carriers while the P-type has holes (represented as positive charges) as the current carriers. N-type and P-type materials exchange charges at the junction of the two materials, creating a thin depletion region that acts as an insulator. Diode leads are identified as the anode lead (connected to the P-type material) and the cathode lead (connected to the N-type material).

The most important operating characteristic of a diode is that it allows current in one direction and blocks current in the opposite direction. When placed in a DC circuit, the diode will either allow

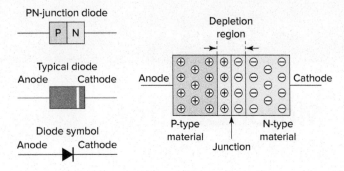

Figure 9-1 PN-junction diode.

or prevent current flow, depending on the polarity of the applied voltage. Figure 9-2 illustrates two basic operating modes of a diode: forward bias and reverse bias. A **forward-bias** voltage forces the positive and negative current carriers to the junction and collapses the depletion region to allow current flow. A **reverse-bias** voltage widens the depletion region so the diode does not conduct. In other words, the diode conducts current when the anode is positive with respect to the cathode (a state called forward-biased) and blocks current when the anode is negative with respect to the cathode (a state called reverse-biased).

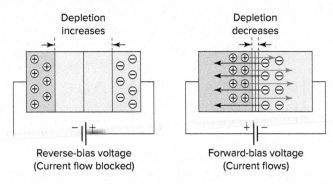

Figure 9-2 Diode forward and reverse biasing.

Rectifier Diode

Rectification is the process of changing AC to DC. Because diodes allow current to flow in only one direction, they are used as rectifiers. There are several ways of connecting diodes to make a rectifier to convert AC into DC. Figure 9-3 shows the schematic for a single-phase, **half-wave rectifier** circuit. The operation of the circuit can be summarized as follows:

- The AC input is applied to the primary of the transformer; the secondary voltage supplies the rectifier and load resistor.
- During the positive half-cycle of the AC input wave, the anode side of the diode is positive.
- The diode is then forward-biased, allowing it to conduct a current to the load. Because the diode acts as a closed switch during this time, the positive half-cycle is developed across the load.
- During the negative half-cycle of the AC input wave, the anode side of the diode is negative.
- The diode is now reverse-biased; as a result, no current can flow through it. The diode acts as an open switch during this time, so no voltage is produced across the load.
- Thus, applying an AC voltage to the circuit produces a pulsating DC voltage across the load.

Diodes can be tested for short-circuit or open-circuit faults with an ohmmeter. It should show continuity when the ohmmeter leads are connected to the diode in one direction but not in the other. If it does not show continuity in either direction, the diode is **open.** If it shows continuity in both directions, the diode is **short-circuited.**

Inductive loads, such as the coils of relays and solenoids, produce a high transient voltage at turnoff. This inductive voltage can be particularly damaging to sensitive circuit components such as transistors and integrated

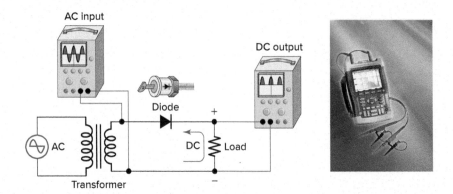

Figure 9-3 Single-phase, half-wave rectifier circuit.
Photo courtesy of Fluke, www.fluke.com. Reproduced with permission.

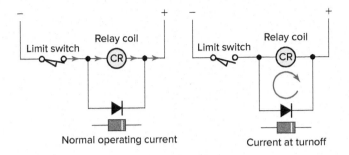

Figure 9-4 Diode connected to suppress inductive voltage.

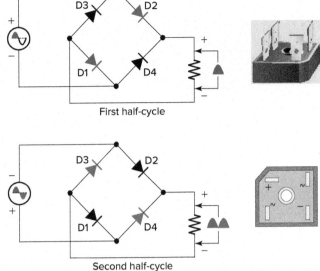

Figure 9-5 Single-phase, full-wave bridge rectifier circuit.
Photo courtesy of Fairchild Semiconductor, www.fairchildsemi.com

circuits. A **diode-clamping,** or despiking, circuit connected in parallel across the inductive load can be used to limit the amount of transient voltage present in the circuit. The diode-clamping circuit of Figure 9-4 illustrates how a diode can be used to suppress the inductive voltage of a relay coil. The operation of the circuit can be summarized as follows:

- The diode acts like a one-way valve for current flow.
- When the limit switch is closed, the diode is reverse-biased.
- Electric current can't flow through the diode so it flows through the relay coil.
- When the limit switch is opened, a voltage opposite to the original applied voltage is generated by the collapsing magnetic field of the coil.
- The diode is now forward-biased and current flows through the diode rather than through the limit switch contacts, bleeding off the high-voltage spike.
- The faster the current is switched off, the greater is the induced voltage. Without the diode, the induced voltage could reach several hundreds or even thousands of volts.
- It is important to note that the diode must be connected in reverse bias relative to the DC supply voltage. Operating the circuit with the diode incorrectly connected in forward bias will create a short circuit across the relay coil that could damage both the diode and the switch.

The half-wave rectifier makes use of only half of the AC input wave. A less-pulsating and greater average direct current can be produced by rectifying both half-cycles of the AC input wave. Such a rectifier circuit is known as a **full-wave rectifier.** A bridge rectifier makes use of four diodes in a bridge arrangement to achieve full-wave rectification. This is a widely used configuration, both with individual diodes and with single-component bridges where the diode bridge is wired internally. Bridge rectifiers are used on DC injection braking of AC motors to change the AC line voltage to DC, which is then applied to the stator for braking purposes. The schematic for a single-phase, full-wave bridge rectifier circuit is shown in Figure 9-5. The operation of the circuit can be summarized as follows:

- During the positive half-cycle, the anodes of D1 and D2 are positive (forward-biased), whereas the anodes of D3 and D4 are negative (reverse-biased). Electron flow is from the negative side of the line, through D1, to the load, then through D2, and back to the other side of the line.
- During the next half-cycle, the polarity of the AC line voltage reverses. As a result, diodes D3 and D4 become forward-biased. Electron flow is now from the negative side of the line through D3, to the load, then through D4, and back to the other side of the line. Note that during this half-cycle the current flows through the load in the same direction, producing a full-wave pulsating direct current.

Some types of direct current loads such as motors, relays, and solenoids will operate without problems on pulsating DC, but other electronic loads will not. The pulsations, or ripple, of the DC voltage can be removed by a **filter circuit.** Filter circuits may consist of capacitors, inductors, and resistors connected in different configurations. The schematic for a simple half-wave capacitor filter circuit is shown in Figure 9-6. Filtering is accomplished by alternate charging and discharging of the

Part 1 Semiconductor Diodes

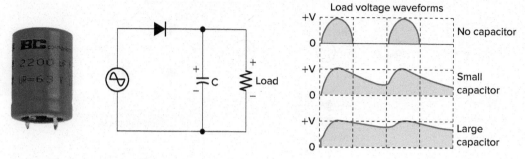

Figure 9-6 Capacitor filter.
Photo courtesy of Vishay Intertechnology, www.vishay.com

capacitor. The operation of the circuit can be summarized as follows:

- The capacitor is connected in parallel with the DC output of the rectifier.
- With no capacitor, the voltage output is normal half-wave pulsating DC.
- With the capacitor installed, on every positive half-cycle of the AC supply, the voltage across the filter capacitor and load resistor rises to the peak value of the AC voltage.
- On the negative half-cycle, the charged capacitor provides the current for the load to provide a more constant DC output voltage.
- The variation in the load voltage, or ripple, is dependent upon the value of the capacitor and load. A larger capacitor will have less voltage ripple.

For heavier load demands, such as those required for industrial applications, the DC output is supplied from a three-phase source. Using three-phase power, it is possible to obtain a low-ripple DC output with very little filtering. Figure 9-7 shows a typical **three-phase full-wave** bridge rectifier circuit. The operation of the circuit can be summarized as follows:

- The six diodes are connected in a bridge configuration, similar to the single-phase rectifier bridge to produce DC.
- The cathodes of the upper diode bank connect to the positive DC output bus.
- The anodes of the lower diode bank connect to the negative DC bus.
- Each diode conducts in succession while the remaining two are blocking.
- Each DC output pulse is 60 degrees in duration.
- The output voltage never drops below a certain voltage level.

Diodes can be tested using the digital multimeter (DMM) diode test function, as illustrated in Figure 9-8. This test is used to check the forward and reverse bias of a diode. Typically, when the diode is connected in forward

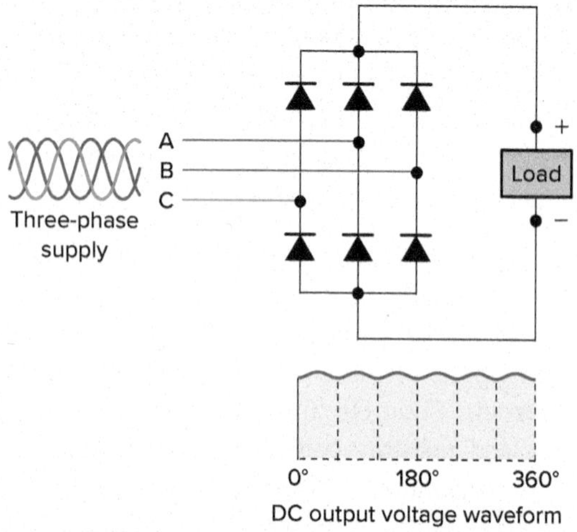

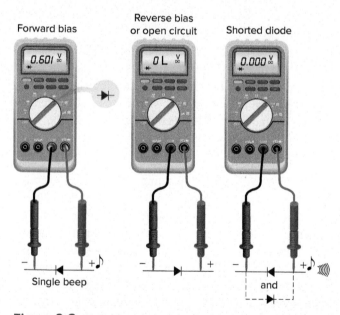

Figure 9-7 Three-phase, full-wave bridge rectifier.

Figure 9-8 Digital multimeter (DMM) diode test function.

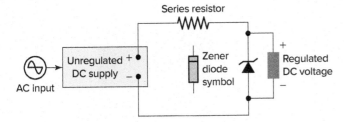

Figure 9-9 Typical zener diode regulator circuit.

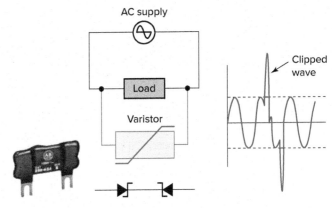

Figure 9-10 Suppressing AC voltage transients.
Photo courtesy of Rockwell Automation, www.rockwellautomation.com

bias, the meter displays the forward voltage drop across it and beeps briefly. When connected in reverse bias or open circuit, the meter displays OL. If the diode is shorted, the meter displays zero and emits a continuous tone when connected in forward or reverse bias.

Zener Diode

A zener diode is a type of diode that permits current to flow in the forward direction like a standard diode, but also in the reverse direction if the voltage is larger than the breakdown voltage, known as the **zener voltage.** This reverse-bias current would destroy a standard diode, but the zener diode is designed to handle it. The specified zener voltage rating of a zener diode indicates the voltage at which the diode begins to conduct when reverse-biased.

Zener diodes are used to provide a fixed reference voltage from a supply voltage that varies. They are commonly found in motor control feedback systems to provide a fixed level of reference voltage in regulated power supply circuits. Figure 9-9 shows a typical **zener diode regulator** circuit. The operation of the circuit can be summarized as follows:

- Input voltage must be higher than the specified zener voltage.
- The zener diode is connected in series with the resistor to allow enough reverse-biased current to flow for the zener to operate.
- Voltage drop across the zener diode will be equal to the zener diode's voltage rating.
- Voltage drop across the series resistor is equal to the difference between the zener voltage and the input voltage.
- The zener diode's voltage remains constant as the input voltage changes within a specified range.
- The change in input voltage appears across the series resistor.

Two back-to-back zener diodes can suppress damaging voltage transients on an AC line. A **metal oxide varistor (MOV)** surge suppressor functions in the same manner as back-to-back zener diodes. The circuit of Figure 9-10 is used to suppress AC voltage transients. The varistor module shown is made to be easily mounted directly across the coil terminals of contactors and starters with 120 V or 240 V AC coils. The operation of the circuit can be summarized as follows:

- Each zener diode acts as an open circuit until the reverse zener voltage across it exceeds its rated value.
- Any greater voltage peak instantly makes the zener diode act like a short circuit that bypasses this voltage away from the rest of the circuit.
- It is recommended that you locate the suppression device as close as possible to the load device.

Light-Emitting Diode

The **light-emitting diode (LED)** is another important diode device. An LED contains a PN junction that emits light when conducting current. When forward-biased, the energy of the electrons flowing through the resistance of the junction is converted directly into light energy. Because the LED is a diode, current will flow only when the LED is connected in forward bias. The LED must be operated within its specified voltage and current ratings to prevent irreversible damage. Figure 9-11 illustrates a simple LED circuit. The LED is connected in series with a resistor that limits the voltage and current to the desired value.

The main advantages of using an LED as a light source rather than an ordinary light bulb are a much lower power consumption, a much higher life expectancy, and high speed of operation. Conventional silicon diodes convert energy into heat. Gallium arsenide diodes convert energy into heat and infrared light. This type of diode is called an **infrared-emitting diode (IRED).** Infrared light is not visible to the

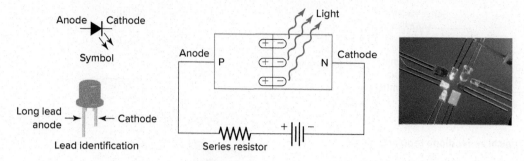

Figure 9-11 Light-emitting diode (LED).
Photo courtesy of Gilway/International Light, www.intl-lighttech.com/products/light-sources/leds

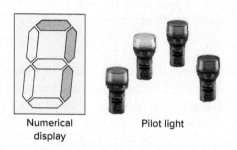

Figure 9-12 LED light sources.
Photo courtesy of Automation Systems Interconnect, www.asi-ez.com.

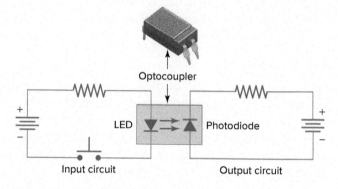

Figure 9-13 An optocoupler circuit.

human eye. By doping gallium arsenide with various materials, LEDs with visible outputs of red, green, yellow, or blue light can be manufactured. Light-emitting diodes are used for pilot lights and digital displays. Figure 9-12 shows a typical seven-segment LED numerical display. By energizing the correct segments, the numbers 0 through 9 can be displayed.

Photodiodes

Photodiodes are PN-junction diodes specifically designed for light detection. They produce current flow when they absorb light. Light energy passes through the lens that exposes the junction. The **photodiode** is designed to operate in the reverse-bias mode. In this device, the reverse-bias leakage current increases with an increase in the light level. Thus, a photodiode exhibits a very high resistance with no light input and less resistance with light input.

There are many situations where signals and data need to be transferred from one piece of equipment to another, without making a direct electrical connection. Often this is because the source and destination are at very different voltage levels, like a microprocessor that is operating from 5 V DC but is being used to control a circuit that is switching 240 V AC. In such situations, the link between the two must be an isolated one, to protect the microprocessor from overvoltage damage.

The circuit of Figure 9-13 uses a photodiode as part of an **optocoupler** (also known as an optoisolator) package containing an LED and a photodiode. Optocouplers are used to electrically isolate one circuit from another. The only thing connecting the two circuits is light, so they are electrically isolated from each other. Optocouplers typically come in a small integrated circuit package and are essentially a combination of an optical transmitter LED and an optical receiver such as a photodiode. The operation of the circuit can be summarized as follows:

- The LED is forward-biased, while the photodiode is reverse-biased.
- With the push button open, the LED is off. No light enters the photodiode, and no current flows in the input circuit.
- The resistance of the photodiode is high, so little or no current flows through the output circuit.
- When the input push button is closed, the LED is forward-biased and turns on.
- Light enters the photodiode so its resistance drops, switching on current to the output load.

Inverters

An **inverter,** is an electronic device or circuitry that changes direct current (DC) to alternating current (AC). Figure 9-14 shows a typical symbol for an inverter. An inverter and a rectifier perform opposite functions in electronic circuits. Both act as electric power converters; a rectifier changes current from AC to DC, while an inverter converts DC to AC. Inverters are used for a variety of electronic applications that include:

- Adjustable speed motor drives
- Uninterruptible power supplies
- Solar power systems
- Inverter welding units

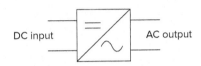

Figure 9-14 Inverter symbol.

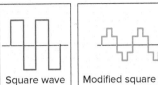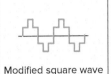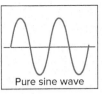

Figure 9-15 Common inverter output waveforms.

Inverters are classified according to their output waveforms (Figure 9-15). The three most common types of inverters made for powering AC loads are:

- **Pure sine wave** inverter used for general applications.
- **Modified square wave** inverter used for resistive, capacitive, and inductive loads.
- **Square wave** inverter used for certain resistive loads. It is the simplest type, but not as popular because of low power quality.

Part 1 Review Questions

1. Compare the type of current carriers associated with N-type and P-type semiconductor materials.
2. State the basic operating characteristic of a diode.
3. How is a diode tested using an ohmmeter?
4. What determines whether a diode is forward- or reverse-biased?
5. Under what condition is a diode considered to be connected in forward bias?
6. What is the function of a rectifier diode?
7. Explain the process by which a single-phase half-wave rectifier changes AC to DC.
8. What is the purpose of a clamping, or despiking, diode?
9. A single-phase half-wave rectifier is replaced with a full-wave bridge type. In what ways will the DC output change?
10. How does a capacitor filter operate to smooth the amount of pulsation (ripple) associated with rectifier circuits?
11. What advantages are gained by using three-phase rectifiers over single-phase types?
12. In what way is the operation of a zener diode different from that of a standard diode?
13. Zener diodes are commonly used in voltage regulation circuits. What operating characteristic of the zener diode makes it useful for this type of application?
14. State the basic operating principle of an LED.
15. How many LEDs are integrated into a single LED numerical display?
16. Explain how a photodiode is designed to detect light.
17. A diode is tested using an ohmmeter and is found to have low resistance in both directions. What type of fault, if any, does this indicate?
18. What is the function of an inverter?
19. List the three common types of inverter output waveforms.

PART 2 TRANSISTORS

The **transistor** is a three-terminal semiconductor device commonly used to amplify a signal or switch a circuit on and off. Amplification is the process of taking a small signal and increasing the signal size. Transistors are used as switches in electric motor drives to control the voltage and current applied to motors. Transistors are capable of extremely fast switching with no moving parts. There are two general types of transistors in use today: the bipolar transistor (often referred to as a bipolar junction transistor, or BJT) and the field-effect transistor (FET). Another common use of transistors is as part of integrated circuits (ICs). As an example, a computer microprocessor chip may contain over a billion transistors.

Bipolar Junction Transistor (BJT)

A bipolar junction transistor (BJT) consists of a three-layer sandwich of doped semiconductor materials consisting of an **emitter** (E), a **base** (B), and a **collector** (C). The emitter layer is medium in size, the base is very thin, and the collector is large. There are two major types of bipolar transistors, called **NPN** and **PNP,** as illustrated in Figure 9-16. The letters refer to the layers of semiconductor material used to make the transistor. The functional difference between an NPN transistor and a PNP transistor is the proper biasing (polarity) of the junctions when operating. For any given state of operation, the current directions and voltage polarities for each kind of transistor are exactly opposite each other. Bipolar transistors are so named because the controlled current must go through two types of semiconductor material, P and N.

The BJT is a **current amplifier** in that a small current flow from the base to the emitter results in a larger flow from the collector to the emitter. This, in effect, is current amplification, with the current gain known as the **beta** of the transistor. The circuit shown in Figure 9-17 illustrates the way in which a BJT is used as a current amplifier

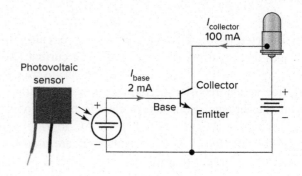

Figure 9-17 BJT current amplification.
Photo courtesy of All Electronics, www.allelectronics.com.

to amplify the small current signal from a photovoltaic sensor. The operation of the circuit can be summarized as follows:

- The transistor is connected to two different DC voltage sources: the supply voltage and the voltage generated by the photovoltaic sensor when exposed to light.
- These voltage supplies are connected so that the base-emitter junction is forward-biased and the base-collector junction is reverse-biased.
- Current in the base lead is called the base current, and current in the collector lead is called the collector current.
- It is called the common-emitter transistor configuration because both the base and collector circuits share the emitter lead as a common connection point.
- The amount of base current determines the amount of collector current.
- With no base current—that is, no light shining on the photovoltaic sensor—there is no collector current (normally off).
- A small increase in base current, generated by the photovoltaic sensor, results in a much larger increase in collector current; thus, the base current acts to control the amount of collector current.
- The current amplification factor, or gain, is the ratio of the collector current to the base current; in this case 100 milliamps (mA) divided by 2 mA, or 50.

When a transistor is used as a switch, it has only two operating states, on and off. Bipolar transistors cannot directly switch AC loads, and they are not usually a good choice for switching higher voltages or currents. In these cases, a relay in conjunction with a low-power transistor is often used. The transistor switches current to the relay coil while the coil contacts switch current to the load. The

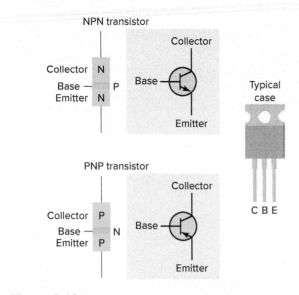

Figure 9-16 Bipolar junction transistor (BJT).
Photo courtesy of Fairchild Semiconductor, www.fairchildsemi.com

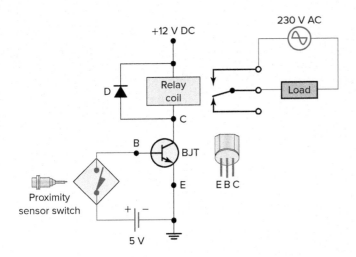

Figure 9-18 BJT switching of an AC load.

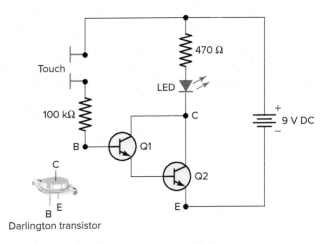

Figure 9-19 Darlington transistor as part of a touch-switch circuit.

circuit shown in Figure 9-18 illustrates the way in which a BJT is used to control an AC load. The operation of the circuit can be summarized as follows:

- A low-power transistor is used to switch the current for the relay's coil.
- With the proximity switch open, no base or collector current flows, so the transistor is switched off. The relay coil will be de-energized and voltage to the load will be switched off by the normally open relay contacts.
- When the transistor is in the off state, the collector current is almost zero, the voltage drop across the collector and emitter is 12 V, and the voltage across the relay coil is 0 V.
- The proximity sensor switch, on closing, establishes a small base current that drives the collector fully on to the point where it is said to be saturated, as it cannot pass any more current.
- The relay coil is energized and its normally open contacts close to switch on the load.
- When the transistor is in the on state, collector current is at its maximum value and the voltage across the collector and emitter drops to near zero while that across the relay coil increases to approximately 12 V.
- The clamping diode prevents the induced voltage at turnoff from becoming high enough to damage the transistor.

The **Darlington transistor** (often called a Darlington pair) is a semiconductor device that combines two bipolar transistors in a single device so that the current amplified by the first transistor is amplified further by the second. The overall current gain is equal to the two individual transistor gains multiplied together. Figure 9-19 shows the Darlington transistor as part of a resistance touch-switch circuit. The operation of the circuit can be summarized as follows:

- The Darlington pairs are packaged with three legs, like a single transistor.
- The base of transistor Q1 is connected to one of the electrodes of the touch switch.
- Placing your finger on the touch plate allows a small amount of current to pass through the skin and establish current flow through the base circuit of Q1 and drive it into saturation.
- The current amplified by Q1 is amplified further by Q2 to switch the LED on.

Like junction diodes, bipolar junction transistors are light-sensitive. **Phototransistors** are designed specifically to take advantage of this fact. The most common phototransistor is an NPN bipolar transistor with a light-sensitive collector-base PN junction. When this junction is exposed to light, it creates a control current flow that switches the transistor on. Photodiodes also can provide a similar function, but at a much lower gain. Figure 9-20 shows a phototransistor employed as part of an optical isolator found in a programmable logic controller (PLC) AC input module circuit. The operation of the circuit can be summarized as follows:

- When the push button is closed, 120 V AC is applied to the bridge rectifier through resistors R1 and R2.
- This produces a low-level DC voltage that is applied across the LED of the optical isolator.
- The zener diode (Z_D) voltage rating sets the minimum level of voltage that can be detected.

Part 2 Transistors 253

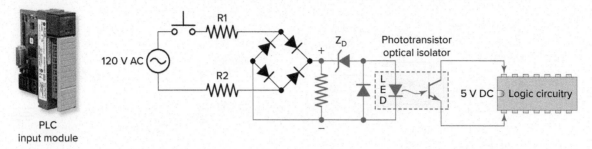

Figure 9-20 Phototransistor employed as part of an optical isolator found in a programmable logic controller (PLC) AC input module circuit.
Photo courtesy of Rockwell Automation, www.rockwellautomation.com

- When light from the LED strikes the phototransistor, it switches into conduction and the status of the push button is communicated in logic, or low-level DC voltage, to the processor.
- The optical isolator not only separates the higher AC input voltage from the logic circuits but also prevents damage to the processor by line voltage transients.

Bipolar transistors can be tested in a fashion similar to that of a diode, using a the digital multimeter (DMM) diode test function. The assumption made when testing transistors in this manner is that a transistor is analogous to a pair of connected diodes, as illustrated in Figure 9-21. When the diode function is used, it will be found that the emitter-base junction possesses a slightly greater forward voltage drop than the collector-base junction. This forward voltage difference is due to the disparity in doping concentration between the emitter and collector regions of the transistor: The emitter is a much more heavily doped piece of semiconductor material than the collector, causing its junction with the base to produce a higher forward voltage drop.

Field-Effect Transistor

The bipolar junction transistor is a current-controlled device, while the **field-effect transistor (FET)** is a voltage-controlled device. The field-effect transistor uses basically no input current. Instead, output current flow is controlled by a varying *electric field*, which is created through the application of a voltage. This is the origin of the term *field effect*. The field-effect transistor was designed to get around the two major disadvantages of the bipolar junction transistor: low switching speed and high drive power, which are imposed by the base current.

The junction field-effect transistor (JFET) is shown in Figure 9-22.

- It is constructed with a bar of N-type material and a gate of P-type material. Because the material in the channel is N type, the device is called an N-channel JFET.
- JFETs have three connections, or leads: source, gate, and drain. These correspond to the emitter, base, and collector of the bipolar transistor, respectively. The names of the terminals refer to their functions.
- The **gate** may be thought of as controlling the opening and closing of a physical gate. This gate permits electrons to flow through or blocks their passage by creating or eliminating a channel between the **source** and the **drain.**

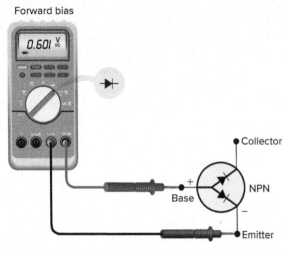

Figure 9-21 Digital multimeter (DMM) bipolar junction transistor test.

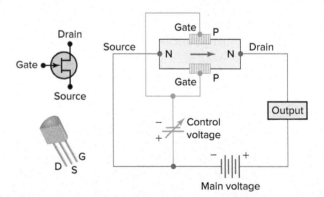

Figure 9-22 Junction field-effect transistor (JFET).

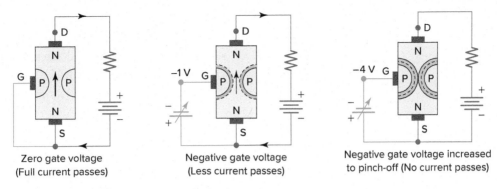

Zero gate voltage
(Full current passes)

Negative gate voltage
(Less current passes)

Negative gate voltage increased
to pinch-off (No current passes)

Figure 9-23 Gate voltage control of current.

- Field-effect transistors are unipolar; their working current flows through only one type of semiconductor material. This is in contrast to bipolar transistors, which have current flowing through both N-type and P-type regions.
- There are also P-channel JFETs that use P-type material for the channel and N-type material for the gate.
- The main difference between the N and P types is that the polarities of voltage they are connected to are opposite.

The JFET operates in the **depletion mode,** which means it is normally on. If a source of voltage is connected to the source and drain, and no source of voltage is connected to the gate, current is free to flow through the channel. Figure 9-23 illustrates the gate voltage control of current through an N-channel JFET. The operation of the circuit can be summarized as follows:

- The normal polarities for biasing the N-channel JFET are as indicated. Note that the JFET is normally operated with the control voltage connected to the source and gate junction in reverse bias. The result is very high input impedance.
- If a source voltage is connected to the source and drain, and no control voltage is connected to the gate, electrons are free to flow through the channel.
- A negative voltage connected to the gate increases the channel resistance and reduces the amount of current flow between the source and the drain.
- Thus, the gate voltage controls the amount of drain current, and the control of this current is almost powerless.
- Continuing to increase the negative gate voltage to the pinch-off point will reduce the drain current to a very low value, effectively zero.

Metal Oxide Semiconductor Field-Effect Transistor (MOSFET)

The metal oxide semiconductor field-effect transistor (MOSFET) is by far the most common field-effect transistor. Field-effect transistors do not require any gate current for operation so the gate structure can be completely insulated from the channel. The gate of a JFET consists of a reverse-biased junction, whereas the gate of a MOSFET consists of a metal electrode insulated by metal oxide from the channel. The insulated gate of a MOSFET has much higher input impedance than the JFET, so it is even less of a load on preceding circuits. Because the oxide layer is extremely thin, the MOSFET is susceptible to destruction by electrostatic charges. Special precautions are necessary when handling or transporting MOS devices.

The MOSFET can be made with a P channel or an N channel. The action of each is the same, but the polarities are reversed. In addition, there are two types of MOSFETs: depletion-mode MOSFETs, or **D type,** and enhancement-mode MOSFETs, or **E type.** Symbols used for N- and P-channel depletion-mode MOSFETs are shown in Figure 9-24. The channel is represented as a solid line to signify that the circuit between the drain and the source is normally complete and that the device is normally on.

Gate voltage in a MOSFET circuit can be of either polarity, since a diode junction is not used. This makes the enhancement mode of operation possible. The enhancement-mode MOSFET is normally off, which means that if a voltage is connected to the drain and source and no voltage is

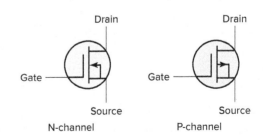

Figure 9-24 Depletion-mode MOSFETs.

Part 2   Transistors   255

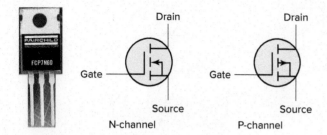

Figure 9-25 Enhancement-mode MOSFETs.
Photo courtesy of Fairchild Semiconductor, www.fairchildsemi.com

connected to the gate, there will be no current flow through the device. The proper gate voltage will attract carriers to the gate region and form a conductive channel.

Thus the channel is considered to be "enhanced," or aided by gate voltage. Figure 9-25 shows the schematic symbols used for enhancement-mode MOSFETs. Note that, unlike the depletion-mode symbols, the line from the source to drain is broken. This implies that the device is normally off.

Figure 9-26 shows an enhancement-mode MOSFET used as part of an off-delay cube timer circuit. Because the gate current flow is negligible, a broad range of time-delay periods, from minutes to hours, is possible. The operation of the circuit can be summarized as follows:

- With the switch initially open, a voltage is applied between the drain and the source but no voltage is applied between the gate and the source. Therefore, no current flows through the MOSFET and the relay coil will be de-energized.
- Closing the switch results in a positive voltage being applied to the gate, which triggers the MOSFET into conduction to energize the relay coil and switch the state of its contacts.
- At the same time the capacitor is charged to 12 V DC.

- The circuit remains in this state with the relay coil energized as long as the switch remains closed.
- When the switch is opened, the timing action begins.
- The positive gate circuit to the 12 V source is opened.
- The stored positive charge of the capacitor keeps the MOSFET switched on.
- The capacitor begins to discharge its stored energy through R1 and R2 while still maintaining a positive voltage at the gate.
- The MOSFET and relay coil continue to conduct a current for as long as it takes the capacitor to discharge.
- The discharge rate, and thus the off-delay timing period, is adjusted by varying the resistance of R2. Increasing the resistance will slow the rate of discharge and increase the timing period. Decreasing the resistance will have the opposite effect.

Power MOSFETs that are designed to handle larger amounts of current are used in some electronic DC motor speed controllers. For this type of application, the MOSFET is used to switch the applied DC voltage on and off very rapidly. The speed of a DC motor is directly proportional to the voltage applied to the armature. By switching the DC motor line voltage, it is possible to control the average voltage applied to the motor armature.

Figure 9-27 shows an enhancement-mode power MOSFET used as part of a chopper circuit. In this circuit the source voltage is chopped by the MOSFET to produce an average voltage somewhere between 0 and 100 percent of the DC supply voltage. The operation of the circuit can be summarized as follows:

- The DC supply and motor field voltage is fixed and the voltage applied to the motor armature is varied by the MOSFET using a technique called pulse-width modulation (PWM).
- Pulse-width modulation works by applying a series of modulated square wave pulses to the gate.

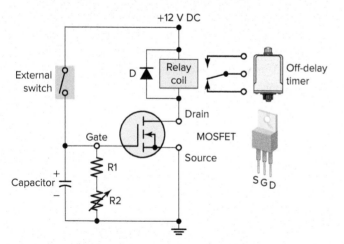

Figure 9-26 MOSFET off-delay timer circuit.

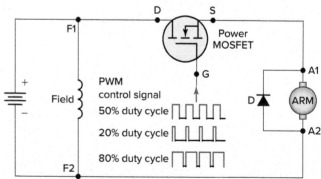

Figure 9-27 Power MOSFET used as part of a chopper circuit.

256　　Chapter 9　Motor Control Electronics

- Motor speed is controlled by driving the motor with short pulses. These pulses vary in duration to change the speed of the motor. The longer the pulses, the faster the motor runs and vice versa.
- By adjusting the duty cycle of the gate signal (modulating the width of the pulse), the time fraction that the MOSFET is on can be varied, along with the average voltage to the motor armature and hence the motor speed.
- The diode (sometimes called a freewheeling diode) is connected in reverse bias to provide a discharge path for the collapsing magnetic field when voltage to the motor armature is switched off.

Field-effect transistors are more difficult to test than bipolar junction transistors. Special care must be taken not to damage the device due to possible static charge buildup. MOSFETs can be tested with a low-voltage ohmmeter, set to the highest possible range. D-MOSFETs that are in a good working order exhibit some continuity between the source and the drain. However, there should be no resistance between the gate to drain and the gate to source terminals. E-MOSFETs that are working properly show no continuity between any two terminals.

Insulated-Gate Bipolar Transistor (IGBT)

The insulated-gate bipolar transistor (**IGBT**) is a cross between a bipolar transistor and a MOSFET in that it combines the positive attributes of both. BJTs have lower on resistance, but have longer switching times, especially at turn off. MOSFETs can be turned on and off much faster, but their on-state resistance is higher. IGBTs have lower on-state power loss in addition to faster switching speeds, allowing the electronic motor drive to operate at much higher switching frequencies and control more power.

The two different schematic symbols used to represent an N-type IGBT and its equivalent circuit are shown in Figure 9-28. Notice that the IGBT has a gate like a MOSFET yet it has an emitter and a collector like a BJT. The equivalent circuit is depicted by a PNP transistor, where the base current is controlled by a MOS transistor. In essence, the IGFET controls the base current of a BJT, which handles the main load current between collector and emitter. This way, there is extremely high current gain (since the insulated gate of the IGFET draws practically no current from the control circuitry), but the collector-to-emitter voltage drop during full conduction is as low as that of an ordinary BJT.

As applications for IGBT components have continued to expand rapidly, semiconductor manufacturers have responded by providing IGBTs in both discrete (individual) and modular packages. Figure 9-29 shows a power electronic module that houses two insulated-gate bipolar

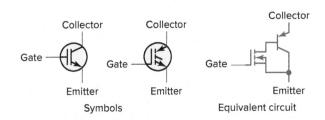

Figure 9-28 Insulated-gate bipolar transistor (IGBT)—N type.
Photo courtesy of Fairchild Semiconductor, www.fairchildsemi.com

power transistors and corresponding diodes. This package provides an easy way to cool the devices and to connect them to the other circuit.

Figure 9-30 illustrates how IGBTs are used in an electronic variable-frequency AC motor drive. A variable-frequency drive controls the speed of the AC motor by varying the **frequency** supplied to the motor. In addition, the drive also regulates the output **voltage** in proportion to the output frequency to provide a relatively constant ratio of voltage to frequency (V/Hz), as required by the characteristics of the AC motor to produce **adequate torque.** The six IGBTs are capable of very high switching speeds and may be required to switch the voltage to the motor thousands of times per second. The operation of the circuit can be summarized as follows:

- The input section of the drive is the converter. It contains six diodes, arranged in an electrical bridge. The diodes convert the three-phase AC power into DC power.

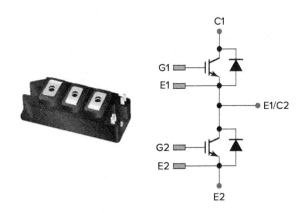

Figure 9-29 Insulated-gate bipolar transistor (IGBT) power module.
Photo courtesy of Fairchild Semiconductor, www.fairchildsemi.com

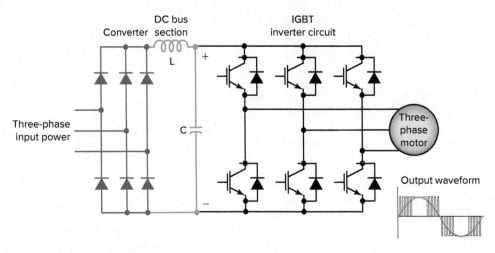

Figure 9-30 IGBTs used in an electronic motor drive.

- The next section—the DC bus—sees a fixed DC voltage.
- The inductor (L) and the capacitor (C) work together to filter out any AC component of the DC waveform. The smoother the DC waveform, the cleaner the output waveform from the drive.
- The DC bus feeds the inverter, the final section of the drive. As the name implies, this section inverts DC voltage back to AC. But it does so in a variable voltage and frequency output.
- Fairly involved control circuitry coordinates the switching of the IGBT devices, typically through a logic control board that dictates the firing of power components in their proper sequence.

PART 2 Review Questions

1. Explain the two main functions of a transistor.
2. What are the two general types of transistor?
3. Give a brief explanation of how current flow through a bipolar junction transistor is controlled.
4. State the two types of bipolar transistors.
5. What names are used to identify the three leads of a BJT?
6. Explain the term *current gain* as it applies to a BJT.
7. Describe the makeup of a Darlington transistor.
8. Give a brief explanation of how current flow through a phototransistor is controlled.
9. How is current flow through a field-effect transistor controlled?
10. State two advantages of field-effect transistors over bipolar types.
11. What names are used to identify the three leads of a junction field-effect transistor?
12. What is the main difference between the operation of N-channel and P-channel JFETs?
13. Why are field-effect transistors said to be unipolar?
14. JFETs are characterized as being "normally on." What does this mean?
15. Compare the gate structure of a JFET and a MOSFET.
16. In what way is the operation of a depletion-mode MOSFET different from that of an enhancement-mode MOSFET?
17. Give a brief explanation of how a power MOSFET is operated as a chopper in a DC electronic motor drive.
18. How is current flow through an insulated-gate bipolar transistor controlled?
19. What names are used to identify the three leads of an IGBT?
20. What features of the IGBT make it the transistor of choice for power electronic motor control applications?
21. What are the advantages of modular packaging of power electronic devices?
22. Give a brief explanation of how a power IGBT is operated as an inverter in an AC electronic motor drive.

PART 3 THYRISTORS

Thyristor is a generic term for a broad range of semiconductor components used as an electronic switch. Like a mechanical switch, it has only two states: on (conducting) and off (not conducting). Thyristors have no linear in-between state as transistors have. In addition to switching, they can also be used to adjust the amount of power applied to a load.

Thyristors are mainly used where high currents and voltages are involved. They are often used to control alternating currents, where the change of polarity of the current causes the device to automatically switch off. The silicon-controlled rectifier and triac are the most frequently used thyristor devices.

Silicon-Controlled Rectifiers (SCRs)

Silicon-controlled rectifiers (SCRs) are similar to silicon diodes except for a third terminal, or gate, which controls, or turns on, the SCR. Basically the **SCR** is a four-layer (PNPN) semiconductor device composed of an anode (A), cathode (K), and gate (G), as shown in Figure 9-31. Common SCR case styles include stud-mounted, hockey puck, and flexible lead. SCRs function as switches to turn on or off small or large amounts of power. High-current SCRs that can handle load currents in the thousands of amperes have provisions for some type of heat sink to dissipate the heat generated by the device.

In function, the SCR has much in common with a diode. Like the diode, it conducts current in only one direction when it is forward-biased from anode to cathode. It is unlike the diode because of the presence of a gate (G) lead, which is used to turn the device on. It requires a momentary positive voltage (forward-biased) applied to the gate to switch it on. When turned on, it conducts like

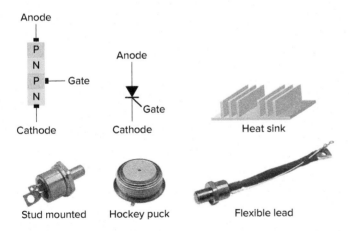

Figure 9-31 Silicon-controlled rectifier (SCR).
Photos courtesy of Vishay Intertechnology, www.vishay.com

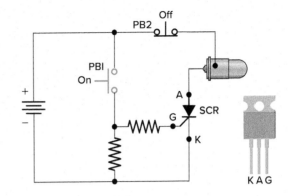

Figure 9-32 SCR operated from a DC source.
Photo courtesy of Fairchild Semiconductor, www.fairchildsemi.com

a diode for one polarity of current. If not triggered on, it will not conduct a current regardless of whether it is forward-biased.

The schematic of an SCR switching circuit that is operated from a DC source is shown in Figure 9-32. The operation of the circuit can be summarized as follows:

- The anode is connected so that it is positive with respect to the cathode (forward-biased).
- Momentarily closing pushbutton PB1 applies a positive current-limited voltage to the gate of the SCR, which switches the anode-to-cathode circuit into conduction, thus turning the lamp on.
- Once the SCR is on, it stays on, even after the gate voltage is removed. The only way to turn the SCR off is to reduce the anode-cathode current to zero by removing the source voltage from the anode-cathode circuit.
- Momentarily pressing pushbutton PB2 opens the anode-to-cathode circuit to switch the lamp off.
- It is important to note that the anode-to-cathode circuit will switch on in just one direction. This occurs only when it is forward-biased with the anode positive with respect to the cathode and a positive voltage is applied to the gate.

The problem of SCR turn off does not occur in AC circuits. The SCR is automatically shut off during each cycle when the AC voltage across the SCR nears zero. As zero voltage is approached, anode current falls below the holding current value. The SCR stays off throughout the entire negative AC cycle because it is reverse-biased.

The schematic of an SCR switching circuit that is operated from an AC source is shown in Figure 9-33. Because the SCR is a rectifier, it can conduct only one-half of the AC input wave. The maximum output delivered to the load, therefore, is 50 percent; its shape is that of a half-wave

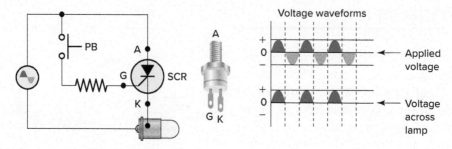

Figure 9-33 SCR operated from an AC source.

pulsating DC waveform. The operation of the circuit can be summarized as follows:

- The anode-cathode circuit can be switched on only during the half-cycle when the anode is positive (forward-biased).
- With the push button open, no gate current flows, so the anode-cathode circuit remains off.
- Pressing the push button continuously closed causes the gate-cathode and anode-cathode circuits to be forward-biased at the same time. This produces a half-wave pulsating direct current through the lamp load.
- When the push button is released, the anode-cathode current is automatically shut off when the AC voltage drops to zero on the sine wave.

When the SCR is connected to an AC source, it can also be used to vary the amount of power delivered to a load by **phase angle control**. Figure 9-34 shows the circuit of a single-phase fully controlled SCR bridge rectifier circuit. The main purpose of this circuit is to supply a variable DC output voltage to the load. The operation of the circuit can be summarized as follows:

- A pulse trigger is applied to the gate at the instant that the SCR is required to turn on. This pulse is relatively short and would typically be applied to the gate via a pulse transformer.
- The circuit has two pairs of SCRs with SCR-1 and SCR-4 forming one pair and SCR-2 and SCR-3 the other pair.
- During the positive half of the AC input waveform, SCR-1 and SCR-4 can be triggered into conduction.
- During the negative half of the AC input waveform, SCR-2 and SCR-3 can be triggered into conduction.
- Power is regulated by advancing or delaying the point at which each pair of SCRs is turned on within each half-cycle.
- Even though the direction of current through the source alternates from one half-cycle to the other half-cycle, the current through the load remains in the same direction.

SCRs are often employed in solid-state reduced-voltage starters to reduce the amount of voltage delivered to an AC motor on starting. Figure 9-35 shows a typical solid-state reduced-voltage power circuit made up of two contactors: a start contactor and a run contactor. The SCR is a unidirectional device in that it can conduct current in one direction only. In this application, bidirectional operation is obtained by connecting two SCRs in an antiparallel (also known as reverse parallel) connection. Using the antiparallel connection, with a suitable triggering circuit for each gate, both positive and negative halves of a sine wave may be

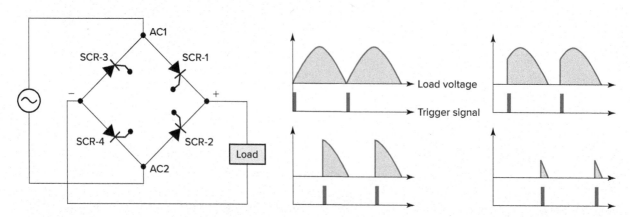

Figure 9-34 Single-phase, fully controlled SCR bridge rectifier circuit.

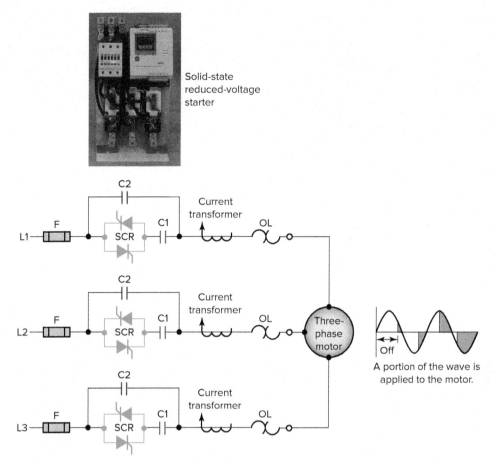

Figure 9-35 Solid-state reduced-voltage motor starter.
General Electric

controlled in conduction. The operation of the circuit can be summarized as follows:

- When the motor is first started, the start contacts (C1) close and reduced voltage is applied to the motor through the antiparallel-connected SCRs.
- Triggering of the SCRs is controlled by logic circuits that chop the applied sine-wave system power so that only a portion of the wave is applied to the motor.
- The logic circuits can be programmed to respond to any of several sensors to control the voltage: internal time ramp, current sensor feedback, or tachometer feedback.
- The voltage is increased until the SCR is being triggered at the zero crossing point and the motor is getting full line voltage.
- At this point, the run or bypass contacts (C2) close and the motor is connected directly across the line and runs with full power applied to the motor terminals.

SCRs usually **fail shorted** rather than open. Shorted SCRs can usually be detected with an ohmmeter check. Measure the anode-to-cathode resistance in both the forward and reverse directions; a good SCR should measure near infinity in both directions.

Small and medium-size SCRs can also be gated on with an ohmmeter. Forward bias the SCR with the ohmmeter by connecting its positive lead to the anode and its negative lead to the cathode. Momentarily touch the gate lead to the anode; this will provide a small positive turn-on voltage to the gate and the cathode-to-anode resistance reading will drop to a low value. Even after you remove the gate voltage, the SCR will stay conducting. Disconnecting the meter leads from the anode or cathode will cause the SCR to revert to its nonconducting state. In this test, the meter resistance acts as the SCR load. On larger SCRs, the unit may not latch on because the test current is not above the SCR holding current. Special testers are required for larger SCRs in order to provide an adequate value of gate voltage and load the SCR sufficiently to latch on. Hockey puck SCRs must be compressed in a heat sink (to make up the internal connections to the semiconductor) before they can be tested or operated.

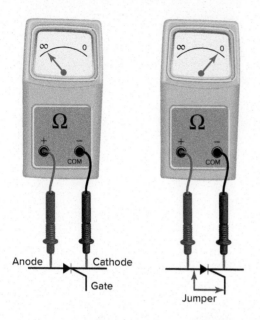

Figure 9-36 Testing an SCR using an ohmmeter.

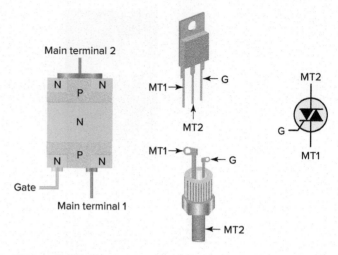

Figure 9-37 Triac.

The SCR can be tested using an ohmmeter, as illustrated in Figure 9-36.

- The positive (red) test lead of the meter is connected to the anode of the SCR, and the negative (black) test lead is applied to the cathode.
- The instrument should show an infinite high resistance.
- A jumper wire can be used to trigger the SCR.
- Without disconnecting the meter, use the jumper to short-circuit the gate terminal of the SCR with the positive lead of the meter.
- The SCR should exhibit a great decrease of resistance.
- When the jumper is disconnected, the device may continue to conduct or may turn off. This depends on the properties of both the SCR and the meter.
- If the holding current of the SCR is small, the ohmmeter could be capable of supplying enough current to keep it turned on. However, if the holding current of the SCR is high, the device will turn off upon disconnection of the jumper.

Triac

The **triac** is a three-terminal device essentially equivalent to two SCRs joined in reverse parallel (paralleled but with the polarity reversed) and with their gates connected together. The result is a bidirectional electronic switch that can be used to provide load current during both halves of the AC supply voltage. Triac connections, shown in Figure 9-37, are called main terminal 1 (MT1), main terminal 2 (MT2),

and gate (G). The leads are designated this way since the triac acts like two opposing diodes when it is turned on and neither lead always acts like a cathode or an anode. Gate current is used to control current from MT1 to MT2. From terminal MT1 to MT2 the current must pass through an NPNP series of layers or a PNPN series of layers. The gate is connected to the same end as MT1, which is important to remember when you connect the triac control circuit. Terminal MT1 is the reference point for measurement of voltage and current at the gate terminal.

The triac may be triggered into conduction by either a positive or a negative voltage applied to its gate electrode. Once triggered, the device continues to conduct until the current through it drops below a certain threshold value, such as at the end of a half-cycle of the alternating current (AC) mains power. This makes the triac very convenient for switching AC loads. The triac is an almost ideal component for controlling AC power loads with a high (on/off) duty cycle. Using a triac eliminates completely the contact sticking, bounce, and wear associated with conventional electromechanical relays. The schematic of a triac switching circuit is shown in Figure 9-38. Maximum

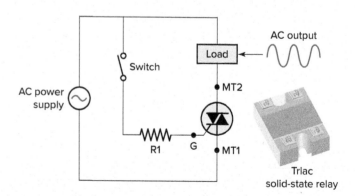

Figure 9-38 Triac switching circuit.

Photo courtesy of Picker Components, www.pickercomponents.com

output is obtained by utilizing both half-waves of the AC input voltage. The operation of the circuit can be summarized as follows:

- The circuit provides random (anywhere in half-cycle) fast turn-on of AC loads.
- When the switch is closed, a small control current will trigger the triac to conduct. Resistor R1 is provided to limit gate current to a small control current value.
- When the switch is then opened, the triac turns off when the AC supply voltage and holding current drop to zero, or reverse polarity.
- In this way large currents can be controlled even with a small switch, because the switch will have to handle only the small control current needed to turn on the triac.

The output module of a programmable logic controller (PLC) serves as the link between the PLC's microprocessor and field load devices. Figure 9-39 shows a triac used to switch AC high voltage and current, controlling the on/off state of the lamp. The optical isolator separates the output signal from the PLC processor circuit from the field load devices. The operation of the circuit can be summarized as follows:

- As part of its normal operation, the processor sets the outputs on or off according to the logic program.
- When the processor calls for the lamp to be on, a small voltage is applied across the LED of the optical isolator.
- The LED emits light, which switches the phototransistor into conduction.
- This in turn switches the triac into conduction to turn on the lamp.

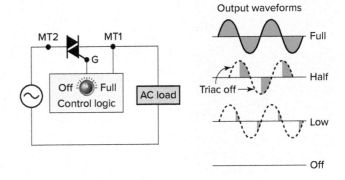

Figure 9-40 Triac variable AC control circuit.

The schematic circuit of Figure 9-40 illustrates how a triac can be used to control the amount of power applied to an AC load. When used for this type of application, a control logic triggering circuit is needed to ensure that the triac conducts at the proper time. The operation of the circuit can be summarized as follows:

- The trigger circuit controls the point on the AC waveform at which the triac is switched on. It proportionally turns on a percentage of each power line half-cycle.
- The resulting waveform is still alternating current, but the average current value is adjustable.
- Since the trigger can cause it to trigger current in either direction, it is an efficient power controller from essentially zero to full power.
- In universal motor circuits, varying the current will change the speed of the motor.

The **diac** is a two-terminal device that behaves like two-series diodes connected in opposite directions. Current flows through the diac (in either direction) when the voltage across it reaches its rated *breakover voltage*. The current pulse produced when the diac changes from a nonconducting to a conducting state is used for SCR and triac gate triggering.

Typically, **light dimmers** are manufactured with a triac as the power control device. A light dimmer works by essentially chopping parts out of the AC voltage. This allows only parts of the waveform to pass to the lamp. The brightness of the lamp is determined by the power transferred to it, so the more the waveform is chopped, the more it dims. A simplified triac/diac incandescent lamp dimmer circuit is shown in Figure 9-41. The operation of the circuit can be summarized as follows:

- With the variable resistor at its lowest value (minimum resistance), the capacitor will charge rapidly at the beginning of each half-cycle of the AC voltage.

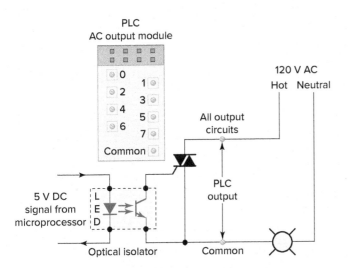

Figure 9-39 Triac switching of PLC output module.

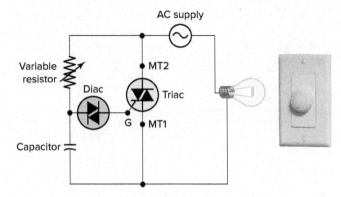

Figure 9-41 Triac/diac lamp dimmer.
Thinkstock Images/Stockbyte/Getty images

- When the voltage across the capacitor reaches the breakover voltage of the diac, the capacitor voltage discharges through the gate of the triac.
- Thus, the triac conducts early in each half-cycle and remains on to the end of each half-cycle.
- As a result, current will flow through the lamp for most of each half-cycle and produce maximum lamp brightness.
- If the resistance of the variable resistor is increased, the time required to charge the capacitor to the breakover voltage of the diac increases.
- This causes the triac to fire later in each half-cycle. So, the length of time current flows through the lamp is reduced, and less light is emitted.
- The diac prevents any gate current until the triggering voltage has reached a certain repeatable level in either direction.

The average current and voltage rating for triacs is much smaller than that for SCRs. Also, triacs are designed to operate at much lower switching frequencies than SCRs and have more difficulty switching power to highly inductive loads.

Triacs, like SCRs, usually fail shorted rather than open. Shorted triacs can usually be detected with an ohmmeter check. Measure the MT1 to MT2 resistance in both directions; a good triac should measure near infinity in both directions. Like SCRs, triacs can fail in other—possibly peculiar—ways, so substitution or bypassing may be necessary to rule out all possibilities.

The procedure for testing a triac is essentially the same as testing an SCR and is illustrated in Figure 9-42.

- The positive test lead of the ohmmeter is connected to MT2, and the negative test lead is applied to MT1.
- The ohmmeter should indicate an infinite high resistance.

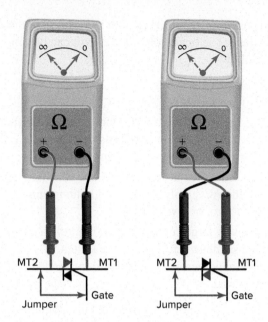

Figure 9-42 Testing a triac using an ohmmeter.

- A jumper wire can be used to trigger the triac into conduction.
- Using the jumper, the gate terminal is momentarily connected to MT2, resulting in a positive triggering pulse being applied to the gate.
- The resistance should go low and remain low even when the gate connection is broken. This indicates that one of the SCRs in the pair is functioning properly.
- Next, the test leads of the ohmmeter are reversed with respect to the anode and the cathode.
- With no gate signal, the ohmmeter should exhibit an infinite high resistance.
- Using the jumper, the gate terminal is momentarily connected to MT2, resulting in a negative triggering pulse being applied to the gate.
- The resistance should go low and remain low even when the gate connection is broken. This indicates that the second SCR in the pair is functioning properly.

Electronic Motor Control Systems

The requirements for an electronic motor control system depend on the type of motor used, process application, level of control required, and amount of system monitoring required. The basic building blocks of the control system are shown in Figure 9-43 and are summarized as follows:

Controller, typically a microcontroller, takes command of motor starting, stopping, speed, and protection.

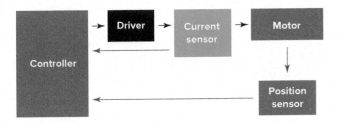

Figure 9-43 Basic building blocks of an electronic control system.

Driver is required to amplify the signals generated by the controller in order to deliver sufficient power to the motor.

Sensors provide signals/information to the controller. The information provided can be in a variety of formats. Digital and analog are two of the most common formats when dealing with sensors. Sensors are often referred to as input devices, as they are normally wired to the input side of a controller.

Part 3 Review Questions

1. In what way does the operation of a thyristor differ from that of a transistor?
2. What are the two most popular types of thyristor?
3. What are the similarities and differences between an SCR and diode?
4. Compare the way in which the control of an SCR is different when operated from an AC supply than when it is operated from a DC supply.
5. Name three common SCR case styles.
6. How are two SCRs connected in antiparallel and what purpose does this connection serve?
7. State the most common failed state (short or open) for both SCRs and triacs. How can an ohmmeter be used to test for this defective condition?
8. SCRs are unidirectional devices while triacs are bidirectional. What does this mean?
9. In what way is the gate triggering of an SCR different from that of a triac?
10. List some of the advantages of switching AC loads with a solid-state triac relay over switching with electromechanical types.
11. Give a brief explanation of how a triac is operated to control the amount of power applied to an AC load.
12. When will a diac conduct current?
13. List some of the limitations triacs have compared to SCRs.
14. An SCR is tested using an ohmmeter and found to have low resistance from anode to cathode at all times. What type of fault, if any, does this indicate?
15. State the function of each of each of the following blocks of an electronic motor control system:
    a. Controller
    b. Driver
    c. Sensors

PART 4 INTEGRATED CIRCUITS (ICS)

Fabrication

An integrated circuit (IC), sometimes called a **chip,** is a semiconductor wafer on which thousands or millions of tiny resistors, capacitors, and transistors are fabricated. IC chips provide a complete circuit function in one small semiconductor package with input and output pin connections, as illustrated in Figure 9-44. Most integrated circuits provide the same functionality as "discrete" semiconductor circuits at higher levels of reliability and at a fraction of the cost. Usually, discrete-component circuit construction is favored only when voltage and power dissipation levels are too high for integrated circuits to handle.

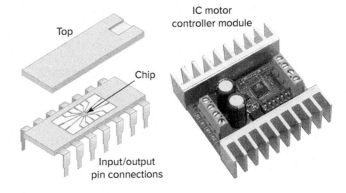

Figure 9-44 Integrated circuit (IC).
Photo courtesy of Dimension Engineering, www.dimensionengineering.com

Integrated circuits may be categorized as either digital or analog, depending on their intended application. **Digital ICs** operate with on/off switch-type signals that have only two different states, called low (logic 0) and high (logic 1). **Analog ICs** contain amplifying-type circuitry and signals capable of an unlimited number of states. The analog and digital processes can be seen in a simple comparison between a light dimmer and light switch. A light dimmer involves an analog process, which varies the intensity of light from off to fully on. The operation of a standard light switch, on the other hand, involves a digital process; the switch can be operated only to turn the light off or on.

Operational Amplifier ICs

Operational amplifier ICs (often called **op-amps**) take the place of amplifiers that formerly required many discrete components. These amplifiers are often used in conjunction with signals from sensors connected in control circuits. An op-amp is basically a high-gain amplifier that can be used to amplify low-level AC or DC signals. The schematic symbol for an op-amp is a triangle, shown in Figure 9-45. The triangle symbolizes direction and points from input to output. The connections associated with an op-amp can be summarized as follows:

- The op-amp has two inputs and a single output. The inverting input (−) produces an output that is 180° out of phase with the input. The second input, called the noninverting input (+), produces an output that is in phase with the input.

- The DC power supply terminals are identified as +V and −V. All op-amps need some type of power supply but some diagrams will not show the power supply terminals, as it is assumed that they are always there. The supply power will be determined by the type of output the op-amp is required to

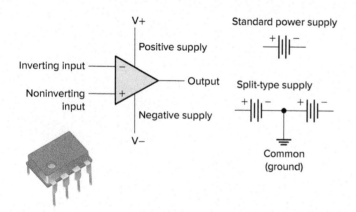

Figure 9-45 Operational amplifier (op-amp).
Photo courtesy of Digi-Key Corporation, www.digikey.com

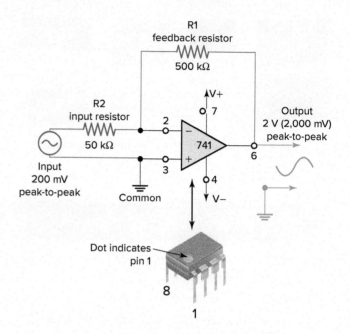

Figure 9-46 741 op-amp voltage amplifier circuit.
Photo courtesy of Digi-Key Corporation, www.digikey.com

produce. For example, if the output signal needs to produce both positive and negative voltages, then the power supply will need to be a split, or differential, type with both positive and negative voltages and a common ground. If the op-amp needs to produce only positive voltages, then the power supply be the standard, or traditional, DC type.

The op-amp is connected in a number of ways to perform different functions. Figure 9-46 shows a schematic diagram of a 741 op-amp circuit configured as an **inverting amplifier**. A split-type power supply, consisting of a positive supply and an equal and opposite negative supply, is used to operate the circuit. The operation of the circuit can be summarized as follows:

- Two resistors, R1 and R2, set the value of the voltage gain of the amplifier.
- Resistor R2 is called the input resistor, and resistor R1 is called the feedback resistor.
- The ratio of the resistance value of R1 to that of R2 sets the voltage gain of the amplifier.
- The op-amp amplifies the input voltage it receives and inverts its polarity.
- The output signal is 180° out of phase with the input signal.
- The op-amp gain for the circuit is calculated as follows:

$$\text{Op-amp gain} = \frac{R1}{R2} = \frac{500 \text{ k}\Omega}{50 \text{ k}\Omega} = 10$$

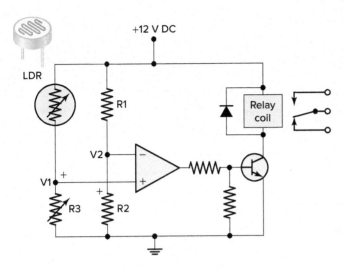

Figure 9-47 Operational amplifier circuit configured as a voltage comparator.

Figure 9-47 shows the schematic diagram of an operational amplifier circuit configured as a **voltage comparator** with its output signal being the difference between the two input signals or voltages, V2 and V1. A light-dependent resistor (LDR) is used to sense the light level. When LDR is not illuminated, its resistance is very high, but once illuminated with light, its resistance drops dramatically. The circuit is operated with a single DC power supply and without a feedback circuit. The operation of the circuit can be summarized as follows:

- The resistor combination R1 and R2 form a fixed reference voltage input V2, set by the ratio of the two resistors.
- The resistor combination LDR and R3 form the variable voltage input V1.

- When the light level sensed by the LDR drops and the variable output voltage V1 falls below the reference voltage at V2, the output from the op-amp changes, activating the relay and switching the connected load.
- Likewise, as the light level increases, the output will switch back, turning the relay off.
- The preset resistor R3 value can be adjusted up or down to increase or decrease resistance; in this way it can make the circuit more or less sensitive.

555 Timer IC

The 555 timer IC is used as a timer in circuits requiring precision timing as well as an oscillator to provide pulses needed to operate digital circuits. Figure 9-48 shows the pinout functional block diagram of the **555 chip.** The internal circuitry of the chip is made up of a complex maze of transistors, diodes, and resistors.

The light switch interval timer shown in Figure 9-49a is an ideal application for lights that are needlessly left on or forgotten. The dial is set for the amount of time the lights are to be left on, from 1 minute to 18 hours. Time intervals are initiated on the closure of the switch. After the preset time has elapsed, the lights will go off automatically regardless of the state (open or closed) of the switch. Figure 9-49b shows a 555 timer version of an interval timer. The operation of the circuit can be summarized as follows:

- The time period is determined by the value of the two external timing components R and C.
- When the switch opens, the external capacitor is held discharged (short-circuited) by a transistor inside the timer.

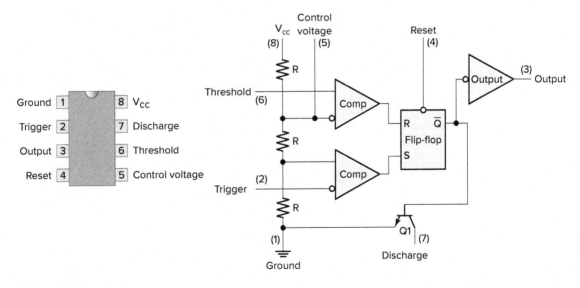

Figure 9-48 The 555 IC timer.

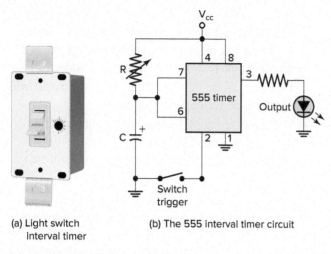

(a) Light switch interval timer

(b) The 555 interval timer circuit

Figure 9-49 A 555 IC interval timer.

- When the switch is closed, it releases the short circuit across the capacitor and triggers the LED into conduction. At this point the timing period starts.
- Capacitor C starts to charge through resistor R.
- When the charge on the capacitor reaches two-thirds of the source voltage, the timing period ends and the LED is automatically switched off.
- At the same time the capacitor discharges to ready for the next triggering sequence.

A common method of DC motor speed control is **pulse-width modulation (PWM).** Pulse-width modulation is the process of switching the power to a device on and off at a given frequency, with varying on and off times. This relation between on and off times is referred to as the **duty cycle.** Figure 9-50 shows the 555 timer used as a pulse width modulator oscillator that controls the speed of a small permanent-magnet DC motor. While most DC motor drives use a microcontroller to generate the required PWM signals, the 555 PWM circuit shown will help you to understand how this type of motor control operates. The operation of the circuit can be summarized as follows:

- The voltage applied across the armature is the average value determined by the length of time that transistor Q2 is turned on compared to the length of time it is turned off (duty cycle).
- Potentiometer R1 controls the length of time the output of the timer will be turned on, which in turn controls the speed of the motor.
- If the wiper of R1 is adjusted to a higher positive voltage, the output will be turned on for a longer period of time than it will be turned off.

Microcontroller

An electronic motor controller can include means for starting and stopping the motor, selecting forward or reverse rotation, selecting and regulating the speed, regulating or limiting the torque, and protecting against overloads and faults. As integrated circuitry evolved, all of the components needed for a controller were built right onto one microcontroller chip. A microcontroller (also called a **digital-signal controller** or **DSC**) is an ultra-large-scale integrated circuit that functions as a complete computer on a chip, containing a processor, memory, and input/output functions. Figure 9-51 shows a typical microcontroller IC that can be used in a variety of control applications.

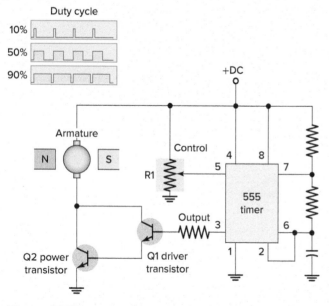

Figure 9-50 The 555 DC motor speed controller.

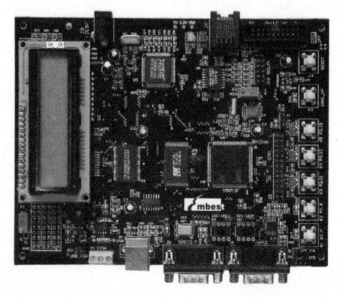

Figure 9-51 Microcontroller used in motor control applications.
Photo courtesy of Embest Technology Co., Ltd.

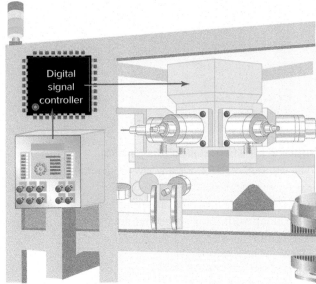

Figure 9-52 Embedded microcontroller.
Photo courtesy of Mosaic Industries, www.mosaic-industries

Microcontrollers are most often **"embedded,"** or physically built into, the devices they control, as shown in Figure 9-52. An embedded microcontroller is designed to do some specific task, rather than to be a general-purpose computer for multiple tasks. The software written for embedded systems is often called firmware and is stored in read-only memory or flash memory chips rather than a disk drive. Microcontrollers often run with limited computer hardware resources: small or no keyboard and screen, and little memory.

The block diagram of Figure 9-53 shows a microcontroller used to control the operation of the inverter section of a variable-speed AC motor drive. The rotating speed of AC induction motors is determined by the frequency of the AC applied to the stator, not the applied voltage. However, stator voltage must also drop to prevent excessive current flow through the stator at low frequencies. The microcontroller controls the voltage and frequency and sets the proper stator voltage for any given input frequency. Two

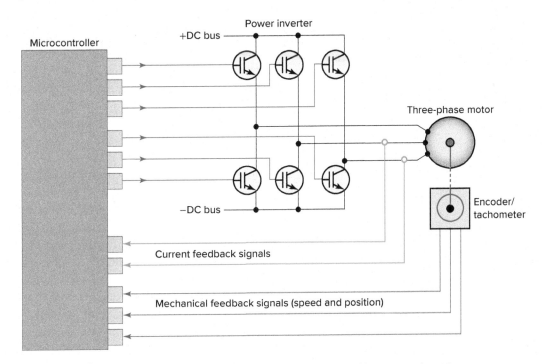

Figure 9-53 Microcontroller used to control the operation of the inverter section of a variable-speed AC motor drive.

Part 4 Integrated Circuits (ICs)

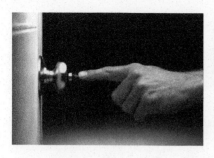

Figure 9-54 Electrostatic discharge (ESD).
Photo courtesy of RTP Company, www.rtpcompany.com.

phase currents are measured and returned to the microcontroller along with rotor speed and angular-position information from the encoder/tachometer.

Electrostatic Discharge (ESD)

Static charge is an unbalanced electrical charge at rest. Typically, it is created by insulator surfaces rubbing together or pulling apart. One surface gains electrons, while the other surface loses electrons. This results in an unbalanced electrical condition known as static charge. When a static charge moves from one surface to another, it becomes **electrostatic discharge (ESD).** Figure 9-54 shows a common example of ESD. When a person (negatively charged) contacts a positively charged or a grounded object, electrons will move from one to the other. ESD, which is a bit annoying but certainly harmless to humans, can be lethal to sensitive electronic devices.

All integrated circuits are sensitive to electrostatic discharge to some degree. If static discharge occurs at a sufficient magnitude, some damage or degradation (the IC is weakened and will often fail later) will usually occur. Damage is mainly due to the current flowing through ICs during discharge. Basically what happens is that a relatively large amount of heat is generated in a localized volume significantly faster than it can be removed, leading to a temperature in excess of the material's safe operating limits.

Figure 9-55 shows an antistatic wrist strap band used to prevent a static charge from building up on the body by safely grounding a person working on sensitive electronic equipment. The wrist strap is connected to ground through a coiled retractable cable and resistor. An approved grounding wrist band has a resistance built into it, so it discharges static electricity but prevents a shock hazard when the wearer is working with de-energized electronic circuits or components. Other precautions that should be taken when working on integrated circuits include:

- Never handle sensitive ICs by their leads.
- Keep your work area clean, especially of common plastics.
- Handle printed circuit boards by their outside corners.
- Always transport and store sensitive ICs and control boards in antistatic packaging.

Digital Logic

Logic circuits perform operations on digital signals. In digital or binary logic circuits there are only two values, 0 and 1. Logically, we can use these two numbers or we can specify that:

$$0 = \text{false} = \text{no} = \text{off} = \text{open} = \text{low}$$
$$1 = \text{true} = \text{yes} = \text{on} = \text{closed} = \text{high}$$

Using the binary two-value logic system, every condition must be either true or false; it cannot be partly true or partly false. While this approach may seem limited, it can be expanded to express very complex relationships and interactions among any number of individual conditions. One reason for the popularity of digital logic circuits is that they provide stable electronic circuits that can switch back and forth between two clearly defined states, with no ambiguity attached. Integrated circuits are the least expensive way to make logic gates in large volumes. They are used in programmable controllers to solve complex logic.

Logic is the ability to make decisions when one or more different factors must be taken into account before an action is taken. A **logic gate** performs a logical operation on one or more logical inputs and produces a single logical output. There are two generally accepted methods of drawing logic gate symbols; the distinctive shape method and the rectangular shape method. The **distinctive shape** method uses different shapes for different logic functions and a bubble (a small circle) to indicate a logical inversion. With the **rectangular shape** method, all functions are shown in rectangular form with the logic function indicated by standard notation inside the rectangle. Logical inversion is indicated by a half arrowhead.

Figure 9-55 Antistatic wrist strap band.
Photo courtesy of Electronix Express, www.elexp.com

270　　Chapter 9　Motor Control Electronics

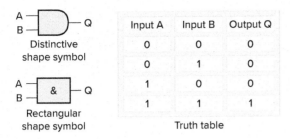

Figure 9-56 Two-input AND gate.

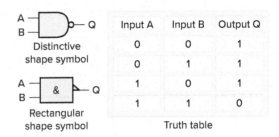

Figure 9-59 Two-input NAND gate.

The **AND gate** is a logic circuit that has two or more inputs and a single output. Figure 9-56 shows the symbols used for a two-input AND gate. The operation of the AND gate is summarized by the table. Such a table, called a truth table, shows the output for each possible input. The basic logic that applies is that if all inputs are 1, the output will be 1. If any input is 0, the output will be 0.

An **OR gate** produces a 1 output if any of its inputs are 1s. The output is 0 if all the inputs are 0s. An OR gate can have two or more inputs; its output is true if at least one input is true. Figure 9-57 shows the symbols used for a two-input OR gate along with its truth table.

The simplest logic circuit is the **NOT circuit.** It performs the function called inversion, or complementation, and is commonly referred to as an inverter. The purpose of the inverter is to make the output state the opposite of the input state. Figure 9-58 shows the symbols used for a NOT function along with its truth table. Unlike the AND and OR gate functions, the NOT function can have only one input. If a 1 is applied to the input of an inverter, a 0 appears on its output. The input to an inverter is labeled A and the output is labeled $\bar{A}$ (read "NOT A"). The bar over the letter indicates the complement of A. Because the inverter has only one input, only two input combinations are possible.

A **NAND gate** is a combination of an inverter and an AND gate. It is called a NAND gate after the NOT-AND function it performs. Figure 9-59 shows the symbols used for a two-input NAND gate along with its truth table. The bubble or half arrowhead on the output end of the symbol means to invert the AND function. Notice that the output of the NAND gate is the complement of the output of an AND gate. A NAND gate can have two or more inputs. Any 0 in the inputs yields a 1 output. The NAND gate is the most commonly used logic function. This is because it can be used to construct an AND gate, OR gate, inverter, or any combination of these functions.

A **NOR gate** is a combination of an inverter and an OR gate. Its name is derived from its NOT-OR function. Figure 9-60 shows the symbols used to represent a two-input NOR gate along with its truth table. The output of the NOR gate is the complement of the output of the OR-function output. The output Q is 1 if NOR inputs A or B are 0. A NOR gate can have two or more inputs, and its output is 1 only if no inputs are 1.

The term *hard-wired* logic refers to logic control functions that are determined by the way devices are interconnected. Hard-wired logic is fixed in that it is changeable only by changing the way the devices are connected. In contrast, programmable logic, such as that used in programmable logic controllers (PLCs), is based on basic

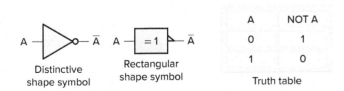

Figure 9-58 The NOT function.

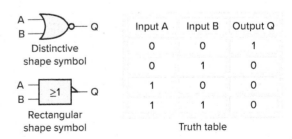

Figure 9-60 Two-input NOR gate.

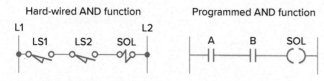

Figure 9-61 Hard-wired circuit and the equivalent PLC logic program for an AND logic control function.

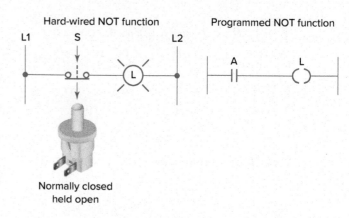

Figure 9-63 Hard-wired circuit and the equivalent PLC logic program for a NOT logic control function.

logic functions, which are easily changed by modifying the program. PLCs use logic functions either singly or in combination to form instructions that will determine if a device is to be switched on or off.

Figure 9-61 shows a hard-wired circuit and the equivalent PLC logic program for an AND logic control function. Both normally open input limit switches (LS1 *and* LS2) must be closed to energize the output solenoid valve. This control logic is implemented in the hard-wired circuit by connecting the two limit switches and the solenoid valve all in series. The PLC program uses the same inputs (LS1 and LS2) and output (SOL) devices connected to the PLC, but implements the logic via the program rather than the wiring.

Figure 9-62 shows a hard-wired circuit and the equivalent PLC logic program for an OR logic control function. Either one of the two normally open push buttons (PB1 *or* PB2) is closed to energize the contactor coil (C). This control logic is implemented in the hard-wired circuit by connecting the two input push buttons in parallel with each other to control the output coil. The PLC program uses the same inputs (PB1 and PB2) and output (C) devices connected to the PLC, but implements the logic via the program rather than the wiring.

AND and OR logic use normally open input devices that must be closed to supply the signal that energizes the loads. NOT logic energizes the load when the control signal is off. Figure 9-63 shows a hard-wired circuit and the equivalent PLC logic program for a NOT logic control function. This example is that of the NOT function used to control the interior lamp of a refrigerator. When the door is opened, the lamp automatically switches on. The switch controlling the lamp is a normally closed type that is held open by the shut door. When the door is opened, the switch returns to its normal closed state and the load (lamp) is energized. For the load to remain energized, there must *not* be a signal from the switch input. To keep the lamp on, the normally closed contact must not change its normally closed state.

Digital sensor signals are often referred to as a discrete signal. A digital signal in its simplest of forms is one piece or **bit** of information. This single piece of information (bit) has one of two values. **Nibbles, bytes,** and **words,** communicate more complex information or data, as illustrated in Figure 9-64. The values increase/decrease in very specific abrupt steps.

Three-wire **PNP** sensors are sometimes called **sourcing sensors.** This refers to the theory of operation of the electronics inside the sensor and type of transistors used, as illustrated in Figure 9-65. For the PNP sensor, when it senses a target, it will connect the output load to the positive source supply.

Word			
Byte		Byte	
Nibble	Nibble	Nibble	Nibble
0 0 0 0	0 0 0 0	0 0 0 0	0 0 0 0

0 Bit

Figure 9-64 Digital sensor signals.

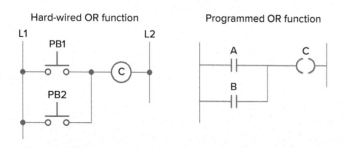

Figure 9-62 Hard-wired circuit and the equivalent PLC logic program for an OR logic control function.

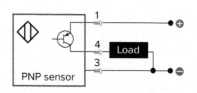

Figure 9-65 PNP sourcing sensor connections.

Chapter 9 Motor Control Electronics

Three-wire **NPN** sensors are sometimes called **sinking** sensors. The difference between these two configurations is simply the direction in which current flows, as illustrated in Figure 9-66. For the NPN sensor, when it senses a target, it will connect the output load to the negative supply. The selection of a PNP sensor verses an NPN sensor is determined by the nature of the circuit the device is to be used in.

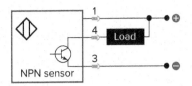

Figure 9-66 NPN sinking sensor connections.

Part 4 Review Questions

1. Describe the makeup of an integrated circuit.
2. What advantages does an IC have over discrete-component circuit construction?
3. What type of circuit is not suited to integrate onto a chip?
4. Compare the functioning of digital and analog integrated circuits.
5. What is an operational amplifier?
6. An operational amplifier is connected as a voltage amplifier. How is the gain of the circuit determined?
7. Explain the operation of an op-amp when configured as a voltage comparator.
8. What are the two major applications for the 555 timer IC?
9. Give a brief explanation of how a 555 interval timer circuit works.
10. A 555 timer is configured as a pulse-width modulator to vary the speed of a DC motor. How exactly does it operate to change the speed of the motor?
11. List some of the control tasks that a microcontroller designed for control of an electric motor may be required to perform.
12. What does the term *embedded* microcontroller refer to?
13. Explain the role of a microcontroller when used to control the inverter section of a variable-speed AC motor drive.
14. In what way can electrostatic discharge damage an integrated circuit?
15. List some of the precautions that should be taken when handling sensitive IC chips.
16. What can the terms logic 0 and logic 1 represent?
17. What makes digital circuitry so popular?
18. It is desired to have a lamp turn on when any one of three switches is closed. What control logic function would be used?
19. What control logic function energizes the load when the control signal is off?
20. What control logic function is used to implement five inputs connected in series with the requirement that all five must be closed to energize the load?
21. Compare the way in which hard-wired logic differs from programmable logic.
22. Compare the difference in the switching action of PNP and NPN three-wire sensors when each senses a target.

Troubleshooting Scenarios

1. One of the diodes of a single-phase full-wave rectifier is mistakenly connected backward in the bridge configuration. What effect will this have on the resultant DC output voltage and current flow through the diodes?
2. The insulation resistance of a motor operated by an electronic drive is to be tested using a megger. What precaution should you take? Why?
3. If the triac in the internal circuit of a lamp dimmer were to fail shorted, what effect would this most likely have on the operation of the circuit?
4. In what ways would the approach to troubleshooting for integrated circuits be different from that for a circuit constructed with discrete components?
5. The toggle switch controlling a 12-light chandelier is to be replaced with an electronic lamp dimmer. Each of the 12 chandelier lamps is rated for 60 W. Dimmer switches rated for 300 W, 600 W, and 1000 W are available for the job. What is the minimum size dimmer that should be used for this installation? Why?

Discussion Topics and Critical Thinking Questions

1. What circuitry would have to be incorporated into an LED pilot light module rated to be operated directly by a 240 V AC source?
2. The best source to verify correct operation of modular electronic components is the operating manual. Why?
3. What are the advantages of using a triac rather than a rheostat for lamp dimming applications?
4. What faults do heat sinks used in power electronic devices guard against?
5. Compare the way analog and digital signals are used to transmit information.

CHAPTER TEN

Adjustable-Speed Drives and PLC Installations

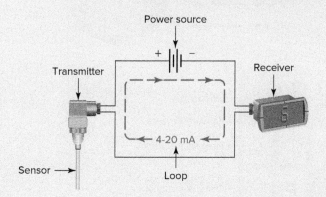

CHAPTER OBJECTIVES

This chapter will help you:

- Understand the operation, installation, and setup for an alternating current (AC) motor drive.
- Understand the operation, installation, and setup for a direct current (DC) motor drive.
- Understand the operation, installation, and setup for PLC motor control.

The two most important emerging technologies associated with motor control are adjustable-speed drives and programmable logic controllers. Adjustable-speed drives (also known as variable-speed drives) allow motor-driven loads to operate within a wide range of speeds. Matching motor speeds to load requirements can increase both the efficiency and performance of a motor installation. A programmable logic controller (PLC) is a type of computer commonly used in motor control applications. The traditional motor control circuit is hard-wired, while PLC control is program-based. This chapter deals with the unique installation requirements of these electronic control systems.

PART 1 AC MOTOR DRIVE FUNDAMENTALS

The primary function of any electronic adjustable-speed drive is to control the **speed, torque, acceleration, deceleration,** and **direction of rotation** of a machine. Unlike constant-speed systems, the adjustable-speed drive permits the selection of an infinite number of speeds within its operating range.

The use of adjustable-speed drives in pump and fan systems can greatly increase their efficiency. Outdated technology most often used throttles or dampers to interrupt the flow as a means to control it. The fluid or air was held back by the throttle or damper, but the

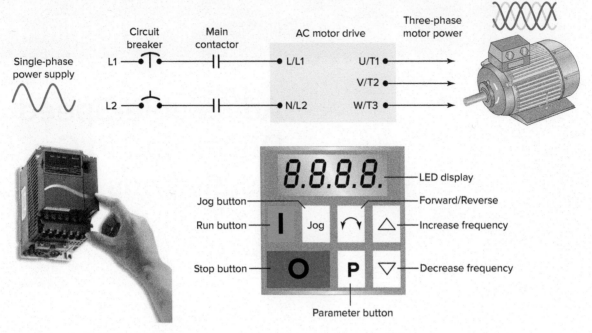

Figure 10-1 Variable-speed AC motor drive used for lower-power applications.
Photo courtesy of Delta Products Corporation, www.delta-americas.com

energy used to move the fluid or air was then dissipated uselessly. That wasted energy was still accounted for and paid for. Running a system this way is like driving a car with the accelerator pressed to the floor while controlling speed with the brake. An electronic adjustable-speed drive, on the other hand, allows precise control of motor output. In the case of centrifugal fans and pumps, there is a significant saving in the power required to handle the load.

Figure 10-1 shows a **variable-speed AC motor drive used for lower-power applications.** Power wiring consists of the conductors supplying power to the drive (L/L1 and N/L2) and the conductors supplying power to the motor (U/T1, V/T2, and W/T3). The North American designation for load conductors is T1, T2, and T3; the European designation for load conductors is U, V, and W. While single-phase supplies power to the drive, the output to the motor is three-phase. Control wiring consists of inputs and outputs connected to the control terminal strip. Various control wiring configurations are used, depending on the make of controller and specific application. Branch circuit protection via circuit breaker or disconnect switch and fuses must be provided to comply with the National Electrical Code (NEC) and all local codes.

Variable-Frequency Drives (VFDs)

Squirrel-cage induction motors are the most common three-phase motors used in commercial and industrial applications. The preferred method of speed control for squirrel-cage induction motors is to alter the frequency of the supply voltage. Since the basis of the drive's operation is to vary the frequency to the motor in order to vary the speed, the best-suited name for the system is the variable-frequency drive (VFD). However, other names used to reference this type of drive include adjustable-speed drive (ASD), adjustable-frequency drive (AFD), variable-speed drive (VSD), and frequency converter (FC).

A VFD controls the speed, torque, and direction of an AC induction motor. It takes fixed voltage and frequency AC input and converts it into a variable voltage and frequency AC output. Figure 10-2 shows the block diagram of a typical three-phase variable-frequency drive controller. The function of each block is as follows:

- **Converter:** A full-wave rectifier that converts the applied AC into DC.

- **DC bus:** Also referred to as a DC link, connects the rectifier output to the input of the inverter. The DC bus functions as a filter to smooth the uneven, rippled output to ensure that the rectified output resembles as closely as possible pure DC.

- **Inverter:** The inverter takes the filtered DC from the DC bus and converts it into a pulsating AC waveform. By controlling the output of the inverter, the pulsating AC waveform can simulate an AC sine wave at different frequencies.

- **Control logic:** The control logic system generates the necessary pulses used to control the firing of the power semiconductor devices such as SCRs and

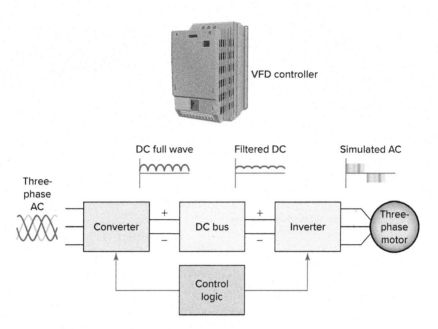

Figure 10-2 Block diagram of a typical three-phase variable-frequency drive.

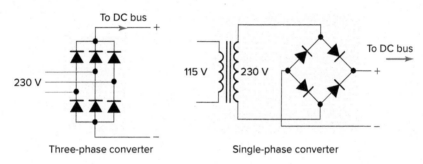

Figure 10-3 Three-phase and single-phase converter input connections.

transistors. Fairly involved control circuitry coordinates the switching of power devices, typically through a control board that dictates the firing of power components in the proper sequence. An embedded microprocessor is used for all internal logic and decision requirements.

Sometimes called the front end of the VFD, the converter is commonly a three-phase, full-wave bridge rectifier. However, one of the advantages of variable-frequency drives is being able to operate a three-phase AC motor from a single-phase AC supply. The key to this process is the **rectification** of the AC input to a DC output. At this rectification point, the DC voltage has no phase characteristics; the VFD is simply producing a filtered pulsating DC waveform. The drive **inverts** the DC waveform into three different pulse-width modulated waveform signatures that duplicate an AC three-phase waveform. Figure 10-3 shows three-phase and single-phase converter input connections. AC input voltage levels that are different from that required to operate the motor require the converter section to raise or lower the voltage to the proper operating level of the motor. As an example, an electric motor drive supplied with 115 V AC that must deliver 230 V AC to the motor requires a transformer capable of stepping up the input voltage.

The VFD offers an alternative to other forms of power conversion in areas where three-phase power is unavailable. Since it converts incoming AC power into DC, the VFD really doesn't care if its source is single or three phase. Regardless of the input power, its output will always be three phase. Drive sizing, however, is a factor since it must be capable of rectifying the higher-current, single-phase source. As a rule of thumb, most manufacturers recommend doubling the normal three-phase capacity of a drive that will be operating on a single-phase input. Single-phase operation is limited to smaller-horsepower motors. Some manufacturers offer some models for single-phase input only and others that are fully rated for both single-phase and three-phase input.

Part 1 AC Motor Drive Fundamentals

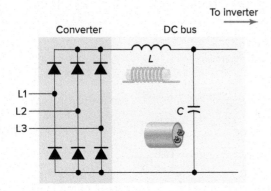

Figure 10-4 Inductor and capacitor connections within the DC bus.

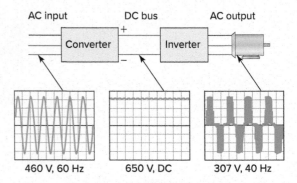

Figure 10-5 Typical VFD voltage measurements.

After full-wave rectification of an AC supply into a VFD, the DC output passes through a DC bus. Figure 10-4 shows the inductor (L) and capacitor (C) connections within the DC bus.

- They work together to filter out any ripple component of the DC waveform.
- The principal energy storage element is the bus capacitors.
- Any ripple that is not smoothed out will show up as distortion in the motor output waveform.
- Most VFD manufacturers provide a special terminal block or parameter setting for DC bus voltage measurement.
- With a 460 V AC input, you should read an average DC bus voltage of about 650 to 680 V DC, as illustrated in Figure 10-5.
- The DC value is calculated by taking the root-mean-square (RMS) value of the line voltage and multiplying it by 1.414.
- Ripple voltage readings of more than 4 V AC on the bus may indicate a possible capacitor filtering problem or a problem with the diode bridge converter section.

The **inverter** is the final output section of a VFD. This is the point where the DC bus voltage is switched on and off at specific intervals. In doing so, the DC energy is changed into three channels of AC energy that an AC motor uses to operate. Today's inverters use insulated-gate bipolar transistors (IGBTs) to switch the DC bus on and off. Figure 10-6 shows a simplified diagram of the three sections of a variable-speed drive. The control logic and inverter section control the output voltage and frequency to the motor. Six switching transistors are used in the inverter section. The control logic uses a microcontroller to switch the transistors on and off at the proper time. The main objective of the VFD is to vary the speed of the motor while providing the closest approximation to a sine wave for current.

In the simplest circuit implementation, two IGBTs are placed in series across the DC supply and are switched on and off to generate one phase of the three phases for the motor. Two other identical circuits generate the other two phases. Figure 10-7 shows a simplified circuit of a pulse-width modulation (PWM) inverter. Switches are used to illustrate the way that the transistors are switched to produce one phase (A to B) of the three-phase output. The output voltage is switched from positive to negative by opening and closing the switches in

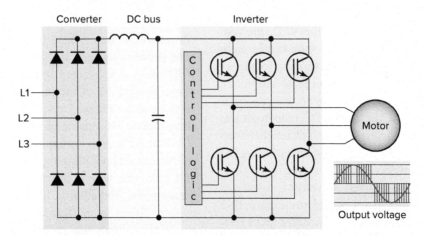

Figure 10-6 The three sections of a variable-speed drive.

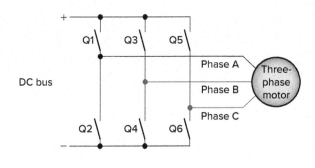

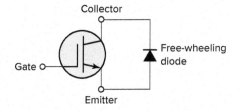

Figure 10-8 **IGBT switch element.**

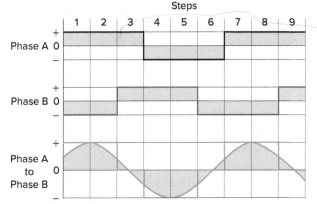

Figure 10-7 **Simplified circuit of a PWM inverter.**

a specific sequence of steps. The operation can be summarized as follows:

- During steps 1 and 2, transistor switches Q1 and Q4 are closed.
- The voltage from phase A to B is positive.
- During step 3, transistor switches Q1 and Q3 are closed.
- The difference in voltage between phase A and phase B is zero, resulting in zero output voltage.
- During steps 4 and 5, transistor switches Q2 and Q3 are closed.
- This results in a negative voltage between phases A and B.
- The other steps continue in a similar manner.
- Output voltage is dependent on the state of the switches (open or closed), and the frequency is dependent on the speed of switching.

Insulated-gate bipolar transistors (IGBTs) are capable of high switching speeds necessary for inverter operation. When a positive voltage is applied to the gate, the IGBT turns on. When the IGBT is turned on, current flows between the collector and the emitter. This is similar to closing a switch. Typically, each switch element is made up of an IGBT and **free-wheeling diode,** as illustrated in Figure 10-8. The free-wheeling diode protects the transistor by conducting current when it turns off and the motor's collapsing magnetic field causes a current surge. This is also the path that is used when the motor is connected to an overhauling load where the inertia of the load is greater then that of the motor.

Figure 10-9 shows the sine-wave (AC) line voltage, superimposed on the pulsed inverter output, or simulated AC. Notice that the pulses are the same height for each pulse. This is because the DC bus voltage the drive uses to create these pulses is constant. Output voltage is varied by changing the width and polarity of the switched pulses. Output frequency is adjusted by changing the switching cycle time. The resulting current in an inductive motor simulates a sine wave of the desired output frequency. Most true RMS-measuring multimeters are fast enough to measure the RMS value of the PWM voltage and current.

There are two frequencies associated with a PWM variable-frequency drive: the fundamental frequency and the carrier frequency. The **fundamental** frequency is the variable frequency a motor uses to vary speed. In a typical VFD, the fundamental frequency will vary from a few hertz up to a few hundred hertz. The inductive reactance of an AC magnetic circuit is directly proportional to the frequency ($X_L = 2fL$). Therefore, when the frequency applied to an induction motor is reduced, the applied voltage must also be reduced to limit the current drawn by the motor at reduced frequencies. The microprocessor control adjusts the output voltage waveform to simultaneously change the voltage and frequency to maintain the constant volts/hertz ratio.

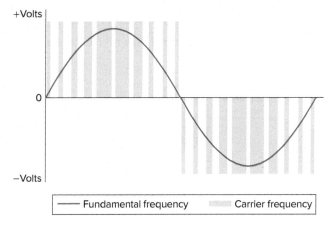

Figure 10-9 **Sine-wave line voltage superimposed on the PWM inverter output.**

Part 1 **AC Motor Drive Fundamentals**

Figure 10-10 Inverter motor and control.
Photo courtesy of Siemens, www.siemens.com

The **carrier** frequency (also known as the switch frequency) is the frequency at which the pulses in pulse-width modulation switch at. The carrier frequency is a fixed frequency substantially higher than the fundamental frequency. This high switching speed produces the classic whine associated with variable-frequency drives. Higher carrier frequency allows a better approximation to the sinusoidal form of the output current. However, higher switch frequencies decrease the efficiency of the drive because of increased heat in the power transistors. The carrier frequency for VFDs is in the 2 to 16 kHz range. Adjusting the carrier frequency automatically in accordance with the changing load and temperature will result in quieter operation.

The **inverter-duty motor** shown in Figure 10-10 is designed for optimized performance to operate in conjunction with a variable-frequency drive. An inverter duty motor can withstand the higher voltage spikes produced by all VFDs and can run at very slow speeds without overheating.

A fundamental advantage of an AC motor drive is that it provides virtually infinite speed control of a standard induction motor. Speed control can be open loop or closed loop. **Open-loop control** uses no feedback of the actual motor speed. If the drive output voltage and frequency are held constant when motor load is increased, shaft rpm will decrease due to additional required torque. This rpm decrease is not monitored, and no compensation takes place. Speed regulation is therefore only determined by the slip characteristics of the connected motor.

Closed-loop control is required if better speed regulation is needed. Closed-loop speed control requires a speed **feedback device**, such as a tachometer or encoder mounted to the shaft of the connected motor. The drive logic compares the reference frequency to the feedback frequency, as illustrated in Figure 10-11. The feedback value is subtracted from the reference value, and a error signal is produced. It is this error signal, acted upon by the control logic, that is used to calculate a new output frequency, which is applied to the motor.

In general, AC drives control motor speed by varying the frequency of the current supplying the motor. Although frequency can be varied in different ways, the two most common speed control methods in use today are volts per hertz (V/Hz) and flux vector.

Volts per Hertz Drive

Of the speed control methods, volts per hertz technology is the most economical and easiest to apply. The **V/Hz drive** controls shaft speed by varying the voltage and frequency of the signal powering the motor. Volts per hertz control in its simplest form takes a speed reference command from an external source and varies the voltage and frequency applied to the motor. By maintaining a **constant V/Hz ratio,** the drive can control the speed of the connected motor. Volts per hertz drives work well on applications in which the load is predictable and does not change quickly, such as fan and pump loads.

In order to prevent overheating, the voltage applied to the motor must be decreased by the same amount as the frequency. V/Hz control runs in open loop without a feedback device. The ratio between voltage and frequency is called volts per hertz (V/Hz). To find the volts per hertz

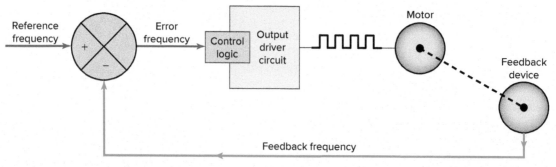

Figure 10-11 Closed-loop control system.

ratio, simply divide the rated nameplate voltage by the rated nameplate frequency. For example the volt per hertz ration for a 460 Volt, 60 Hz motor is calculated as follows:

$$\text{V/Hz} = \frac{\text{Voltage}}{\text{Frequency}} = \frac{460 \text{ V}}{60 \text{ Hz}} = 7.67 \text{ V/Hz}$$

Volts per hertz control provides a linear (straight-line) voltage ratio to the frequency of a motor from 0 rpm to base speed. This is illustrated in Figure 10-12 using a 460 V AC, 60 Hz motor as an example.

- The volts per hertz ratio of 7.67 is supplied to the motor at any frequency between 0 and 60 Hz.
- If applied frequency is reduced to 30 Hz, the shaft will slow to half its original speed.
- In this situation, a V/Hz drive also halves the voltage (here, to 230 V AC) in order to maintain the 7.67 V/Hz ratio, which allows the motor to continue producing its rated torque.
- Horsepower increases and torque remains constant up to the base speed; however, above base speed (i.e., above 60 Hz frequency), the torque decreases while the horsepower remains constant.
- This can easily be understood by the simple relationship between horsepower, speed, and torque:

$$\text{Horsepower} = \text{Torque} \times \text{Speed} \times K$$

where K is a units constant.

Volts per hertz control of an AC induction motor is based on the principle that to maintain constant magnetic flux in the motor, the terminal voltage magnitude must increase roughly proportionally to the applied frequency. This is only an approximate relationship, and volts per hertz drives designs may include the following refinements:

- **Low-frequency voltage boost** (also referred to as *IR* compensation)—Below 15 Hz the voltage applied to the motor is boosted to compensate for the power losses AC motors experience at low speeds and increase the starting torque capability.
- **Steady-state slip compensation**—Increases frequency on the basis of a current measurement to give better steady-state speed regulation.
- **Stability compensation**—To overcome midfrequency instabilities evident in high-efficiency motors.

Flux Vector Drive

You may recall that a vector quantity has both magnitude and direction. Similarly, a **flux vector drive** controls both the magnitude and direction of the magnetic flux in an AC induction motor (Figure 10-13). Vector control treats an AC machine similar to the way that a DC drive controls a DC machine.

- Current drawn from a motor can be broken down into two components: excitation, magnetization, or flux current (similar to field current of a DC motor) and in-phase current (similar to armature current of a DC motor). The vector sum of these two components equals the total current the motor draws.
- Excitation or flux current is controlled by varying the voltage to the motor. Increasing the voltage increases flux current, and decreasing the voltage decreases the flux current.
- The in-phase current is dependent on the load. The larger the load, the greater the in-phase current.
- Independent control of the voltage means that the microprocessor has direct control of motor flux.
- Increasing flux reduces motor slip and, conversely, decreasing flux increases motor slip.

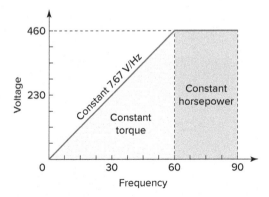

Figure 10-12 V/Hz control of a 460 V AC, 60 Hz motor.

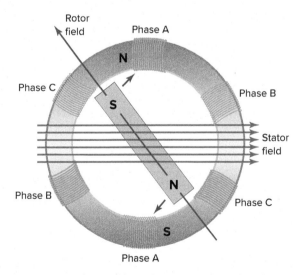

Figure 10-13 Vector drives control both magnitude and direction of the magnetic flux.

Part 1 AC Motor Drive Fundamentals

- Precise control of flux allows slip to be correlated to motor torque.
- If slip is known, the drive can accurately compensate for slip by increasing the drive's output frequency, thereby maintaining very tight speed regulation.

A flux vector drive uses feedback from what is happening at the motor to make changes in the output of the drive. However, it still relies on the basic volts per hertz core for controlling the motor. These combined techniques control not only the magnitude of motor flux but also its orientation, thus the flux vector name. The flux vector method provides more precise motor speed and torque control.

Vector control is popular for applications that need either better speed regulation and/or optimum torque, especially at lower frequencies. Flux vector control drives are available in either open-loop configurations (drives that do not require direct speed feedback) or closed-loop configurations (drives that require an external encoder to provide speed feedback).

Sensorless open-loop vector control involves calculating an estimate of the motor's magnetic flux and torque based on the measured value of voltage and current of the motor. This requires the parameters and characteristics of the induction motor to be programmed into the software. By reading the back EMF current and comparing it to the applied voltage, the VFD calculates the rotor speed and, therefore, slip. Once the speed is known, the drive compensates for the slip by increasing output frequency to get good speed regulation and directly controls the flux current by increasing or decreasing the applied voltage for excellent torque control.

The block diagram of a sensorless flux vector control drive is shown in Figure 10-14. Its operation can be summarized as follows:

- Slip is the difference between the rotor speed and the synchronous speed of the magnetic field and is required to produce motor torque. The *slip estimator* block keeps actual motor rotor speed close to the desired set speed.
- The *torque current estimator* block determines the percent of current that is in phase with the voltage, providing an approximate torque current. This is used to estimate the amount of slip, providing better speed control under load.
- *V angle* controls the amount of total motor current that goes into motor flux enabled by the torque current estimator. By controlling this angle, low-speed operation and torque control are improved over those of the standard V/Hz drive.
- The *flux vector control* retains the V/Hz core and adds additional blocks around the core to improve the performance of the drive.
- The *current resolver* attempts to identify the flux and torque producing currents in the motor and makes these values available to other blocks in the drive.
- The *current limit block* monitors motor current and alters the frequency command when the motor current exceeds a predetermined value.

A **closed-loop vector drive** uses a motor-mounted encoder or similar sensor to give positive shaft position indication back to the microprocessor. The motor's rotor position and speed are monitored in real time via a digital

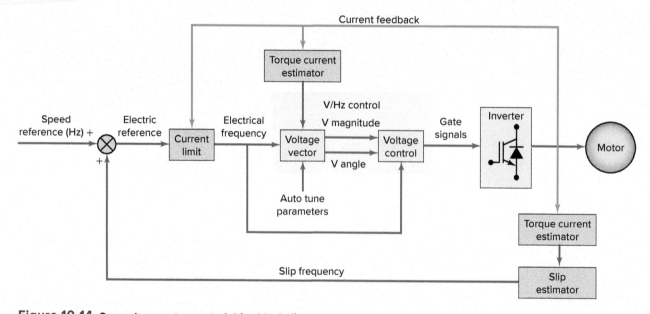

Figure 10-14 Sensorless vector control drive block diagram.

encoder to determine and control the motor's actual speed, torque, and power produced. Figure 10-15 shows a typical flux vector drive and motor-mounted encoder used in AC vector drive applications.

- The encoder operates by sending digital pulses back to the drive, indicating both speed and direction.
- The processor counts the pulses and uses this information, along with information about the motor itself, in order to control the motor torque and associated operating speed.
- The most common encoders deliver 1,024 pulses per revolution.
- To guard against electromagnetic interference (EMI), the cable assembly between the encoder and the drive should be one continuous shielded cable.

A closed-loop vector drive can also make an AC motor develop continuous full torque at zero speed, something that previously only DC drives were capable of. That

Encoder   Flux vector drive

Figure 10-15 Flux vector drive and encoder.
Flux vector drive material and associated copyrights are proprietary to, and used with the permission of, Schneider Electric.

makes them suitable for crane and hoist applications where the motor must produce full torque before the brake is released or else the load begins dropping and cannot be stopped.

Part 1 Review Questions

1. List the common basic control function of an electronic adjustable-speed drive.
2. What is the preferred method for altering the speed of a squirrel-cage induction motor?
3. State the prime function of each of the following parts of a variable-frequency drive: (a) converter, (b) DC bus, (c) inverter, and (d) control logic.
4. Explain how it is possible for a VFD to operate a three-phase motor from a single-phase power source.
5. Calculate the average DC bus voltage for a line voltage of 230 V AC.
6. Which component is the main energy storage element of the DC bus?
7. What types of transistor are currently used in the inverter section of a VFD?
8. How is the output voltage of an inverter varied?
9. How is the output frequency of the inverter varied?
10. What is the difference between the fundamental frequency and the carrier frequency of a variable-frequency drive?
11. In what ways is an inverter-duty motor different from standard motor types?
12. Compare open- and closed-loop motor control.
13. Explain how the output voltage of a VFD is controlled using the volts per hertz speed-control method.
14. Calculate the V/Hz for a 230 volt, 60 Hz motor.
15. How does a flux vector improve on the basic V/Hz control technique?

PART 2 VFD INSTALLATION AND PROGRAMMING PARAMETERS

Careful planning of your VFD installation will help avoid many problems. Follow the instructions from the VFD manufacturer for required and optional installation requirements. Important considerations include temperature and line power quality requirements, along with electrical connections, grounding, fault protection, motor protection, and environmental parameters.

Selecting the Drive

In selecting a drive, consideration must be given to the load characteristics of the driven machinery. The three basic load categories can be summarized as follows:

- **Constant-torque** loads require a constant motor torque throughout the operational speed range.

Loads of these types are essentially friction loads such as traction drives and conveyors.

- **Variable-torque** loads require much lower torque at low speeds than at high speeds. Loads that exhibit variable torque characteristics include centrifugal fans, pumps, and blowers.
- **Shock (impact)** loads require a motor to operate at normal load conditions followed by a sudden, large load applied to the motor. An example would be the sudden shock load that results from engaging a clutch that applies a large load to the motor (as it would during a hard start). This current spike could cause the VFD to trip as a result an excessive motor current fault.

Line and Load Reactors

A VFD reactor, as shown in Figure 10-16, is basically an inductor installed on the input or output of the drive. **Line reactors** stabilize the current waveform on the input side of a VFD, reducing harmonic distortion and the burden on upstream electrical equipment. Harmonics are high-frequency voltage and current distortions within the power system normally caused by nonlinear loads that do not have a constant current draw, but rather draw current in pulses. VFDs create harmonics when they convert AC into DC and DC back to AC.

By absorbing line spikes and filling some voltage sags, line and load reactors can prevent overvoltage and undervoltage tripping problems. **Load reactors,** connected between the VFD and the motor, will dampen overshoot peak voltage and reduce motor heating and audible noise. A load reactor helps to extend the life of the motor and increase the distance that the motor can be from the drive.

Location

Location is an important consideration in installing VFDs as this can have a significant effect on the drive's performance and reliability. Location considerations are summarized as follows:

- Mount the drive near the motor. Excessive cable length between the VFD and the motor can result in extremely high voltage spikes at the motor leads. It is important to verify the maximum cable distance stated in the drive specifications, when you are installing drives onto AC induction motors. Excessive voltages can reduce the expected life of the insulation system, especially non-inverter-duty motors.
- The enclosure surrounding the drive should be well ventilated or in a climate-controlled environment, as the build up of excess heat may damage the VFD components over time. Large fluctuations in ambient temperatures can result in condensation forming inside the drive enclosures and possibly damaging components.
- Locations in dusty, wet, and corrosive environments, constant vibration, and direct sunlight should be avoided.
- The location should have adequate lighting and sufficient working space to carry out maintenance of the drive. NEC Article 110 lists requirements for working space and illumination.

Enclosures

Once a suitable location is chosen, it is then important to select the appropriate NEMA-type enclosure on the basis of use and service. Factory enclosures of VFDs, such as the one shown in Figure 10-17, should have a NEMA rating appropriate to the level of protection for the environment.

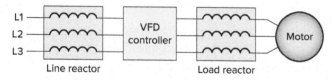

Figure 10-16 VFD reactor.
TCI, LLC

Figure 10-17 VFD mounted within an enclosure.
Photo courtesy of Joliet Technologies, www.joliettech.com

DIN rail and mounting clip

Figure 10-18 VFD mounting technique.
Photo courtesy of Winford Engineering, LLC, www.winford.com

Mounting Techniques

Most small VFDs are mounted in rack slots or on a DIN rail as illustrated in Figure 10-18. The clips for mounting to the DIN rail are usually built into the fins of the heat sink to which the VFD is mounted. This makes them easily installable in control cabinets. Larger VFD units usually have through-hole mounting to accommodate individual fasteners. The fastening method should be adequate to support the weight of the drive and allow the free flow of air across the heat sink; airflow in some applications is aided by a cooling fan.

Operator Interface

The typical VFD operator interface, shown in Figure 10-19, provides a means for an operator to start and stop the motor and adjust the operating speed. Additional operator control functions might include reversing and switching between manual speed adjustment and automatic control from an external process control signal. The operator interface often includes an alphanumeric display and/or indication lights and meters to provide information about the operation of the drive. When mounted within another enclosure, a remote operator keypad and display may be cable-connected and mounted a short distance from the controller.

A communications port is normally available to allow the VFD to be configured, adjusted, monitored, and controlled using a personal computer (PC). PC-based software offers greater flexibility, as more detailed information on the drive parameters can be viewed simultaneously on the monitor. Modes of operation may include program, monitor, and run. Typical data accessible in real time include:

- Frequency output
- Voltage output
- Current output
- Motor rpm
- Motor kilowatts
- DC bus volts
- Parameter settings
- Faults

Electromagnetic Interference

Electromagnetic interference (EMI), also called electrical noise, is the unwanted signals generated by electrical and electronic equipment. EMI-based drive problems range from corrupted data transmission to electric motor drive damage. Modern drives using IGBT switches for motor frequency control are very efficient because of their high switching speed, which unfortunately also results in much higher EMI being generated. All drive manufacturers detail installation procedures that must be followed in order to prevent excessive noise on both sides of the drive. Some of these noise suppression procedures include the following:

- Use a shielded power cable, such as the one shown in Figure 10-20, to connect the VFD to the motor.
- Use a built-in or external EMI filter.
- Use twisted pair control wiring leads to provide a balanced capacitive coupling.
- Use shielded cable to return the noise current flowing in the shield back to the source, instead of through the signal leads.
- Maintain at least 8-inch separation between control and power wires in open air, conduit, or cable trays.
- Use a common-mode choke wound with multiple turns of both signal and shield.
- Use optical isolation modules for control signal communications.

Inherent in all motor cables is line-to-line and line-to-ground capacitance. The longer the cable, the greater is this capacitance. **Electrical spikes** occur on the outputs of PWM drives due to currents charging the cable capacitances. Higher voltages, such as 460 V AC, along with

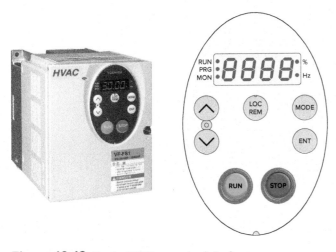

Figure 10-19 Typical VFD operator interface.
Photo courtesy of Toshiba International Corporation, Industrial Division

Figure 10-20 VFD shielded power cable.

higher capacitances, result in larger current spikes. These spikes can shorten the lives of inverters and motors. For this reason cable length must be limited to that recommended by the manufacturer.

Grounding

Figure 10-21 illustrates the general grounding requirements for a VFD. **Proper grounding** plays a key role in the safe and reliable operation of the VFD system. All electric motor drives, motors, and related equipment must meet the grounding and bonding requirements of NEC Article 250. The drive's safety ground must be connected to system ground. Ground impedance must conform to the requirements of the NEC in order to provide equal potential between all metal surfaces and a low-impedance path to activate overcurrent devices and reduce electromagnetic interference.

Bypass Contactor

A bypass contactor is intended for use in case of a drive failure for short-time emergency service. A typical diagram of the power circuit connection of a **VFD bypass contactor** is shown in Figure 10-22. The isolation contactor electrically isolates the drive during bypass operation and is mechanically and electrically interlocked with the bypass contactor to ensure that both cannot be closed at the same time. Upon a sensed malfunction of the VFD, the control circuit automatically opens the drive isolation contactor and closes the bypass contactor to keep the motor connected to the source. When automatic transfer to bypass operation

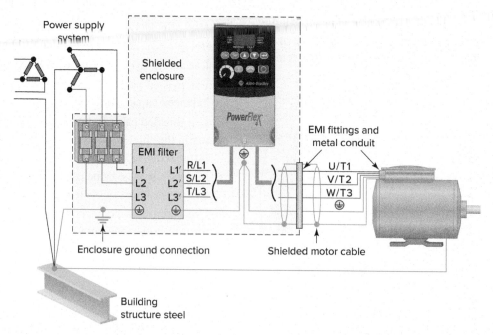

Figure 10-21 General grounding requirements for a variable-frequency drive.
Photo courtesy of Rockwell Automation, www.rockwellautomation.com

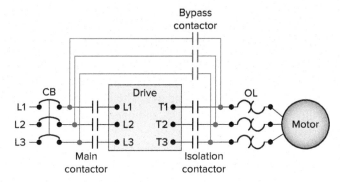

Figure 10-22 Power circuit connection of a VFD bypass contactor.

occurs, the motor continues to operate at full speed. The drive isolation contactor must be opened during closing of the bypass contactor so that AC power is not fed into the output of the VFD, causing damage. The automatic switch to bypass ensures no downtime and no interruption of service to critical loads. For example, in HVAC applications, this allows heating or cooling to be maintained at all times.

Disconnecting Means

Safety in operation and maintenance dictates that all motor-operated equipment has a means of fully disconnecting the power supply. This is a requirement of National Electrical Code (NEC) and Occupational Safety and Health Administration (OSHA) regulations. As with starters, to reduce cost and size, most VFD manufacturers do not provide a disconnect switch as part of their standard drive package. If the optional input disconnect is not specified, a separate switch or circuit breaker must be installed. **Article 430.102** of the NEC includes requirements for disconnecting means for the motor itself and for the motor controller; both sets of requirements must be satisfied. The general rules that apply are as follows:

- Under 1000 V, the controller disconnecting means must be within sight (and less than 50 ft according to definitions) of the motor controller as specified in Article 430.102(A).
- The controller is not required to be in sight from the motor.
- If within site of the motor, the controller disconnecting means shall also be permitted as the motor disconnecting means according to Article 430.102(B).
- The motor disconnecting means shall be in sight of the motor. See exceptions under Article 430.102(B) that would allow motor disconnecting means to be out of sight of the motor. These exceptions, if applicable, would allow one lockable disconnect to serve as both controller and motor disconnect while not in sight of the motor.

Motor Protection

VFDs can operate as motor protection devices along with their role as motor speed controllers. Some VFDs have short-circuit protection (usually in the form of fuses) already installed by the manufacturer, as shown in the VFD package of Figure 10-23. The selection and sizing of these fuses is critical for semiconductor protection in the event of a fault. The manufacturer's recommendations must be followed in installing or replacing fuses for the VFD to assure fast operation of fuses in case of a fault.

In most drive applications, the drive itself provides overload protection for the motor. However, the feeder cable cannot be protected by VFD built-in protection. The motor drive provides protection based on motor nameplate information that is programmed into it. Controllers incorporate many complex protective functions, such as:

- **Stall prevention**
- **Current limitation and overcurrent protection**
- **Short-circuit protection**
- **Undervoltage and overvoltage protection**

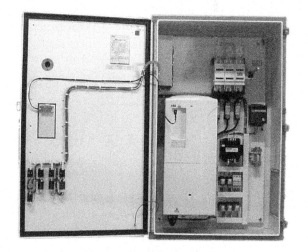

Figure 10-23 Typical VFD package.
Photos courtesy of Joliet Technologies, www.joliettech.com

- Ground fault protection
- Power supply phase failure protection
- Motor thermal protection through sensing of the motor winding temperature

When a VFD is not approved for overload protection, or if multiple motors are fed from the drive, one or more external overload relays must be provided. The most common practice is to use a motor overcurrent relay that will protect all three phases and protect against single phasing.

Braking

With AC motors, there is excessive energy generated when the load drives the motor during deceleration, instead of the motor driving the load. This energy goes back into the drive and will result in an increasing DC bus voltage. If the bus voltage goes too high, the drive will be damaged. Depending on design, a VFD can redirect this excess energy through resistors or back to the AC supply.

When **dynamic braking** or a **chopper** is used, the drive connects a braking resistor across the DC bus, as shown in Figure 10-24, to absorb the excess energy. For smaller-horsepower motors, the resistance is built into the drive. External resistance banks are used for larger-horsepower motors to dissipate the increased heat load.

Regenerative braking is similar to dynamic braking, except the excess energy is redirected back to the AC supply. VFDs designed to use regenerative braking are required to have an active front end to control regenerative current. With this option the diodes in the converter bridge are replaced with IGBT modules. The IGBT modules are switched by the control logic, and operate in both motoring and regenerative modes.

DC-injection braking is a standard feature on a number of variable-frequency drives.

- DC injection braking generates electromagnetic forces in the motor when the controller, in stop mode, injects direct current into the stator windings—after it has cut off alternating current supply to two of the stator phases—thus turning off the normal rotating magnetic field.
- Most DC injection braking systems have the ability to adjust the length of time they will operate and the maximum torque they will apply.
- They generally begin braking when they detect that the motor is no longer receiving its run command and come equipped with hardware to prevent the motor from receiving another run command until the braking is finished.

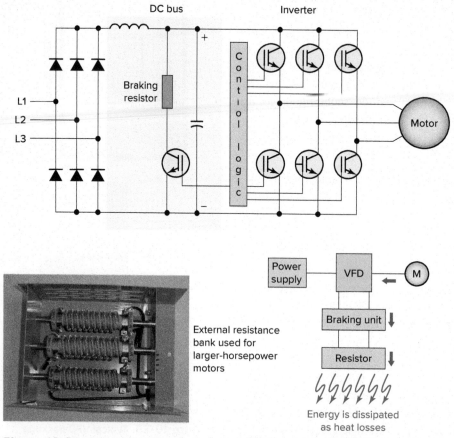

Figure 10-24 Dynamic braking applied to a variable-frequency drive.
Photo courtesy of Transfab TMS, www.transfabtms.com

Ramping

Variable-frequency drives offer many of the same advantages as reduced-voltage and soft start starters. The timed speed ramp-up feature found in VFDs is similar to the soft start function of starters. However, the VFD timed speed ramp-up generally has a much smoother acceleration than soft starting, which is usually done in steps. Soft starting with a VFD reduces the frequency initially supplied to the motor and steps up the frequency over a preset period of time. VFDs with soft start capabilities have replaced many of the older types of reduced-voltage starters. While VFDs offer soft start capabilities, true soft start starters are not considered variable-frequency drives.

Ramping is the ability of a VFD to increase or decrease the voltage and frequency to an AC motor gradually. This accelerates and decelerates the motor smoothly, as shown in Figure 10-25, with less stress on both the motor and the connected load. Ramp-up is generally a smoother acceleration than the stepped increases used in soft starters. The length of time preset for the speed ramp-up can be varied from a few seconds to 120 seconds or more, depending on the drive capabilities.

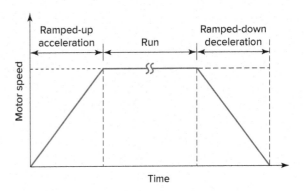

Figure 10-25 Variable-frequency drive ramping functions.

Timed ramp-down is a function of VFDs that provides smooth deceleration, bringing the motor to a full stop in a preset time. Acceleration and deceleration are separately programmable. Depending on drive parameters, ramp-down times can vary from fractions of a second (when used with dynamic braking) to more than 120 seconds. The ramp-down function is applied in processes that require smooth stops, but also require the process to stop within a given period of time.

Control Inputs and Outputs

Figure 10-26 illustrates typical power and control inputs and output termination points found on VFDs. The three-phase

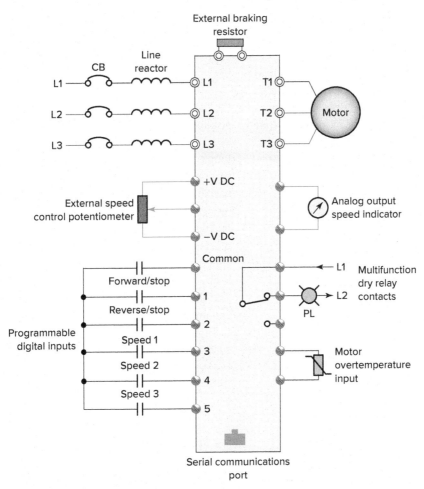

Figure 10-26 Typical VFD control inputs and outputs.

Part 2 VFD Installation and Programming Parameters

power source is connected at the line input terminals L1, L2, and L3 and the motor feed conductors are connected to the output motor terminals T1, T2, and T3. The line and motor terminals pass through electronic circuitry so there is no direct connection between them, as with an across-the-line starter.

Most VFDs contain **control terminal strips** for external connection for both digital and analog inputs and outputs. The number and types of inputs and outputs will vary with the complexity of the drive and serve as a means of comparison between manufacturers of VFDs. Variable frequency drive inputs and outputs are either digital or analog signals. Digital inputs and outputs have two states (either on or off), while analog inputs and outputs have many states that vary across a range of values.

Digital **Digital inputs** are used to interface the drive with devices such as push buttons, selector switches, relay contacts, and PLC digital output modules. Each digital input may have a preset function assigned to it, such as start/stop, forward/reverse, external fault, and preset speed selections. For example, if a motor has to operate at three different speeds, a relay or switch contact could be made to close and send signals to separate digital inputs points that would change the motor speed to the preset value.

Figure 10-27 shows typical digital input connections for two-wire or three-wire control with stop, forward, reverse, and jogging functions. Because VFDs are electronic devices, they can have only one phase rotation output at a time. Therefore, **interlocking,** as required on electromechanical devices, is not required for VFD forward/reverse operations. Inputs can also be programmed for two- or three-wire control to accommodate either maintained or momentary start

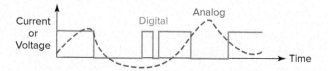

Figure 10-28 Analog and digital input signals.

methods. Note that the control logic is determined and executed by the program within the drive and not by the hard-wiring arrangement of the input control devices.

Digital/relay outputs. Digital/relay outputs are two-position signals (on/off) sent by the VFD to devices such as pilot lights, alarms, auxiliary relays, solenoids, and PLC digital input modules. Digital outputs have a voltage potential (e.g., 24 V DC) coming from them. Relay outputs, which are known as "dry" contacts, switch something external, closing or opening another potential. Relay outputs are normally rated for both AC and DC voltages.

Analog **Analog inputs** are signals of varying amplitude as opposed to a discrete digital signal where each bit is represented by two distinct amplitudes (Figure 10-28). The analog signal takes the form of a current or voltage the magnitude of which is proportional to the value of the measured parameter. Some examples of analog signals are:

- Measurement of temperature and pressure.
- Speed setpoint from a speed control potentiometer.
- Fluid flow rate.

The two most common types of analog input signals are voltage and current. Analog control signals are usually 0–10 V for voltage and 4–20 milliamps (mA) for current loops. This transmitted signal from instruments and sensors in the field represents 0 to 100 percent of a process variable. The **4–20 mA loop** is most commonly used in process control applications. The reason zero is at four milliamps and not zero milliamps is that this allows the receiving instrumentation to differentiate between a zero signal and a broken wire or a dead instrument. This makes it more immune to noise and more reliable over distance. The **0–10 V** systems are often used heating, ventilation, and air conditioning (HVAC) applications.

Figure 10-29 shows the components of a 4–20 mA current loop:

- **Sensor** measures the value of the process variable. The makeup of the sensor depends on what exactly it is intended to measure. Typical measurements include temperature, flow, level, and pressure.
- **Transmitter** takes the senor measurement value and converts it into a meaningful 4–20 mA value. For example, if a level sensor was used to measure

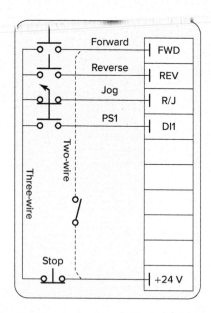

Figure 10-27 Input digital connections for two-wire or three-wire control with stop, forward, reverse, and jogging functions.

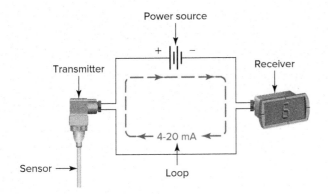

Figure 10-29 Components of a 4–20 mA current loop.

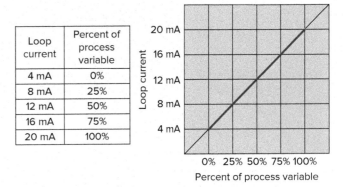

Figure 10-30 Relationship between a 4–20 mA loop input and the resulting percentage of the process range output.

the height of liquid in a 30-foot tank, the transmitter would need to translate an empty-tank as a 4 mA signal, a half-tank as a 12 mA signal, and a full-tank as a 12 mA signal.

- **Power source** provides the power that powers the transmitter. The power supply must output a DC current. Commonly used DC voltages are 9 V, 12 V, and 24 V. Power supply voltage must be at least 10 percent greater than the total voltage drop of the attached components and wiring.
- **Loop** refers to the conductors connecting the sensor to the device receiving the 4–20 mA signal and then back to the transmitter. The current signal on the loop is regulated by the transmitter according to the sensor's measurement. The conductors themselves are a source of resistance that causes a small voltage drop on the system. This it is normally not a problem for a loop that is less than 1,000 feet in length.
- **Receiver** interprets and reacts to the received 4–20 mA signal. The receiver signal is than translated into units that can be easily understood, such as the feet of liquid in a tank or the temperature degrees of a liquid. Digital displays, controllers, actuators, and valves are common devices to incorporate into the loop.

Analog outputs are similar to analog inputs. They are used when you need to control a device that involves a range of values. Like analog inputs, the analog outputs can be a voltage or current signal. Figure 10-30 shows the linear relationship between a 4–20 mA loop input and the resulting percentage of the process span output. Note that for every 4 mA change in input signal there is a corresponding 25 percent change in output. The current in the loop represents the corresponding analog output value. The 4–20 mA signal will vary in an analog fashion as the process changes so that by monitoring the current signal value, the process value will be known.

Verifying a 4–20 mA loop is important in both troubleshooting and calibrating process systems. Calculations of loop current to the measured process value are shown in following examples.

EXAMPLE 10-1

Problem: A 4–20 mA transmitter has an input range of 0–150°C. If the loop current is 8 mA, what temperature in °C should be indicated?

Solution:

$$\text{Temperature} = \frac{\text{Process value high} - \text{Process value low}}{\text{Loop mA value high} - \text{Loop mA value low}}$$
$$\times (\text{Loop current} - \text{Loop mA value low})$$
$$+ \text{Process value low}$$

$$= \frac{150 - 0}{20 - 4} \times (8 - 4) + 0$$

$$= 37.5°C$$

EXAMPLE 10-2

Problem: A 4–20 mA transmitter has an input range of 50–400°F. If the loop current is 8 mA, what temperature in °F should be indicated?

Solution:

$$\text{Temperature} = \frac{\text{Process value high} - \text{Process value low}}{\text{Loop mA value high} - \text{Loop mA value low}}$$
$$\times (\text{Loop current} - \text{Loop mA value low})$$
$$+ \text{Process value low}$$

$$= \frac{400 - 50}{20 - 4} \times (16 - 4) + 50$$

$$= 312.5°F$$

EXAMPLE 10-3

Problem: An electronic temperature transmitter is ranged from 40 to 140°F and has a 4–20 mA output signal. Calculate the current output by this transmitter if the measured temperature is 60°F.

Solution:

Loop current = $\dfrac{\text{Loop mA value high} - \text{loop mA value low}}{\text{Process value high} - \text{Process value low}}$
$\times$ (Process value − Process value low)
+ loop mA value low

$= \dfrac{20 - 4}{140 - 40} \times (60 - 40) + 4$

$= 7.2$ mA

Motor Nameplate Data

Motor specifications are programmed into the VFD to ensure optimum drive performance as well as adequate fault and overload protection. This may include the following items found on the nameplate, as illustrated in Figure 10-31, or derived through measurements:

- **Frequency (hertz)**—Nameplate frequency required by the motor to achieve base speed. The default value is normally 60 Hz.
- **Speed (rpm)**—Nameplate maximum speed at which the motor should be rotated.
- **Full-load current (amperes)**—Nameplate maximum current that the motor may use.
- **Supply voltage (volts)**—Nameplate voltage required by the motor to achieve maximum torque.
- **Power rating (horsepower or kilowatts)**—Nameplate rating of motors manufactured in the United States are generally rated in horsepower (hp). Equipment manufactured in Europe is generally rated in kilowatts (kW). Horsepower can be converted as follows: 1 hp = 0.746 kW.
- **Motor magnetizing current (amperes)**—Current that the motor draws when operating with no load at nameplate rated voltage and frequency. If not specified, it can be measured using a true-RMS clamp-on ammeter.
- **Motor stator resistance (ohms)**—DC resistance of the stator between any two phases. If not specified, it can be measured with an ohmmeter.

Derating

Derating a VFD means using a larger than normal drive in the application. Derating is required when a drive is to be operated outside the normal operating range specified by the manufacturer. Most manufacturers provide derating figures when the drive is to be operated outside the specified temperature, voltage, and altitude. As an example, derating must be considered when the drive is installed at high altitude, greater than 1,000 meters (3,300 ft.). The cooling effect of the drive is deteriorated because of the reduced density of the air at high altitudes.

Induction motor adjustable speed duty

MODEL	5KAF			SERIAL NO.			
POWER	400HP			TYPE	KAF	ENCLOSURE	TEAO
RPM BASE	1200			FRAME	6811		
AMPERES	368	SINE		DRIVE END BRG	SKF 6319		
VOLTS	575	V (MAX.)	575	LUBRICATION	GREASE		
PHASE	3	HERTZ	60	LUBRICANT			
EFF.	0.9567	PF.	0.85	OPP. DRIVE END BRG	SKF 6319		
SERVICE FACTOR	1.15	SINE 1.0	ASD	LUBRICATION	GREASE		
INSULATION CLASS	F			LUBRICANT			
MAX. TEMP. RISE	80	°C AT SF$_{BY}^{STATOR}$		OIL PRESSURE	TO		PSI
TIME RATING	CONTINUOUS			OIL FLOW	TO		GPM/BRG
INVERTER TYPE	IGBT-PWM			AMB. TEMP. (°C)	40	MAX	MIN
ALTITUDE	1000	(M)		MANUFACTURING DATE:			
SUITABLE FOR OPERATION FROM	120	TO	1200	RPM AT CONSTANT TORQUE			
SUITABLE FOR OPERATION FROM	1200	TO	1800	RPM AT CONSTANT POWER			
SUITABLE FOR OPERATION FROM		TO		RPM AT VARIABLE TORQUE			
CONSTANT VOLTS / HERTZ TO 1200 RPM							

Figure 10-31 Entering motor nameplate data.

Types of Variable-Frequency Drives

The evolution of AC drive technology has seen many changes in a relatively small time frame. As a result, newer drives with greater functionality are now available. Most variable-frequency drives manufactured today are pulse-width modulation drives that convert the 60 Hz line power into direct current, then pulse the output voltage for varying lengths of time to mimic an alternating current at the frequency desired. Many older VFDs were distinguished by the type of inverter circuitry used in the drive. Two earlier types of drives were the *voltage-source inverter* and the *current-source inverter*.

Figure 10-32 shows a simplified circuit of one type of **voltage-source inverter (VSI),** also called variable-voltage inverter (VVI).

- This inverter uses a **silicon-controlled rectifier (SCR)** converter bridge to convert the incoming AC voltage into DC.
- The SCRs provide a means of controlling the value of the rectified DC voltage.
- The energy storage in the DC link between the converter and the inverter consists of capacitors.
- The inverter section utilizes six SCR switches.
- The control logic (not shown) uses a microprocessor to switch the SCRs on and off, providing a variable voltage and frequency to the motor.
- This type of switching is often referred to as six-step because it takes six 60-degree steps to complete one 360-degree cycle.
- Although the motor prefers a smooth sine wave, a six-step output can be satisfactorily used.
- The main disadvantage is torque pulsation, which occurs each time an SCR is switched.
- The pulsations can be noticeable at low speeds as speed variations in the motor.
- These speed variations are sometimes referred to as **cogging.**
- The nonsinusoidal current waveform causes extra heating in the motor, requiring a motor derating.
- Voltage-source inverter drives can operate any number of motors up to the total rated horsepower of the drive.

With a **current-source inverter,** the DC power supply is configured as a current source rather than a voltage source.

- These drives employ a **closed-loop** system that monitors the actual speed of the motor and compares it with the preset reference speed, creating an error signal that is used to increase or decrease the current to the motor.
- Figure 10-33 shows the simplified circuit of a current-source inverter (CSI).
- The converter is connected to the inverter through a large series inductor.
- This inductor opposes any change in current and is of a sufficiently high inductance value that direct current is constrained to be almost constant.
- As a result, the output produced is almost a square wave of current.
- Current-source inverters are used for large drives—about 200 hp—because of their simplicity, regeneration braking capabilities, reliability, and lower cost.
- Because current-source inverters monitor the actual motor speed, they can be used to control only a single corresponding motor with characteristics that match the drive.

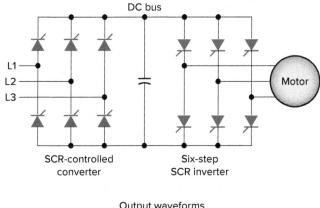
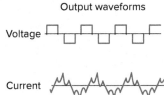

Figure 10-32 Simplified circuit of a voltage-source inverter (VSI).

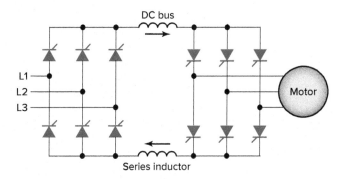

Figure 10-33 Simplified circuit of a current-source inverter (CSI).

PID Control

Most VFD applications require the AC motor to run at a specific speed as set by the keypad, speed potentiometer, or analog input. Some drives provide an alternative option that allows precise process control through a setpoint controller or PID mode of operation. Many variable-frequency drives come equipped with a built-in **proportional-integral-derivative (PID)** controller. The PID loop is used to maintain a process variable, such as speed, as illustrated in Figure 10-34.

- The desired speed, or setpoint, and the actual speed values are input to a summation point.
- These two signals are opposite in polarity and yield a zero error or deviation whenever the desired speed equals the actual speed.
- If the two signals differ in value, the error signal will have a positive or negative value, depending on whether the actual speed is greater or less than the desired speed. This error signal is input to the PID controller.
- The terms *proportional, integral,* and *derivative* describe three basic mathematical functions then applied to the error signal.
- The PID output reacts to the error and outputs a frequency to try to reduce the error value to zero.
- The controller's job is to make the speed adjustments quickly, with a minimum of overshoot or oscillations.
- Tuning the PID controller involves gain and time adjustments designed to improve performance and result in a fast response with a minimum overshoot, allowing the motor to settle in quickly to the new speed.
- Some drives have a PID autotune function designed to ease the tuning process.

Parameter Programming

The main drive program is contained in the processor's firmware and not normally accessible to the VFD user. A **parameter** is a variable associated with the operation of the drive that can be programmed or adjusted. Parameters provide a degree of adjustment so that the user can customize the drive to suit specific motor and driven equipment requirements. The number of parameters can range from 50, for small basic drives, to over 200 for larger, more complex drives. Some VFDs provide upload/download and parameter copy capability. Common adjustable parameters include the following:

- **Preset speeds**
- **Minimum and maximum speeds**
- **Acceleration and deceleration rates**
- **Two- and three-wire remote control modes**
- **Stop modes: ramp, coast, DC injection**
- **Automatic torque boost**
- **Current limit**
- **Configurable input jog**
- **V/Hz settings**
- **Carrier frequency**
- **Program password**

VFDs come with factory default settings for most parameters that are more conservative in nature. The default value settings simplify the start-up procedure. However, parameters for motor nameplate data are not factory-set (unless a matched drive and motor has been purchased) and must be entered in the field. In general there are three types of parameters:

- **Tunable on the fly**—Parameters can be adjusted or changed while the drive is running or stopped.
- **Configurable**—Parameters can be adjusted or changed only while the drive is stopped.
- **Read only**—Parameters cannot be adjusted.

Figure 10-35 shows an integral keypad with an LED display used to program and operate a small drive locally. The display shows either a parameter number or a parameter value. The drive's parameter menu outlines what the parameter number represents and what numerical selections or options for the parameter are available. Parameter menu formats vary between make and model. This drive has two kinds of parameters: program parameters (P-00 through P-64), which configure the drive operation,

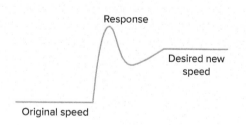

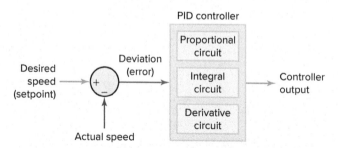

Figure 10-34 PID loop.

Figure 10-35 Integral keypad with an LED display used to program and operate a small drive locally.
Photo courtesy of Rockwell Automation, www.rockwellautomation.com

and display parameters (d-00 through d-64), which display information. Examples of program parameters include:

P-00 minimum speed—Use this parameter to set the lowest frequency the drive will output. Default setting is 0.

P-01 maximum speed—Use this parameter to set the highest frequency the drive will output. Default setting is 60 Hz.

P-02 motor overload current—Set this parameter to the motor nameplate full-load ampere rating. Default setting is 100 percent of the rated drive current.

P-30 acceleration time—Use this parameter to define the time it will take the drive to ramp up from 0 Hz to maximum speed. Default setting is 5.0 seconds.

Examples of display parameters include:

d-00 command frequency—This parameter represents the frequency that the drive is commanded to output.

d-01 output frequency—This parameter represents the output frequency at the motor terminals.

d-02 output current—This parameter represents the motor current.

d-03 bus voltage—This parameter represents the DC bus voltage level.

Diagnostics and Troubleshooting

Most VFDs come equipped with self-diagnostic controls to help trace the source of problems. Always observe the following precautions when troubleshooting the drive:

- Stop the drive.
- Disconnect, tag, and lock out AC power before working on the drive.
- Verify that there is no voltage present at the AC input power terminals (Figure 10-36). It's important to

Figure 10-36 Taking measurements on a VFD.
Photo courtesy of Fluke, www.fluke.com. Reproduced with permission.

remember that DC bus capacitors retain hazardous voltages after input power has been disconnected. Therefore, wait 5 minutes for the DC bus capacitors to discharge once power has been disconnected. Check the voltage with a voltmeter to ensure that the capacitors have discharged before touching any internal components.

Problem indicators may include the following:

LEDs provide a quick indication of problems. Normally, a steady glowing light means everything is running properly. Flashing yellow or red lights indicate a problem with the drive that should be checked. Consult the operator's manual for the specific drive to determine what a particular flashing light means.

Alarms indicate conditions that may affect drive operation or application performance. They are cleared automatically when the condition that caused the alarm is no longer present. **Configurable** alarms alert the operator to conditions that, if left untreated, may lead to a drive fault. The drive continues to operate during the alarm condition, and the alarms can be enabled or disabled by the programmer or operator. **Nonconfigurable** alarms alert the operator of conditions caused by improper programming and prevent the drive from starting until the problem is resolved. These alarms can never be disabled.

Fault parameters settings indicate conditions within the drive that require immediate attention. The drive responds to a fault by coasting to a stop and turning off output power to the motor. Autoreset faults reset automatically if, after a preset time, the condition that caused the fault is no longer present. The drive then restarts. Nonresettable faults may require drive or motor repair; the fault must be corrected before it can be cleared. User-configurable faults can be enabled and disabled to enunciate or ignore a fault condition.

Fault queues normally retain a history of faults. Typically, queues hold only a limited number of entries; therefore, when the queue is full, older faults are discarded when new faults occur. The system typically assigns a timestamp to the fault so that programmers or operators can determine when a fault occurred relative to the last drive power-up.

A complete listing of all the different types of faults and the appropriate corrective actions can typically be found in the operator's manual for a specific drive. The following are examples of typical fault codes and corrective action:

Display Code	Fault Description	Fault Cause	Corrective Action
CF	Undervoltage	Low input line. Temporary loss of input line.	Check input line to verify voltage is within operating specifications.
OL	Motor overload	Excessive driven load.	Reduce the load.
J1, J2, or J3	Ground short	Phase A, B, or C.	Verify output wiring is correct. Verify output phase is not grounded. Verify motor is not damaged.
OH	Overtemperature	Operating environment is too hot. Fan is blocked or not operating. Excessive driven load.	Verify the ambient temperature is less than 50°C. Verify clearance above and below drive. Check for fan obstruction. Reduce the carrier frequency. Reduce the load.

Part 2 Review Questions

1. Compare the torque characteristics of constant, variable, and shock loads.
2. Explain the function of reactors that are connected in series with the line and load side of a VFD.
3. List the factors that need to be taken into account in selecting the location for an electric motor drive.
4. What is the purpose of a VFD enclosure?
5. What is the function of a drive's operator interface?
6. How are the effects of electromagnetic interference minimized in a VFD installation?
7. In addition to line losses, why must cable lengths between a PWM drive and a motor be kept to a minimum?
8. Why is proper grounding required for the safe and reliable operation of a VFD system?
9. A bypass contactor working in conjunction with an isolation contactor is utilized in certain VFD installations. What purpose is served by this combination of contactors, and how do they work together to achieve this?
10. Outline the basic code requirement for a drive's controller and motor disconnecting means.
11. Summarize the types of built-in protective functions that can be programmed into a variable-speed drive.
12. Compare the operation of dynamic, regenerative, and DC-injection VFD braking systems.
13. Ramped acceleration and deceleration are two key features of VFDs. Give a brief explanation of how each is achieved.
14. Compare digital and analog input and output drive control signals.
15. List typical preset functions that digital inputs may have assigned to them.
16. Why is forward/reverse interlocking not required on VFDs?
17. Compare the manner in which the control logic is executed in a hard-wired electromagnetic starter and a VFD.
18. Explain the term *dry contact* as it relates to a drive's relay output.

19. List typical motor nameplate data that may be required to be programmed into the VFD.
20. Explain the term *derating* as it applies to a VFD and list factors that might require a drive to be derated.
21. Compare PWM, VVI, and CSI variable-frequency drives.
22. Outline how setpoint speed is maintained in a PID control loop.
23. Explain the term *parameter* as it applies to the VFD.
24. List several common adjustable parameters associated with AC drives.
25. What is the difference between program and display parameters?
26. What potential safety hazard is associated with DC bus capacitors?
27. When LEDs are used for problem indicators, what does a steady glow indicate compared to a flashing one?
28. Compare the operation of autoreset, nonresettable, and user-configurable fault parameters.
29. Explain the term *fault queues* as it applies to VFDs.
30. Name the two main types of analog input control signals?
31. A 4–20 mA input signal loop represents 0 to 100 percent of the process variable. What is the reason zero percent of the process variable is represented by four milliamps instead of zero milliamps?
32. State the basic function of each of the following components of a 4–20 mA current loop.
    a. Sensor
    b. Transmitter
    c. Power source
    d. Loop
    e. Receiver
33. A 4–20 mA pressure transmitter has a range of 0 to 100 psi. If loop current is 12 mA, what value of psi pressure should be indicated?
34. A pressure sensor transmitter is ranged from 50 to 400 psi and has a 4–20 mA output signal. Calculate what the milliamp current output of the transmitter should be for a pressure reading of 250 psi.

PART 3 DC MOTOR DRIVE FUNDAMENTALS

Applications

DC drive technology is the oldest form of electrical speed control. The speed of a DC motor is the simplest to control, and it can be varied over a very wide range. These drives are designed to handle applications such as:

Winders/coilers—In motor winder operations, maintaining tension is very important. DC motors are able to operate at rated current over a wide speed range, including low speeds.

Crane/hoist—DC drives offer several advantages in applications that operate at low speeds, such as cranes and hoists. Advantages include low-speed accuracy, short-time overload capacity, size, and torque providing control. Figure 10-37 shows a typical DC hoist motor and drive used on hoisting applications where an overhauling load is present. Generated power from the DC motor is used for braking, and excess power is fed back into the AC line. This power helps reduce energy requirements and eliminates the need for heat-producing dynamic braking resistors. Peak current of at least 250 percent is available for short-term loads.

Mining/drilling—The DC motor drive is often preferred in the high-horsepower applications required in the mining and drilling industry. For this type of application, DC drives offer advantages in size and cost (Figure 10-38). They are rugged, dependable, and industry-proven.

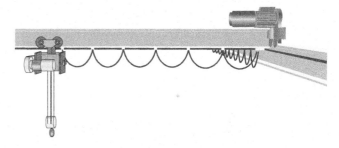

Figure 10-37 Typical DC hoist motor and drive.

DC Drives—Principles of Operation

DC electronic variable-speed drives vary the speed of DC motors with greater efficiency and speed regulation than resistor control circuits. Since the speed of a DC motor is directly proportional to armature voltage and inversely proportional to field current, either armature voltage or field current can be used to control speed. To change the direction of rotation of a DC motor, either the armature polarity can be reversed (Figure 10-39) or the field polarity can be reversed.

The block diagram of a DC drive system made up of a DC motor and an electronic drive controller is shown in Figure 10-40.

Figure 10-38 Electric mining shovel—DC drive.
Courtesy of Joy Global

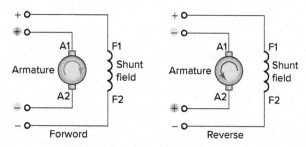

Figure 10-39 Changing direction by reversing the polarity of the armature.

- The shunt motor is constructed with armature and field windings.
- A common classification of DC motors is by the type of field excitation winding.
- Shunt-wound DC motors are the most commonly used type for adjustable-speed control. In most instances the shunt field winding is excited, as shown, with a constant-level voltage from the controller.
- The SCR (silicon controller rectifier), also known as a thyristor, of the power conversion section converts the fixed-voltage alternating current (AC) of the power source into an adjustable-voltage, controlled, direct current (DC) output, which is applied to the armature of a DC motor.
- Speed control is achieved by regulating the armature voltage to the motor.
- Motor speed is directly proportional to the voltage applied to the armature.

The main function of a DC drive is to convert the fixed applied AC voltage into a variable rectified DC voltage. SCR switching semiconductors provide a convenient method of accomplishing this.

- They provide a controllable power output by phase angle control.
- The firing angle, or point in time where the SCR is triggered into conduction, is synchronized with the phase rotation of the AC power source, as illustrated in Figure 10-41.
- The amount of rectified DC voltage is controlled by timing the input pulse current to the gate.
- Applying gate current near the beginning of the sine-wave cycle results in a higher average voltage applied to the motor armature.
- Gate current applied later in the cycle results in a lower average DC output voltage.
- The effect is similar to a very high speed switch, capable of being turned on and off at an infinite number of points within each half-cycle.
- This occurs at a rate of 60 times a second on a 60 Hz line, to deliver a precise amount of power to the motor.

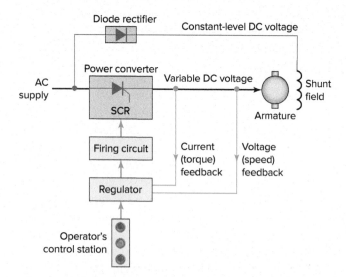

Figure 10-40 Block diagram of a DC drive system.

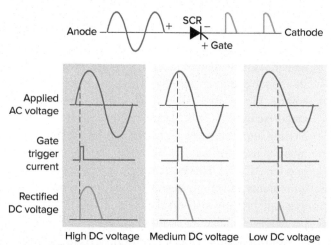

Figure 10-41 SCR conversion from AC to variable DC.

298   **Chapter 10** Adjustable-Speed Drives and PLC Installations

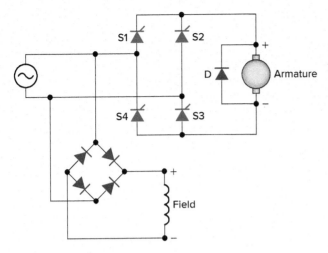

Figure 10-42 Fully controlled SCR bridge rectifier powered by a single-phase AC supply.

Single-Phase Input—DC Drive

Armature voltage-controlled DC drives are constant-torque drives, capable of rated motor torque at any speed up to rated motor base speed. Fully controlled rectifier circuits are built with SCRs. Figure 10-42 shows a fully controlled SCR bridge rectifier powered by a single-phase AC supply.

- The SCRs rectify the supply voltage (changing the voltage from AC to DC) as well as controlling the output DC voltage level.
- In this circuit, silicon-controlled rectifiers S1 and S3 are triggered into conduction on the positive half of the input waveform and S2 and S4 on the negative half.
- Freewheeling diode D (also called a suppressor diode) is connected across the armature to provide a path for release of energy stored in the armature when the applied voltage drops to zero.
- A separate diode bridge rectifier is used to convert the alternating current into a constant direct current required for the field circuit.

Single-phase controlled bridge rectifiers are commonly used in the smaller-horsepower DC drives such as the one shown in Figure 10-43. The terminal diagram shows the

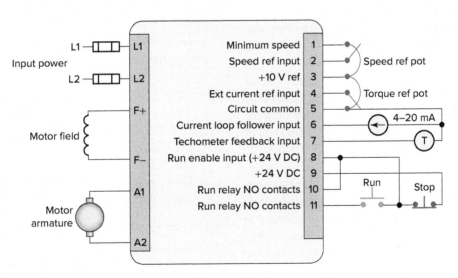

Figure 10-43 Smaller-horsepower-rated DC drive.
Photo Courtesy of Emerson Industrial Automation, Control Techniques Americas LLC, www.controltechniques.com

Part 3 DC Motor Drive Fundamentals

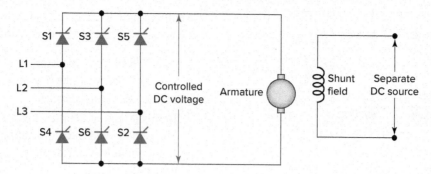

Figure 10-44 Three-phase fully controlled rectifier.

input and output power and control terminations available for use with the drive. Features include:

- **Speed or torque control**
- **Tachometer input**
- **Fused input**
- **Speed or current monitoring (0–10 V DC or 4–20 mA)**

Three-Phase Input—DC Drive

Controlled bridge rectifiers are not limited to single-phase designs. In most commercial and industrial control systems, AC power is available in three-phase form for maximum horsepower and efficiency. Typically six SCRs are connected together, as shown in Figure 10-44, to make a **three-phase fully controlled rectifier.** This three-phase bridge rectifier circuit has three legs, each phase connected to one of the three phase voltages. It can be seen that the bridge circuit has two halves, the positive half consisting of the SCRs S1, S3, and S5 and the negative half consisting of the SCRs S2, S4, and S6. At any time when there is current flow, one SCR from each half conducts.

The variable DC output voltage from the rectifier supplies voltage to the motor armature in order to run it at the desired speed. The gate firing angle of the SCRs in the bridge rectifier, along with the maximum positive and negative values of the AC sine wave, determines the value of the motor armature voltage. The motor draws current from the three-phase AC power source in proportion to the amount of mechanical load applied to the motor shaft. Unlike AC drives, bypassing the drive to run the motor is not possible.

Larger-horsepower three-phase drive panels often consist of a power module mounted on a chassis with line fuses and disconnect. This design simplifies mounting and makes connecting power cables easier as well. Figure 10-45 shows a three-phase input DC drive with the following drive power specifications:

- Nominal line voltage for three-phase—230/460 V AC
- Voltage variation—+15%, –10% of nominal

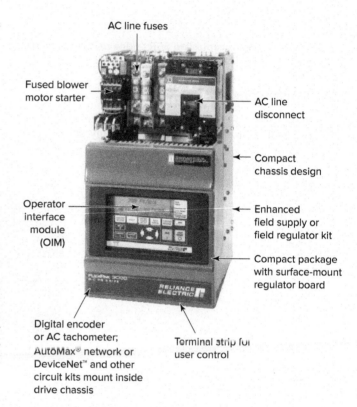

Figure 10-45 Three-phase input DC drive.
Photo courtesy of Rockwell Automation, www.rockwellautomation.com

- Nominal line frequency—50 or 60 cycles per second
- DC voltage rating 230 V AC line: Armature voltage 240 V DC; field voltage 150 V DC
- DC voltage rating 460 V AC line: Armature voltage 500 V DC; field voltage 300 V DC

Field Voltage Control

To control the speed of a DC motor below its base speed, the voltage applied to the armature of the motor is varied while the field voltage is held at its nominal value. To control the speed above its base speed, the armature is supplied with its rated voltage and the field is weakened. For this reason, an additional variable-voltage field regulator, as illustrated in Figure 10-46 is needed for DC drives with field voltage control.

- Field weakening is the act of reducing the current applied to a DC motor shunt field.
- This action weakens the strength of the magnetic field and thereby increases the motor speed.
- The weakened field reduces the counter EMF generated in the armature; therefore, the armature current and the speed increase.
- **Field loss detection must be provided for all DC drives to protect against excessive motor speed due to loss of motor field current.**

DC drives with motor field control provide coordinated automatic armature and field voltage control for extended speed range and constant-horsepower applications. The motor is armature-voltage-controlled for constant-torque, variable-horsepower operation to base speed, where it is transferred to field control for constant-horsepower, variable-torque operation to motor maximum speed.

Nonregenerative and Regenerative DC Drives

Nonregenerative DC drives, also known as single-quadrant drives, rotate in one direction only and they have no inherent braking capabilities.

- Stopping the motor is done by removing voltage and allowing the motor to coast to a stop.
- Typically, nonregenerative drives operate high-friction loads such as mixers, where the load exerts a strong natural brake.
- In applications where supplemental quick braking and/or motor reversing is required, dynamic braking

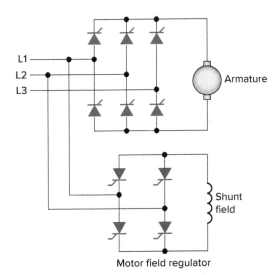

Figure 10-46 DC drive motor field regulator.

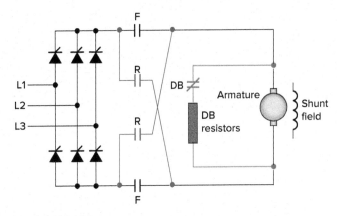

Figure 10-47 Nonregenerative DC drive with external dynamic braking and reversing contactors.

and forward and reverse circuitry, such as shown in Figure 10-47, may be provided by external means.

- Dynamic braking (DB) requires the addition of a DB contactor and DB resistors that dissipate the braking energy as heat.
- The addition of an electromechanical (magnetic) reversing contactor or manual switch permits the reversing of the controller polarity and, therefore, the direction of rotation of the motor armature.
- Field contactor reverse kits can also be installed to provide bidirectional rotation by reversing the polarity of the shunt field.

All DC motors are DC generators as well. The term *regenerative* describes the ability of the drive under braking conditions to convert the generated energy of the motor into electrical energy, which is returned (or regenerated) to the AC power source. Regenerative drives are also known as **four-quadrant drives.** They are capable of controlling not only the speed and direction of motor rotation, but also the direction of motor torque, as illustrated in Figure 10-48.

- **Quadrant I**—Drive delivers forward torque, motor rotating forward (motoring mode of operation). This is the normal condition, providing power to a load similar to that of a motor starter.

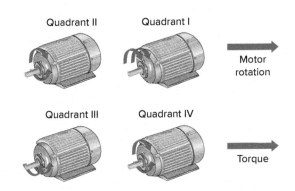

Figure 10-48 Regenerative four-quadrant drive control.

- **Quadrant II**—Drive delivers reverse torque, motor rotating forward (generating mode of operation). This is a regenerative condition, where the drive itself is absorbing power from a load, such as an overhauling load or deceleration.
- **Quadrant III**—Drive delivers reverse torque, motor rotating reverse (motoring mode of operation). Basically the same as in quadrant I and similar to a reversing starter.
- **Quadrant IV**—Drive delivers forward torque with motor rotating in reverse (generating mode of operation). This is the other regenerative condition, where again, the drive is absorbing power from the load in order to bring the motor toward zero speed.

A single-quadrant nonregenerative DC drive has one power bridge with six SCRs used to control the applied voltage level to the motor armature. The nonregenerative drive can run in only motoring mode and would require physically switching armature or field leads to reverse the torque direction.

A four-quadrant regenerative DC drive will have two complete sets of power bridges, with 12 controlled SCRs connected in inverse parallel as illustrated in Figure 10-49. One bridge controls forward torque, and the other controls reverse torque. During operation, only one set of bridges is active at a time. For straight motoring in the forward direction, the forward bridge would be in control of the power to the motor. For straight motoring in the reverse direction, the reverse bridge is in control.

Cranes and hoists use DC regenerative drives to hold back **overhauling loads** such as a raised weight (Figure 10-50) or a machine's flywheel. Whenever the inertia of the motor load is greater than the motor rotor inertia, the load will be driving the motor and is called an overhauling load. Overhauling load results in generator action within the motor, which will cause the motor to send current into the drive. Regenerative braking is summarized as follows:

- During normal forward operation, the forward bridge acts as a rectifier, supplying power to the motor. During this period, gate pulses are withheld from reverse bridge so that it is inactive.
- When motor speed is reduced, the control circuit withholds the pulses to the forward bridge and simultaneously applies pulses to reverse bridge. During this period, the motor acts as a generator, and the reverse bridge conducts current through the armature in the reverse direction back to the AC line. This current reverses the torque, and the motor speed decreases rapidly.

Figure 10-50 Motor overhauling load.

Both regeneration and dynamic braking slow down a rotating DC motor and its load. However, there are significant differences in stopping time and controllability during stopping, and safety issues depending on how one defines what should happen under emergency conditions. Regenerative braking will stop the load smoothly and faster than a dynamic brake for fast stop or emergency stop requirements. In addition, regenerative braking will regenerate power to the supply if the load is overhauling.

Parameter Programming

Programming parameters associated with DC drives are extensive and similar to those used in conjunction with AC drives. An operator's panel is used for programming of control setup and operating parameters for a DC drive (Figure 10-51).

Speed Setpoint This signal is derived from a closely regulated fixed-voltage source applied to a potentiometer, as illustrated in Figure 10-52. The potentiometer has the capability of accepting the fixed voltage and dividing it down to any value, for example, 10 to 0 V, depending on where it is set. A 10 V input to the drive from the speed potentiometer corresponds to maximum motor speed and

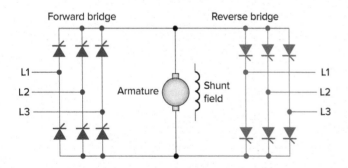

Figure 10-49 Four-quadrant regenerative DC drive.

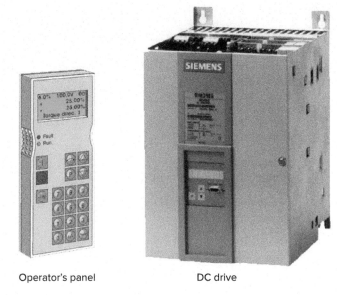

Operator's panel   DC drive

Figure 10-51 Operator's panel used for programming of control setup and operating parameters for a DC drive.
Photos courtesy of Siemens, www.siemens.com

0 V corresponds to zero speed. Similarly, any speed between zero and the maximum can be obtained by adjusting the speed control to the appropriate setting.

Speed Feedback Information In order to "close the loop" and control motor speed accurately, it is necessary to provide the control with a feedback signal related to motor speed. The standard method of doing this in a simple control is by monitoring the armature voltage and feeding it back into the drive for comparison with the input setpoint signal. The armature voltage feedback system is generally known as a **voltage-regulated drive.**

A second and more accurate method of obtaining the motor speed feedback information is from a motor-mounted tachometer. The output of this tachometer is directly related to the speed of the motor. When tachometer feedback is used, the drive is referred to as a **speed-regulated drive.**

In some newer high-performance digital drives, the feedback can come from a motor-mounted encoder that feeds back voltage pulses at a rate related to motor speed.

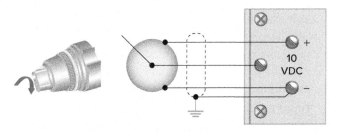

Figure 10-52 Speed potentiometer.

These pulses are counted and processed digitally and compared to the setpoint, and an error signal is produced to regulate the armature voltage and speed.

Current Feedback Information The second source of feedback information is obtained by monitoring the motor armature current. This is an accurate indication of the torque required by the load. The current feedback signal is used to eliminate the speed droop that normally would occur with increased torque load on the motor and to limit the current to a value that will protect the power semiconductors from damage. The current-limiting action of most controls is adjustable and is usually called **current limit or torque limit.**

Minimum Speed In most cases, when the controller is initially installed, the speed potentiometer can be turned down to its lowest point and the output voltage from the controller will go to zero, causing the motor to stop. There are, however, situations where this is not desirable. For example, there are some applications that may need to be kept running at a minimum speed and accelerated up to operating speed as necessary. The typical minimum speed adjustment is from 0 to 30 percent of motor base speed.

Maximum Speed The maximum speed adjustment sets the maximum speed attainable. In some cases, it's desirable to limit the motor speed (and machine speed) to something less than would be available at this maximum setting. The maximum adjustment allows this to be done.

IR Compensation Although a typical DC motor presents a mostly inductive load, there is always a small amount of fixed resistance in the armature circuit. *IR* compensation is a method used to adjust for the drop in a motor's speed due to armature resistance. This helps stabilize the motor's speed from a no-load to full-load condition. *IR* compensation should be applied only to voltage-regulated drives.

Acceleration Time As its name implies, the acceleration time adjustment will extend or shorten the amount of time for the motor to go from zero speed up to the set speed. It also regulates the time it takes to change speeds from one setting (e.g., 40 percent) to another setting (e.g., 80 percent).

Deceleration Time The deceleration time adjustment allows loads to be slowed down over an extended period of time. For example, if power is removed from the motor and it stops in 3 seconds, then the deceleration time adjustment would allow you to adjust this time typically within a 0.5- to 30-second range.

Part 3 DC Motor Drive Fundamentals

Part 3 Review Questions

1. List three types of operations where DC drives are commonly found.
2. How can the speed of a DC motor be varied?
3. What are the two main functions of the SCR semiconductors used in a DC drive power converter?
4. Explain how SCR phase angle control operates to vary the DC output from an SCR.
5. Armature-voltage-controlled DC drives are classified as constant-torque drives. What does this mean?
6. Why is three-phase AC power, rather than single-phase, used to power most commercial and industrial DC drives?
7. List what input line and output load voltage information must be specified for a DC drive.
8. How can the speed of a DC motor be increased above that of its base speed?
9. Why must field loss protection be provided for all DC drives?
10. Compare the braking capabilities of nonregenerative and regenerative DC drives.
11. A regenerative DC drive requires two sets of power bridges. Why?
12. Explain what is meant by an overhauling load.
13. What are the advantages of regenerative braking versus dynamic braking?
14. How is the desired speed of a drive normally set?
15. List three methods used by DC drives to send feedback information from the motor back to the drive regulator.
16. What functions require monitoring of the motor armature current?
17. Under what operating condition would the minimum speed adjustment parameter be utilized?
18. Under what operating condition would the maximum speed adjustment parameter be utilized?
19. *IR* compensation is a parameter found in most DC drives. What is its purpose?
20. What, in addition to the time it takes for the motor to go from zero to set speed, does acceleration time regulate?

PART 4 PROGRAMMABLE LOGIC CONTROLLERS (PLCS)

PLC Sections and Configurations

A **programmable logic controller (PLC)** is an industrial-grade computer that is capable of being programmed to perform control functions. The programmable controller has eliminated much of the hard wiring associated with conventional relay control circuits. Other benefits PLCs provide include easy programming and installation, fast control response, network compatibility, troubleshooting and testing convenience, and high reliability. Programmable logic controllers are now the most widely used industrial process control technology.

Figure 10-53 shows the major sections of a programmable logic controller system. Their basic functions are summarized as follows:

Power supply The power supply of a PLC system converts either AC line voltage or, in some applications, a DC source voltage, into low-voltage DC required by the processor and I/O modules. In addition to voltages required for the internal operation of these components, the power supply in certain applications may provide low-voltage DC to external loads as well. Power supplies are available for different input voltages including 120 V AC, 240 V AC, 24 V AC, and 24 V DC. The required output current rating of the power supply is based on the type of processor, the number and types of input/output (I/O) modules, and any external loads that may be required to be connected to the power supply.

Processing unit The central processing unit (CPU), also called processor, and associated memory form the intelligence of a PLC system. Unlike other modules that simply direct input and output signals, the CPU evaluates the status of inputs, outputs, and other data as it executes a stored program. The CPU then sends signals to update the status of outputs. Processors are rated as to their available memory and I/O capacity, as well as the different types and number of available programming instructions.

Input module Input modules enable the PLC to sense and control the system it is operating. The prime function of an input module is to take the input signals from

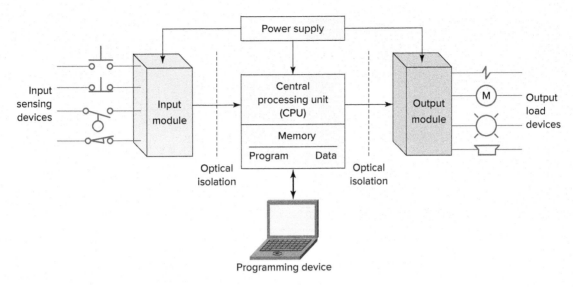

Figure 10-53 The major sections of a PLC system.

the field devices switches or sensors and convert them into logic signals that can be used by the CPU. In addition, the input module provides electrical isolation between the input field devices and the PLC. The types of input modules required depend on the types of input devices used. Some input modules respond to digital inputs, also called discrete inputs, which are either on or off. Other input modules respond to analog signals that represent conditions as a range of voltage or current.

Output module Output modules control the system by operating motor starters, contactors, solenoids, and the like. They convert control signals from the CPU into digital or analog values that can be used to control various output field devices (loads). They also provide electrical isolation between the output field devices and the PLC.

Programming device The programming device is used to enter or change the PLC's program or to monitor or change stored values. Once entered, the program is stored in the CPU. A personal computer (PC) is the most commonly used programming device and communicates with CPU via a communications port.

Small-size **fixed PLCs,** such as the Micro PLC shown in Figure 10-54, are stand-alone, self-contained units. A fixed controller consists of a power supply, processor (CPU), and fixed number of input/outputs (I/Os) in a single unit. They are constructed in one package with no separate, removable units. The number of available I/O points varies and usually can be increased by the addition of expansion modules. Fixed controllers are small and less expensive but limited to smaller, less complex applications.

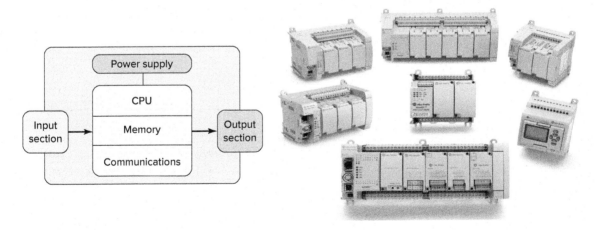

Figure 10-54 Fixed programmable controller.
Photo courtesy of Rockwell Automation, www.rockwellautomation.com

Part 4 Programmable Logic Controllers (PLCs)

Figure 10-55 Modular programmable controller.
Photo courtesy of Rockwell Automation,
www.rockwellautomation.com

A **modular PLC,** such as the one shown in Figure 10-55, is made up of several different physical components. It consists of a rack or chassis, power supply, processor (CPU), and I/O modules. The chassis is divided by compartments into which the separate modules can be plugged. The complete assembly provides all of the control functions required for a particular application. This feature greatly increases your options and the system's flexibility. You can choose from a variety of modules available from the manufacturer and mix them any way you desire.

Ladder Logic Programming

A *program* is a user-developed series of instructions that directs the PLC to execute actions. A **programming language** provides rules for combining the instructions so that they produce the desired actions. Relay ladder logic (RLL) is the standard programming language used with PLCs. Its origin is based on electromechanical relay control. The ladder logic program graphically represents rungs of contacts, coils, and special instruction blocks. The program, Figure 10-56 is contained in the PLC's memory and takes the place of much of the external wiring that would normally be required for control purposes. Hardwiring, though, is still required to connect input and output field devices.

Figure 10-57 shows the traditional electrical diagram for a hard-wired motor start/stop circuit.

- The diagram consists of two vertical supply lines with a single rung.
- Every rung contains at least one load device as well as the conditions that control the device.
- The rung is said to have electrical continuity whenever a current path is established between L1 and L2.
- Pressing the start push button results in electrical continuity to energize the starter coil (M) and close the seal-in contact (M1).

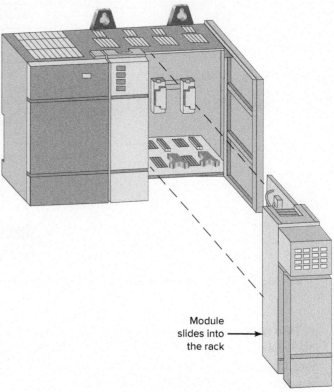

Module slides into the rack

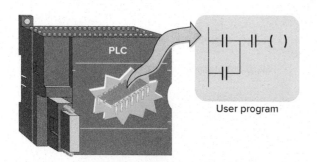

Figure 10-56 PLC relay ladder logic program.

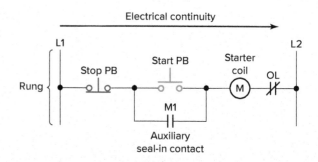

Figure 10-57 Hard-wired start/stop circuit.

306 Chapter 10 Adjustable-Speed Drives and PLC Installations

Instruction	Symbol	State
XIC Examine if closed	─┤ ├─	If the input device is open, the instruction is false. If the input device is closed, the instruction is true.
XIO Examine if open	─┤/├─	If the input device is open, the instruction is true. If the input device is closed, the instruction is false.
OTE Output energize	─( )─	If the rung has logic continuity, the output is energized. If the rung doesn't have logic continuity, the output is de-energized.

Figure 10-58 Basic PLC instructions.

- After the start button is released, electrical continuity is maintained by the seal-in contact.
- When the stop push button is pressed, electrical continuity is lost, and the starter coil de-energizes.

On an electrical diagram, the symbols represent real-world devices. In the electrical diagram, the electrical states of the devices are described as being open/closed or off/on. In the ladder logic program, instructions are either false/true or binary 0/1. The three basic PLC instructions (Figure 10-58) are:

XIC—Examine if closed instruction

XIO—Examine if open instruction

OTE—Output energize instruction

Each of the input and output connection points on a PLC has an **address** associated with it. This address will indicate what PLC input is connected to what input device and what PLC output will drive what output device. Types of addressing formats include **rack/slot-based** and **tag-based** formats. Addressing formats may vary between PLC modules manufactured by the same company. Also no two different PLC manufacturers have identical addressing formats. Understanding the addressing scheme used is of prime importance when it comes to programming and wiring. PLCs with fixed I/O typically have all their input and output locations predefined.

Figure 10-59 illustrates how the basic PLC instructions are applied in a programmed start/stop motor circuit, which looks and acts much like the electric hard-wired circuit. The operation of the program can be summarized as follows:

- The normally closed stop push button is closed, making the stop instruction (I1) true.
- Closing the start push button makes the start instruction (I2) true and establishes logical continuity of the rung.
- Rung logic continuity energizes the motor starter coil.
- The starter auxiliary contact M1 closes, making its instruction (I3) true.
- After the start button is released, electrical continuity is maintained by the true I3 instruction.

Figure 10-60 illustrates typical PLC wiring, with input/output notation, designed to implement the motor start/stop control. A fixed controller with eight predefined fixed inputs (I1 to I8) and four predefined fixed relay outputs (Q1 to Q4) is used to control and monitor the motor start/stop operation. The job is completed as follows:

- The power supply is connected to terminals L1 and L2 of the controller.
- Q1 normally open output relay contact, M starter coil, and the OL relay contact are hard-wired in series between L1 and L2.
- The stop push button, start push button, and M1 auxiliary seal-in contact inputs are connected to I1, I2, and I3, respectively.

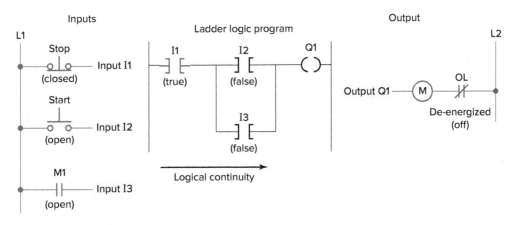

Figure 10-59 Programmed start/stop circuit.

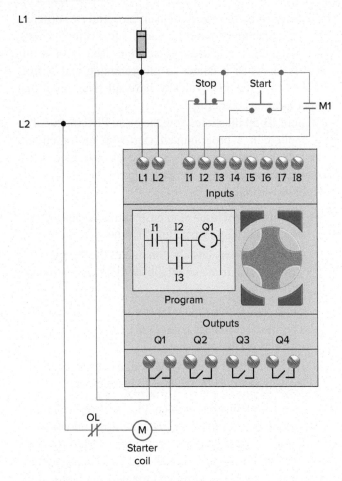

Figure 10-60 PLC wiring designed to implement the motor start/stop control.

- The ladder logic program is entered using the front keypad and LCD display or a personal computer connected via the communications port.
- Power is applied and the PLC is placed in the run mode to operate the system.

Figure 10-61 illustrates the original motor start/stop program modified to include remote standby and run pilot lights. The operation of the pilot lights is summarized as follows:

- Both examine-if-open and examine-if-closed Q1 programmed contacts are referenced to the starter output coil address Q1.
- When output Q1 is not energized, the examine-if-open Q1 instruction will be true, establishing rung continuity and energizing output Q2 to switch on the standby pilot light.
- In addition, when output Q1 is not energized, the examine-if-closed Q1 instruction will be false. No rung continuity exists, so the run pilot light connected to output Q3 will be de-energized.
- When output Q1 is energized, the Q1 examine-if-open Q1 instruction becomes false and the Q1 examine-if-closed instruction becomes true. This results in the standby pilot light switching off and the run pilot light switching on.

The term **hard-wired** refers to logic control functions that are determined by the way the devices are interconnected. Hard-wired logic is fixed; it is changeable only by altering the way in which the devices are connected. In contrast, programmable control is based on logic functions, which are programmable and easily changed. Figure 10-62 shows the wiring changes that would be required, in addition to the program changes, in order to implement the motor control with standby and run pilot lights. All existing wiring remains intact. The only new wiring required is connection from the remote pilot lights to outputs Q2 and Q3.

Safety considerations need be developed as part of the PLC program. One such consideration involves the use of a motor starter seal-in contact in place of a programmed

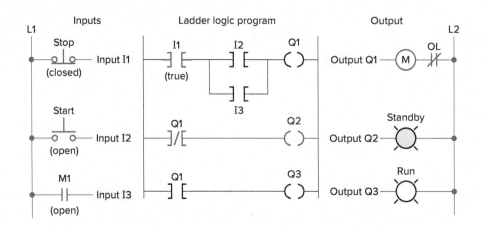

Figure 10-61 Programmed start/stop circuit with remote standby and run pilot lights.

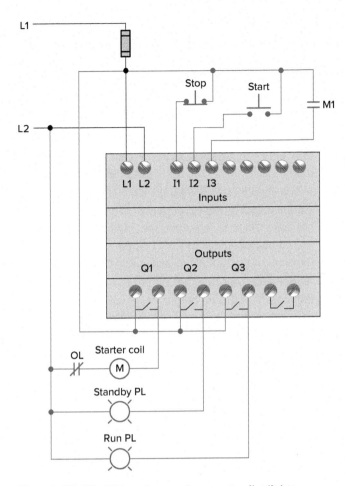

Figure 10-62 Wiring changes for remote pilot lights.

contact referenced to the output coil instruction. The use of the field-generated **starter auxiliary contact** status in the program is safer because it provides positive feedback to the processor about the exact status of the motor. Assume, for example, that the OL contact of the starter opens under overload condition. The motor, of course, would stop operating because power would be lost to the starter coil. If the program were implemented using a normally open contact instruction referenced to the output coil instruction as the seal-in for the circuit, the processor would never know that power had been lost to the motor. When the OL was reset, the motor would restart instantly, creating a potentially unsafe operating condition.

Another safety consideration concerns the wiring of stop buttons. A stop push button is generally considered a safety function as well as an operating function. As such, it should be wired using a **normally closed push button and programmed for an examine-if-closed condition.** Using a normally open push button programmed for an examine-if-open condition will produce the same logic but is not considered as safe. Assume that the latter configuration is used. If, by some chain of events, the circuit between the push button and the input point were to be broken, the stop button could be depressed forever, but the PLC logic could never react to the stop command because the input would never be true. The same would result if power were lost to the stop pushbutton control circuit. If the normally closed wiring configuration is used, the input point receives power continuously unless the stop function is desired. Any faults occurring with the stop circuit wiring, or a loss of circuit power, would effectively be equivalent to an intentional stop.

Programming Timers

The most commonly used PLC instruction, after coils and contacts, is the timer. **Timers** are programming functions that keep track of time and provide various responses depending on the elapsed time. They operate in a manner similar to hard-wired electromechanical timers. Even though each manufacturer may represent timer instructions differently on the ladder diagram, most operate in the same manner. Common PLC timer instructions include:

On-delay timer (TON) is a programming instruction typically used to delay the start of a machine or process for a set period of time.

Off-delay timer (TOF) is a programming instruction typically used to delay the shutdown of a machine or process for a set period of time.

Retentive timer (RTO) is a programming instruction typically used to track the length of time a machine has been operating or to shut down a process after an accumulated time period of recurring faults.

Timers in a programmable controller are output instructions. Figure 10-63 shows the on-delay timer instruction used by Allen-Bradley SLC-500 controllers. The following are associated with this timer instruction:

- **Type of timer.** TON (on delay).
- **Timer number.** Address T4:0.
- **Time base.** 1.0 second. The time base of the timer determines the duration of each time base interval. Time intervals are accumulated or counted by the timer.

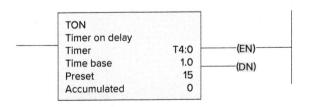

Figure 10-63 On-delay timer instruction.

Part 4 Programmable Logic Controllers (PLCs)

- **Preset time.** 15. Used in conjunction with the time base to set the time-delay period. In this case the time delay period would be 15 seconds (1 second × 15).
- **Accumulated value.** The time that has elapsed since the timer was last reset.
- **(EN)—Enable bit.** Is true whenever the timer instruction is true.
- **(DN)—Done bit.** Changes state whenever the accumulated value reaches the preset value.

The most commonly used timer is the on-delay type. You can use this timer to turn an output on or off after the timer has been on for a preset period of time. Figure 10-64 illustrates the operation of a typical on-delay programmed timer. The operation of the timer is summarized as follows:

- As long as switch input A is true (closed), the timer on delay T4:0 will increment every 1 second toward the preset value of 15 seconds.
- As long as switch input A is true (closed), the timer's enable bit (EN) will be true or set to 1. With rung continuity established, the green pilot light output C will be energized (turned on) at all times that the switch is closed.
- The current number of seconds that have passed will be displayed in the accumulated value portion of the instruction.
- When the accumulated value is equal to the preset value, the timer's done bit (DN) will be true or set to 1. Rung continuity is established, and the red pilot light output B will be energized (turned on).
- The processor resets the accumulated time to zero when the rung condition goes false, regardless of whether the timer timed out or not.

Figure 10-65 shows the wiring for the on-delay timer implemented using the Allen-Bradley SLC 500 modular controller with slot-based addressing. A 16-point (0 to 15) AC input module is plugged into slot 1 and a 16-point (0 to 15) AC output module into slot 2 of the single-rack chassis. The wiring is as shown and the addressing format is as follows:

Address I:1/2 (switch input A). The letter I indicates that it is an input, the 1 indicates that the AC input module is in slot 1, and the 2 indicates the terminal of the module to which it is connected.

Address O:2/3 (red PL output B). The letter O indicates that it is an output, the 2 indicates that the AC output module is in slot 2, and the 3 indicates the terminal of the module to which it is connected.

Address O:2/8 (green PL output C). The letter O indicates that it is an output, the 2 indicates that the AC output module is in slot 2, and the 8 indicates the terminal of the module to which it is connected.

Programming Counters

Most PLC manufacturers provide counters as part of their instruction set. A programmed **counter** can count, calculate, or keep a record of the number of times something happens. One of the most common counter applications is counting the number of items moving past a given point. The two counter types are the count-up (CTU) counter and the count-down (CTD) counter. Count-up instructions are used alone or in conjunction with count-down instructions having the same address.

Counters, like timers, in a programmable controller are output instructions. Other similarities include the following:

- A timer and counter both have an accumulator. For a timer, the accumulator is the number of time base intervals the instruction has counted.

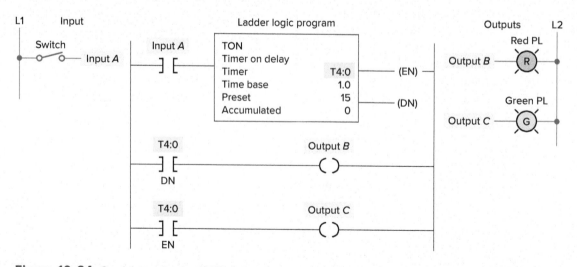

Figure 10-64 On-delay programmed timer.

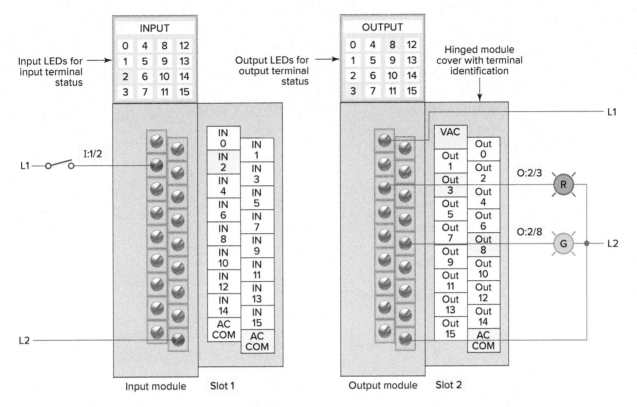

Figure 10-65 Wiring for the on-delay timer using the Allen-Bradley SLC 500 modular controller.

For a counter, the accumulator is the number of false-to-true transitions of its logic rung that have occurred.

- A timer and counter both have a preset value. The preset value is the set point that you enter in the timer or counter instruction. When the accumulated value becomes equal to or greater than the preset value, the done bit (DN) status is set to 1.

Figure 10-66 shows the standard counter instruction used by Allen-Bradley controllers. The following are associated with this counter instruction:

- **Type of counter.** Up-counter (CTU).
- **Counter number.** Address C5:1.
- **Preset value.** 7.
- **Accumulated value.** Initially set at 0.

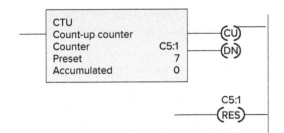

Figure 10-66 Up-counter (CTU) instruction.

- **(CU)—enable bit.** Is true whenever the rung conditions for the counter are true.
- **(DN)—done bit.** Changes state whenever the accumulated value reaches the preset value.
- **(OV)—overflow bit.** Is true whenever the counter counts past its maximum value.
- **(RES)—Reset.** Instruction with the same address as the counter that is being reset is used to return counter accumulator values to zero.

Count-up counters are used when a total count is needed. The number stored in the counter accumulator is incremented each time the logic rung for the counter goes from false to true. It thus can be used to count false-to-true transitions of an input instruction and then trigger an event after a preset number of counts or transitions. The up-counter output instruction will increment by 1 each time the counted event occurs.

Figure 10-67 illustrates the operation of a programmed up-counter used to turn the red pilot light on and the green pilot light off after an accumulated count of 7. Figure 10-68 shows the wiring for the up-counter implemented using the Allen-Bradley SLC 500 modular controller with slot-based addressing. The operation of the program is summarized as follows:

- Operation of PB1 (input I:1/0) provides the off-to-on transition pulses that are counted by the counter.

Part 4 Programmable Logic Controllers (PLCs)   311

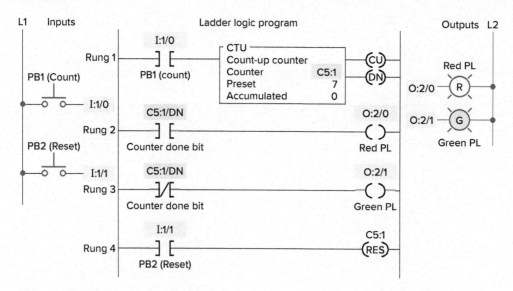

Figure 10-67 Programmed up-counter.

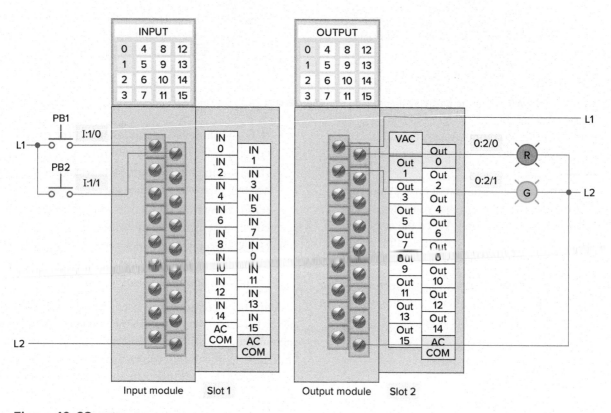

Figure 10-68 Wiring for the up-counter implemented using the Allen-Bradley SLC 500 controller.

- The preset value of the counter is set to 7.
- Each false-to-true transition of rung 1 increases the counter's accumulated value by 1.
- After 7 pulses, or counts, when the preset counter value equals the accumulated counter value, output DN is energized.
- As a result, rung 2 becomes true and energizes output O:2/0 to switch the red pilot light on.
- At the same time, rung 3 becomes false and de-energizes output O:2/1 to switch the green pilot light off.
- The counter is reset by closing PB2 (input I:1/1) and resets the accumulated count to zero.
- Counting can resume when rung 4 goes false again.

The down-counter output instruction will count down or decrement by 1 each time the counted event occurs.

312 Chapter 10 Adjustable-Speed Drives and PLC Installations

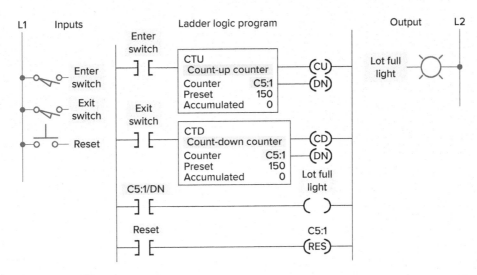

Figure 10-69 Parking garage counter.

Count-down counters (CTDs) are used when a preset number of items (or events) exist and the number must be decreased (or decremented) as items are taken away or events occur. An example of a down-counter application is keeping track of the number of parts leaving a storage bin.

Often the down-counter is used in conjunction with the up-counter to form an up/down-counter. A typical application for an up/down-counter could be to keep count of the cars that enter and leave a parking garage. Figure 10-69 shows a PLC program that could be used to implement this application. The operation of the program is summarized as follows:

- As a car enters, it triggers the up-counter output instruction and increments the accumulated count by 1.
- Conversely, as a car leaves, it triggers the down-counter output instruction and decrements the accumulated count by 1.
- Because both the up- and down-counters have the same address, the accumulated value will be the same in both.
- Whenever the accumulated value equals the preset value, the counter output is energized to light up the lot full sign.
- A reset button has been provided to reset the accumulated count.

Troubleshooting

PLCs are relatively easy to troubleshoot because the control program can be displayed on a monitor and watched in real time as it executes. When a problem occurs, the source of a problem can generally be narrowed down to the processor module, I/O hardware, machine inputs or outputs, or ladder logic program.

Processor Module The **processor module** monitors itself continually for any problems that might cause the controller to execute the user program improperly. An error table includes all of the **error codes** and **error messages** that the panel will display if the listed cause is detected. In addition, LEDs on the processor module may indicate the current operating mode and a detected fault.

Most PLCs incorporate a **watchdog timer** to monitor the scan process of the system. It monitors how long it takes the CPU to complete a scan. If the CPU scan takes too long, a watchdog major error will be declared. PLC user manuals will show how to apply this function. The PLC processor hardware is not likely to fail as they are very reliable when operated within the stated limits of temperature, moisture, and harsh environments.

Input Malfunctions If the controller is operating in the run mode but output devices do not operate as programmed, the faults could be associated with any of the following:

- Input and output wiring between field devices and modules
- Field device or module power supplies
- Input sensing devices
- Output actuators
- PLC I/O modules
- PLC processor

LED status indicators can provide much information about field devices, wiring, and modules. For an input module, a lit LED indicates that the input device is activated and that its signal is present at the module. Narrowing down the problem source can usually be accomplished by comparing the actual status of the suspect input with

Part 4 Programmable Logic Controllers (PLCs)

the status indicator. The circuit of Figure 10-70 illustrates how to check for discrete input malfunctions. The steps taken can be summarized as follows:

- When input hardware is suspected to be the source of a problem, the first check is to see if the status indicator on the input module illuminates when it is receiving power from its corresponding input device (e.g., push button, limit switch).
- If the status indicator on the input module does *not* illuminate when the input device is on, take a voltage measurement across the input terminal to check for the proper voltage level.
- If the voltage level is correct, then the input module should be replaced.
- If the voltage level is not correct, the power supply, wiring, or input device may be faulty.

If the program monitor does not show the correct status indication for a condition instruction, the input module may not be converting the input signal properly to the logic level voltage required by the processor module. In this case, the input module should be replaced. If a replacement module does not eliminate the problem and the wiring is assumed to be correct, then the I/O rack, communication cable, or processor should be suspected. Figure 10-71

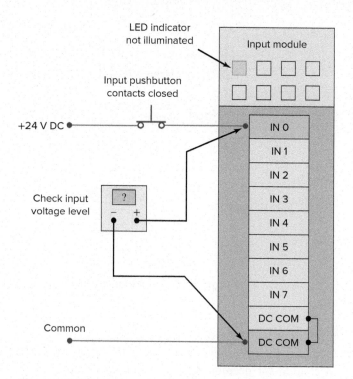

Figure 10-70 Checking for input malfunctions.

shows a typical discrete input device troubleshooting guide. This guide reviews condition instructions and how their true/false status relates to external input devices.

		Input device troubleshooting guide		
Sensor condition	Status indicator	Ladder instructions		Possible faults(s)
		─] [─	─]/[─	
Closed — ON 24 V DC input	ON	True ─] [─	False ─]/[─	None - correct indications
Open — OFF 0 V DC input	OFF	False ─] [─	True ─]/[─	None - correct indications
Closed — ON 24 V DC input	ON	False ─] [─	True ─]/[─	Sensor condition, voltage, status indicator are correct. Ladder instructions have incorrect indications. **Input module or processor fault.**
Closed — ON 0 V DC input	OFF	False ─] [─	True ─]/[─	Status indicator and instructions agree but not with the sensor condition. **Open field device or wiring fault.**
Open — OFF 0 V DC input	OFF	True ─] [─	False ─]/[─	Sensor condition, voltage, status indicator are correct. Ladder instructions have incorrect indications. **Input module or processor fault.**
Open — OFF 24 V DC input	ON	True ─] [─	False ─]/[─	Input voltage, status indicator, and ladder instructions agree but not with sensor condition. **Short circuit in the field device or wiring fault.**

Figure 10-71 Input troubleshooting guide.

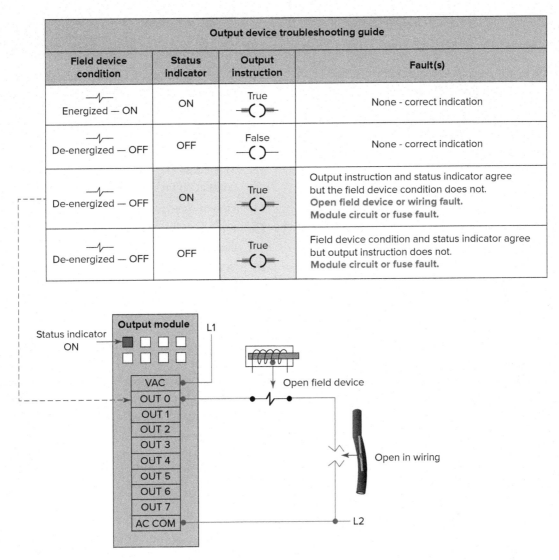

Figure 10-72 Output troubleshooting guide.

Output Malfunctions In addition to the logic indicator, output modules may incorporate a blown fuse or an electronic protection indicator. They are used to provide protection for the modules from short-circuit and overload current conditions. When an output does not energize as expected, first check the output protection indicator. Many output modules have each output protected.

Figure 10-72 shows a typical discrete output module troubleshooting guide. In general, the following items should be noted when troubleshooting discrete output modules:

- If the blown fuse indicator is not illuminated (fuse OK), then check to see if the output device is responding to the LED status indicator.

- An output module's logic status indicator functions similarly to an input module's status indicator. When it is on, the status LED indicates that the module's logic circuitry has recognized a command from the processor to turn on.
- If an output rung is energized, the module status indicator is on, and the output device is not responding, then the wiring to the output device or the output device itself should be suspected.
- If, according to the programming device monitor, an output device is commanded to turn on but the status indicator is off, then the output module or processors may be at fault.
- Check the voltage at the output; if incorrect, the power supply, wiring, or output device may be faulty.

Part 4 Review Questions

1. What is a programmable logic controller, and how is it utilized in motor applications?
2. List the five major components of a PLC system along with the function provided by each.
3. List the advantages of fixed and modular PLCs.
4. Explain the function of a PLC program.
5. Compare the way in which the states of devices are referred to in an electrical diagram and in a ladder logic program.
6. When does logic continuity occur in a rung of a PLC ladder logic program?
7. Compare hard-wired–type control with programmable control.
8. For safety reasons, a programmed start/stop circuit is always wired using normally closed stop push buttons. Why?
9. Compare the way in which electromechanical and programmed timers operate.
10. State what timer instruction(s) is associated with each of the following:
    a. Time that has elapsed since the timer was last reset.
    b. Time-delay period.
    c. Contact that changes state after the timer times out.
11. What would be the correct address for a push button connected to terminal 4, slot 1, of a single-rack SCL 500 modular PLC controller?
12. List three common functions programmed counters perform.
13. What does the accumulator of a programmed counter count?
14. How is a programmed counter reset to zero?
15. Explain the principle of operation of an up/down-counter.
16. What makes programmed PLC control easier to troubleshoot than hard-wired types?
17. Explain the function of the processor module watchdog timer.
18. The status indicator on a discrete input module does not illuminate when the input device is on. How would you proceed to determine the problem?
19. The status indicator on a discrete output module is on, but the output device is not responding. What problem should be suspected?

Troubleshooting Scenarios

1. Most VFD manufacturers provide a special terminal block for DC bus voltage measurement. What conditions could bus voltage measurements be used to check for?
2. For each VFD problem given, list what things in general you would check to determine the problem.
    a. Motor does not start—no output voltage to the motor.
    b. Drive started but motor not rotating—a speed of 0 Hz is displayed.
    c. Motor not accelerating properly.
3. What might be the consequence of setting the acceleration time of a DC drive to a very low value.
4. A normally closed contact-type limit switch is programmed to operate a solenoid valve as part of a motor control system. This limit switch is to be replaced with a normally open contact type. What changes, if any, would be required to have the circuit operate as before with the new limit switch installed?
5. The single-pole switch input to a PLC input module is closed, but the LED status indicator light does not come on. A check of voltage to the input module indicates that no voltage is present. Suggest two possible causes of the problem.

Discussion Topics and Critical Thinking Questions

1. AC adjustable-speed drives are available with a broad selection of features. Do a vendor search on the Internet and prepare a paper on the standard and optional features of a VFD of your choice or one assigned by the instructor.

2. Why might you want to copy your drive parameter settings if this option is available to you?

3. The use of a bypass contactor circuit to operate a DC drive is not an option. Explain why.

4. Three motors are to be started in sequence with a 10-second time delay between them. Design a PLC program that will perform this operation.

5. A parking garage has one entrance and two exits. Design a PLC program that will track the number of vehicles parked in the garage at any one time.

APPENDIX A
ELECTRIC MOTORS AND CONTROL SYSTEMS ACTIVITIES MANUAL

Contents

Preface vii

Part 1
ACTIVITIES MANUAL

Chapter 1
Safety in the Workplace 1

PART 1 Quiz   Protecting against Electric Shock 1
PART 2 Quiz   Grounding—Lockout—Codes 3

Chapter 2
Understanding Electrical Drawings 7

PART 1 Quiz   Symbols—Abbreviations—Ladder Diagrams 7
PART 2 Quiz   Wiring—Single-Line Diagrams—Block Diagrams 10
PART 3 Quiz   Motor Terminal Connections 12
PART 4 Quiz   Motor Nameplate and Terminology 15
PART 5 Quiz   Manual and Magnetic Motor Starters 19

Hands-on Practical Assignments 22

Chapter 3
Motor Transformers and Distribution Systems 29

PART 1 Quiz   Power Distribution Systems 29
PART 2 Quiz   Transformer Principles 34
PART 3 Quiz   Transformer Connections and Systems 37

Hands-on Practical Assignments 40

Chapter 4
Motor Control Devices 45

PART 1 Quiz   Manually Operated Switches 45
PART 2 Quiz   Mechanically Operated Switches 47
PART 3 Quiz   Sensors 49
PART 4 Quiz   Actuators 56

Hands-on Practical Assignments 59

Chapter 5
Electric Motors 70

PART 1 Quiz   Motor Principle 70
PART 2 Quiz   Direct Current Motors 72
PART 3 Quiz   Three-Phase Alternating Current Motors 78
PART 4 Quiz   Single-Phase Alternating Current Motors 84
PART 5 Quiz   Alternating Current Motor Drives 88
PART 6 Quiz   Motor Selection 90
PART 7 Quiz   Motor Installation 92
PART 8 Quiz   Motor Maintenance and Troubleshooting 94

Chapter 6
Contactors and Motor Starters 102

PART 1 Quiz   Magnetic Contactor 102
PART 2 Quiz   Contactor Ratings, Enclosures, and Solid-State Types 108
PART 3 Quiz   Motor Starters 111

Chapter 7
Relays 119

PART 1 Quiz	Electromechanical Control Relays	119
PART 2 Quiz	Solid-State Relays	122
PART 3 Quiz	Timing Relays	125
PART 4 Quiz	Latching Relays	130
PART 5 Quiz	Relay Control Logic	132

Hands-on Practical Assignments   136

Chapter 8
Motor Control Circuits 149

PART 1 Quiz	NEC Motor Installation Requirements	149
PART 2 Quiz	Motor Starting	152
PART 3 Quiz	Motor Reversing and Jogging	158
PART 4 Quiz	Motor Stopping	160
PART 5 Quiz	Motor Speed	162

Hands-on Practical Assignments   164

Chapter 9
Motor Control Electronics 183

PART 1 Quiz	Semiconductor Diodes	183
PART 2 Quiz	Transistors	189
PART 3 Quiz	Thyristors	197
PART 4 Quiz	Integrated Circuits (ICs)	204

Chapter 10
Adjustable-Speed Drives and PLC Installations 211

PART 1 Quiz	AC Motor Drive Fundamentals	211
PART 2 Quiz	VFD Installation and Programming Parameters	215
PART 3 Quiz	DC Motor Drive Fundamentals	220
PART 4 Quiz	Programmable Logic Controllers (PLCs)	223

Part 2
"The Constructor"
SIMULATION LAB MANUAL

Chapter 2
Understanding Electrical Diagrams
Analysis and Troubleshooting Assignments

Chapter 3
Motor Transformers and Distribution Systems
Analysis and Troubleshooting Assignments

Chapter 4
Motor Control Devices
Analysis and Troubleshooting Assignments

Chapter 5
Electric Motors
Analysis and Troubleshooting Assignments

Chapter 6
Contactors and Motor Starters
Analysis and Troubleshooting Assignments

Chapter 7
Relays
Analysis and Troubleshooting Assignments

Chapter 8
Motor Control Circuits
Analysis and Troubleshooting Assignments

INDEX

A

AC contactors, coils, 164
AC generator, 106–107
AC induction motors, 30
    dual-voltage, 32–34
    full-voltage starting of, 218–223
    multispeed, 34–36
    reversing of, 231–234
    single-phase, 30–31
    squirrel-cage, 30, 124–128
    three-phase, 31–32, 124–128
    wound-rotor models, 30, 128–129
AC inverting amplifier, 266
AC load
    BJT and, 252–253
    SSR, 192
AC motor drives, 136–139, 275–283
    flux vector drive, 281–283
    inverter duty motor, 139
    variable-frequency drive, 276–280
    variable-speed, 276
    volts per hertz, 280–281
AC solenoids, 99
Acceleration time, 303
Across-the-line starter, 41, 218–223
Actual speed, AC motors, 123
Actuators, 98–103
    defined, 98
    relays, 98–99
    servo motors, 102–103
    solenoid, 99–100
    solenoid valves, 100–101
    stepper motors, 101–102
Additive polarity, 62–63
Adequate torque, 257
Adjustable-frequency drive (AFD), 276
Adjustable-speed drive (ASD)
    AC, 275–283
    DC, 297–303
    VFD, 284–296
AFD. *See* Adjustable-frequency drive (AFD)
Alarms, VFD, 295
Alternating current (AC) motors
    actual speed, 123
    drives (*See* AC motor drives)
    induction (*See* AC induction motors)
    single-phase, 131–136
    synchronous speed of, 122–123
    three-phase, 122–130
Alternating relays, 204–206
Ambient temperature (AMD/DEG), 38, 142–143, 181
American Wire Gauge (AWG), 212
Ampacity, conductor, 51
Amplification, 252
Analog Hall effect sensors, 92

Analog IC, 266
Analog inputs/outputs, 290–291
Analog temperature sensors, 95
Analog-switching relay, 194
AND gate, 271
AND logic function, 207
Antiplugging, 238–240
Arc blast, 4
Arc burns, 3
Arc chute, 167, 168
Arc flash hazards, 4
Arc suppression, 166–168
Arcing fault, 4
Armature current, 236
Armature reaction, 117
Article 240.6, NEC, 214
Article 430, NEC, 148, 211
    Part II, 212
    Part III, 214
    Part IV, 214
    Part IX, 215
    Part VI, 216
    Part VII, 215
Article 430.102, NEC, 287
Article 430.32, NEC, 214
Article 430.6, NEC, 212
Article 430.72, NEC, 214
Article 430.75, NEC, 217
Article 725.43, NEC, 214
ASD. *See* Adjustable-speed drive (ASD)
Asymmetrical timers, 201
Asynchronous motors, 129
Automatic reset, 178
Automatic starter, 41
Automatic-reset thermal protection, 40, 150
Autotransformer starters, 224–225
Autotransformers, 58, 67
    variable, 67
Auxiliary contact, 41, 161
    blocks, 166
AWG. *See* American Wire Gauge (AWG)

B

Babbitt, 147
Ball bearings, 147, 148
Base (B) section, 252
Base speed, DC motor, 118
Battery-operated latching alarm circuit, 204
BDT. *See* Breakdown torque (BDT)
Bearings, 147–148
    ball, 147, 148
    roller, 147
    sleeve, 147, 148
    thrust, 147
Bellows, 84

I-1

Beta, transistor, 252
Bimetallic type of thermal overload relay, 179
Bipolar junction transistor (BJT), 252–254
BLDC (brushless DC motor), 103, 120–121
Block diagrams, 26–27
Blowout coils, 167, 168
Body resistance values, 2
Bolted fault, 4
Bonding, 10, 53
Bourdon tube, 84
Braking
   DC-injection, 240–241, 288
   dynamic, 240, 288
   electromechanical friction, 241
   parking, 241
   regenerative, 29, 288
   variable-frequency drive, 288
Branch circuits, 51
Break, 190
Break-make push button, 73
Breakdown torque (BDT), 143
Breakover voltage, 263
Bridge rectifier, 247
Brushless DC motor (BLDC), 103
Buddy rule, 152
Built-in thermal protection, 40, 150–151
Burns, 3. *See also* specific types
Bus duct, 51
Busbar, 54
Busways, 51
Bypass contactors, 286–287

C

Cable trays, 51
Canadian Standards Association (CSA), 145
Capacitive loads, 170
Capacitive proximity sensors, 88–89
Capacitor start, 31
Capacitor-start motors, 133–134
   dual-speed, 134
Capacitor-start/capacitor-run motor, 134
Capacitors, 52
   short-circuited, 133
Capillary tube temperature switch, 83
Carrier frequency, 280
CE (Conformité Européene), 145
CEMF. *See* Counter electromotive force (CEMF)
Central processing unit (CPU), 304
Central-station system, 47
Centrifugal switches, 132–133
Chip. *See* Integrated circuit (IC)
Clamp-on ammeter, 152
Closed transition, 227–228
Closed-loop control, 280
Closed-loop mode, servo motors, 102
Closed-loop vector drive, 282–283
Code letter, 39, 140
Cogging, 293
Coilers, 297
Coils
   blowout, 167, 168
   deenergized, 187
   losses, 61
   magnetic contactors, 163–164
   relay, 188
   shading, 165
Cold junction, 93

Collector (C) section, 252
Combination logic functions, 208
Combination starter, 223
Commutating poles, 117
Commutation, 110, 111
Compound DC motors, 114–115
Conductor sizing
   for motor branch circuits, 212
   and motor installation, 149
Conduits, 51
   layout diagram, 24
Configurable alarms, 295
Configurable parameters, 294
Confined spaces, 6–7
Connection diagrams, 41
Consequent pole motors, 127
Constant horsepower, 35
Constant torque, 35
Constant V/Hz ratio, 280
Constant-horsepower loads, 142
Constant-torque loads, 141, 283–284
Contact block, 76
Contactor, 41. *See also* Magnetic contactors
Contacts, control devices, 20
Continuous duty, 143
Control circuit
   defined, 16
   hand-off-auto, 222
   magnetic contactors, 159
   motor, 16–17
   relay, 186
Control inputs/outputs, VFD, 289–292
Control logic, 276–277
Control relay (CR), 85
Control relay jogging circuit, 237
Control section, of DC motor drive, 120
Control terminal strips, 290
Controllers, motor. *See* Motor controllers
Converter, 136, 276
Copper loop, 135
Copper losses, 140
Cores, 51
   losses, 131
   losses, 61
Count-down (CTD) counter, 313
Count-up (CTU) counter, 311–312
Counter electromotive force (CEMF), 116–117, 218
Coupling, by means of gears or pulleys/belts, 146–147
CPU. *See* Central processing unit (CPU)
CR (Control relay), 85
Cranes, 297
CSA (Canadian Standards Association), 145
CSI. *See* Current-source inverter (CSI)
CT. *See* Current transformer (CT)
CTD counter. *See* Count-down (CTD) counter
CTU counter. *See* Count-up (CTU) counter
Current
   armature, 236
   defined, 2
   Eddy, 165
   field, 236
   inrush, 61
   leakage, 195
   magnetizing, 59
   as motor selection criteria, 140
   pathways, 2
   relative magnitude and effect of, 2

transformers, 58–60
Current amplifier, 252
Current feedback information, 303
Current limit, 303
   block, 282
   start, 229
Current rating, 38
Current resolver, 282
Current transformer (CT), 67–68
Current-source inverter (CSI), 293

D

D type, depletion-mode MOSFET, 255–256
Danger tags, 12
Darlington transistor, 253
Dashpot timers, 196
DC (direct-current) motors, 110–121, 120–121, 155
   armature reaction in, 117
   components of, 110
   compound, 114–115
   connections, 28–30
   counter electromotive force, 116–117
   direction of rotation, 115–116
   drives (See DC motor drives)
   performance parameters, 1101
   permanent-magnet, 110–112
   reversing of, 236
   series, 112
   shunt-type, 113–114
   speed, 118–119
   starting of, 229–230
DC brushless servo motor, 103
DC bus, 276
DC contactors, coils, 164
DC filter, 136
DC load, SSR, 192
DC motor drives, 119–120, 297–303
   applications, 297
   block diagram of, 298
   field voltage control, 300–301
   function of, 298
   nonregenerative/regenerative, 301–302
   parameter programming, 302–303
   principles of operation, 297–299
   single-phase input, 299–300
   three-phase input, 300
DC solenoids, 99
DC-injection braking, 240–241, 288
Dead band, 84
Dead front, 53
Deceleration time, 303
Deenergized state, 20
Definite-purpose contactors, 161
Depletion mode
   JFET, 255
   MOSFET, 255–256
Derating, 292
Design letter, 39, 131
Despiking, 247
Diac, 263
Diaphragm, 84
Diffuse scan sensor, 90–91
Digital Hall effect sensors, 92
Digital IC, 266
Digital inputs, 289–292
Digital logic, 270–273
Digital multimeter (DMM), 152

bipolar junction transistor test, 254
   diode test function, 248–249
Digital signals, states of, 207
Digital temperature sensors, 95
Digital-signal controller (DSC), 268
Digital/relay outputs, 290
DIN rail, 189
Diode-clamping, 247
Diodes, semiconductor, 245–251
   light-emitting, 249–250, 295
   operation of, 245–246
   photodiodes, 250
   PN-junction, 245
   rectifier, 246–249
   zener, 249
direct-current (DC) motors, 120–121
Direct-drive motors, 146
Direction of rotation, DC motor, 115–116
Discharge, 6
Disconnecting means, 215–216
   VFD, 287
Distribution systems. See Power distribution systems
DMM. See Digital multimeter (DMM)
Double-break relay, 98
DPDT crossed-wired alternating relays, 205–206
Drilling, 297
Drives, motor. See Motor drives
Dropout voltage, 164
Drum switches, 79
Dry contact, 99
DSC. See Digital-signal controller (DSC)
Dual-in-line package (DIP) switch, 182
Dual-element fuses, 182
Dual-ramp start, 229
Dual-speed capacitor-start motors, 134
Dual-voltage AC motor, connections of, 32–34
Dual-voltage split-phase motor, 133
Duty cycle, 39, 143
Dynamic braking, 29, 288

E

E type, depletion-mode MOSFET, 255–256
Eddy currents, 165
EDM effect. See Electric discharge machining (EDM) effect
Efficiency, 40
   motors, 140–141
   of transformer, 61
EGC (equipment grounding conductor), 10–11
Electric discharge machining (EDM) effect, 149
Electric motors, 105–156
   AC (See Alternating current (AC) motors)
   classification of, 28, 109
   direct-current (See DC (direct-current) motors)
   electromagnetism in, 106
   generators, 106–107
   installation (See Installation, motors)
   magnetism in, 105–106
   maintenance, 151–152
   manual starters, 42–43
   nameplate (See Motor nameplates)
   NEC installation requirements for, 211–217
   overheats, 153
   principle of, 105–109
   protection, 287–288
   reversing of (See Reversing, of motors)
   rotation, 107–109
   selection of (See Selection criteria, motors)

Electric motors (continued)
    servo, 102–103
    starting of (See Motor starting)
    stepper, 101–102
    stopping (See Stopping, of motors)
    terms, abbreviations for, 17
    troubleshooting (See Motor troubleshooting)
Electrical burns, 3
Electrical codes/standards, 12–15
    IEC, 14–15
    IEEE, 15
    NEC, 13
    NEMA, 1
    NFPA, 14
    NRTL, 14
    OSHA, 12–13
Electrical control symbols/markings, 182–184
Electrical drawings, 16–45
    block diagrams, 26–27
    motor abbreviations/terms, 17
    motor ladder diagrams, 17–23
    motor terminal connections, 28–36
    motor terminology, 41–42
    power rise diagram, 27
    single-line diagrams, 26
    symbols, 16–23
    wiring diagrams, 24–26
Electrical interlocking, 232–233
Electrical noise. See Electromagnetic interference (EMI)
Electrical shock, 1–3
Electrical shock protection
    basics of, 6
    grounding/bonding, 9–11
    lockout/tagout, 11–12
    personal protective equipment, 5–7
Electrical short circuits. See Short circuits
Electrical spikes, 285–286
Electrically held contactor, 161
Electrically interlocked circuits, 115
Electromagnetic interference (EMI), 285–286
Electromagnetism, 106
Electromechanical friction brakes, 241
Electromechanical relay (EMR). See also Relays
    applications, 188
    operation of, 186–188
    styles/specifications, 188–190
Electromotive force (EMF) counter, 116–117
Electronic overload relay, 181–182
Electrostatic discharge (ESD), 270
Embedded microcontrollers, 269
Emergency stop (E-stop) switches, 8, 76, 222–223
EMF. See Electromotive force (EMF)
EMI. See Electromagnetic interference (EMI)
Emitter (E) section, 252
EMR. See Electromechanical relay (EMR)
Enclosures
    hazardous-location, 171
    magnetic contactors, 171–172
    of magnetic motor starters, 176
    motor nameplate information, 40
    and motor selection, 143–144
    NEMA types, 74, 75, 171–172
    nonhazardous-location, 171
    for transformers, 57
    variable-frequency drive, 284
Encoder, 96
Energy-efficient motors, 141

Enhancement-mode MOSFETs, 256
Equipment grounding, 10
    bus, 53, 56
Equipment grounding conductor (EGC), 10–11, 54, 149
ESD. See Electrostatic discharge (ESD)
Eutectic, 179
Extended push buttons, 74–75

F
Face shields, 6
Feedback device, 280
Feeders, 50–51
FET. See Field-effect transistor (FET)
Fiber optic sensors, 91
Field current, 236
Field voltage control, 300–301
Field-effect transistor (FET), 254–255
Filter circuits, 247–248
Fire extinguishers, 14
Fire, classes of, 14
Fixed PLC, 305
FLA. See Full-load ampere (FLA)
Float switches, 84–86
Flow measurement sensors, 96–97
Flow switches, 84–86
FLT. See Full-load torque (FLT)
Flush push buttons, 74
Flux vector control, 282
Flux vector drive, 281–283
Fork lever, limit switches, 81
Forward-bias voltage, 246
Foundation, for motor installation, 146
Four-quadrant DC drives, 301–302
Frame sizes, 40, 141
Free-wheeling diode, 279
Frequency, 141, 292
Full voltage starter. See Across-the-line starter
Full-load ampere (FLA), 140, 180, 292
    rating, 212, 214
Full-load current, 60–61
    rating, 212
Full-load speed, 141
Full-load torque (FLT), 143
Full-voltage start, 229
Full-wave rectifier, 247
Fundamental frequency, 279

G
Generators, 106–107, 240
GFCI (ground-fault circuit interrupter), 11
Ground fault, 11
Ground-fault circuit interrupter (GFCI), 11
Ground-fault current path, 11
Ground-fault motor protection, 177
Grounded conductor, 11
Grounding, 153
    defined, 53
    equipment, 10
    and motor installation, 149
    of panelboards, 53
    proper, 286
    requirements for VFD, 286
    shock protection with, 9–11
    system, 10
Grounding electrode conductor, 11

H

Half-shrouded pushbutton operators, 75
Half-wave rectifier, 246
Hall effect sensors, 91–92
Hand-off-auto (HOA) control circuit, 222
Hard-wired logic, 271–272, 308
Hazardous location motors, 144
Hazardous-location enclosures, 171
Heat sink, 173
High-current load circuit, 188
High-inertia loads, 142
High-voltage load circuit, 188
High-voltage primary switchgear, 49
HOA control circuit. *See* Hand-off-auto (HOA) control circuit
Hockey-puck, SSR type, 192
Hoist, 297
Hold-in voltage, 164
Horsepower (HP), 110
    rating, 39
Hot junction, 93
Hot sticks, 6
Hot-spot allowance, 143
Humming noise, 164
Hysteresis zone, 88

I

Ice-cube, SSR type, 192
IEC (International Electrotechnical Commission), 14–15
    magnetic contactors ratings, 170–171
    markings/symbols, 182–184
IEC contactor, 160
IEEE (Institute of Electrical and Electronics Engineers), 15
IGBT. *See* Insulated-gate bipolar transistor (IGBT)
Illuminated pushbutton operator, 75
Impulse relays, 204–206
Inching. *See* Jogging
Inductive loads, 170, 190
Inductive-type proximity sensors, 87–88
Infrared (IR) thermometer, 152
Infrared-emitting diode (IRED), 249–250
Injected DC voltage, 240–241
Input module, 304–305
Inrush current, 61, 140
Installation, motors, 146–151
    bearings, 147–148
    built-in thermal protection, 150–151
    conductor size, 149
    electrical connections, 148
    foundation, 146
    grounding, 149
    motor and load alignment, 146–147
    mounting, 146
    voltage levels/balance, 149–150
Instant-on relays, 194
Institute of Electrical and Electronics Engineers (IEEE), 15
Instrument transformers, 67–68
Insulated-gate bipolar transistor (IGBT), 257–258
Insulation check, 69
Insulation class, 38–39, 143
Integrated circuit (IC), 265–273
    {five}555 timer, 267–268
    analog, 266
    digital, 266
    fabrication of, 265–266
    operational amplifier, 266–267
    temperature sensors, 95

Interlocking, 290
Intermittent duty, 143
International Electrotechnical Commission (IEC), 14–15
    on physical dimensions of motors, 40
Interpoles, 117
Interposing relay, 190–191
Inverse time circuit, 214
Inverter, 136
    PWM, 278
Inverter block, 27
Inverter duty, 35
Inverter-duty motor, 280
Inverters, 251
IR compensation, 281, 303
IRED. *See* Infrared-emitting diode (IRED)
Isolation transformers, 58

J

JFET. *See* Junction field-effect transistor (JFET)
Jogging, 41, 236–237
Junction field-effect transistor (JFET), 254–255

K

kcmil (thousand circular mil), 212
Kick start, 229
Kilovolt-amperes (kVA), 60–61

L

Ladder diagrams, 17–23
    rung numbering in, 21
Laser alignment kit, 146
Latched contactor circuit, 173
Latching relays, 203–206
    alternating relays, 204–206
    applications, 204
    magnetic, 204
    mechanical, 203–204
LDR. *See* Light-dependent resistor (LDR)
Leakage current, 195
LED. *See* Light-emitting diode (LED)
Left-hand conductor rule, 106
Legend plates, 76
Lever type, limit switches, 81
Light dimmers, 263–264
Light-dependent resistor (LDR), 267
Light-emitting diode (LED), 249–250, 295
Limit switches, 80–82, 234
Line frequency, 38
Line reactors, 284
Linear solenoids, 100
Linear speed acceleration, 229
Lines of flux, 106
Load, 20
    alignment, 146–147
    capacitive, 170
    constant-torque, 283–284
    inductive, 170, 190
    motor-selection requirements, 141–142
    nonlinear, 170
    overhauling, 302
    resistive, 170
    shock (impact), 284
    switching, 159–162
    utilization, by magnetic contactors, 170
    variable-torque, 284

Load reactors, 284
Locked-rotor current, 41, 126–127, 140, 218
Locked-rotor torque (LRT), 143
Locking mechanism, mechanical latching relay, 203
Lockout, 6, 11–12
   procedure, steps in, 12
Logic
   defined, 270
   digital, 270–273
   hard-wired, 271–272, 308
   relay control, 207–209
Logic gates, 270
Low-frequency voltage boost, 281
Low-impedance busways, 51
Low-voltage distribution section, 50
Low-voltage protection (LVP), 41, 43–44
Low-voltage release (LVR), 41, 43–44
LRT. See Locked-rotor torque (LRT)
LVP. See Low-voltage protection (LVP)

M

Magnetic contactors, 158–168
   arc suppression with, 166–168
   assemblies, 164–166
   coils, 163–164
   defined, 41, 158
   enclosures, 171–172
   IEC ratings, 170–171
   NEMA ratings, 169–170
   operating mechanisms for, 163
   solid-state, 172–174
   switching loads with, 159–162
   three-pole, 159
Magnetic field, 105–106
Magnetic flowmeters, 97
Magnetic latching relays, 204
Magnetic pickup, 95–96
Magnetic starters, 43–45, 175–176
Magnetism, 105–106
Magnetizing current, 59, 292
Main breaker-type panelboards, 54
Main lug panelboard, 54
Maintained-type push button, 76
Manual motor starters, 42–43, 218–220
Manual reset, 178
Manual-reset thermal protection, 151
Manually operated switches, 72–79
   drum switch, 79
   pilot lights, 77–78
   primary/pilot control devices, 72–73
   pushbutton switches, 73–77
   selector switches, 78–79
   toggle switches, 73
Manufacturer, motor, 37
Maximum speed, 303
MCC. See Motor control center (MCC)
Mechanical functions, representation of, 22–23
Mechanical interlocking, 231–232
Mechanical latching relays, 203–204
Mechanical losses, 140
Mechanical power rating, 140
Mechanically held contactors, 161
Mechanically operated switches, 80–86
   float switch, 84–86
   flow switch, 84–86
   limit switches, 80–82
   pressure sensors, 84

   pressure switches, 83–84
   temperature control devices, 82–83
Megger, 69
Megohmmeter, 152
Melting alloy-type thermal overload relays, 179
Metal oxide semiconductor field-effect transistor (MOSFET), 255–257
   power, 256–257
Metric motors, 144–145
Micro limit switches, 82
Microcontroller, 268–270, 269
Microprocessor-based modular overload relay, 182
Milliampere, 2
Minimum speed, 303
Mining, 297
Modular PLC, 306
Momentary operation, motor, 236
Momentary-type push button, 76
Monitor phase rotation, 35
MOSFET. See Metal oxide semiconductor field-effect transistor (MOSFET)
Motor branch circuit
   conductors, sizing of, 212
   control circuit, providing, 216–217
   controller selection for, 215
   disconnecting means for, 215–216
   elements of, 211
   protection, 212–214
Motor control center (MCC), 54–56
Motor control circuits, 211–243
   design/elements/function of, 216–217
   motor branch circuit (See Motor branch circuit)
   NEC installation requirements, 211–217
   for reversing motors, 231–234
   for starting motors, 218–231
Motor control devices, 72–103
   actuators, 98–103
   disconnecting means for, 215–216
   sensors, 86–97 (See also Sensors)
   switches, 72–86
Motor control electronics, 245–273
   digital logic, 270–273
   electrostatic discharge, 270
   integrated circuits, 265–273
   microcontroller, 268–270, 269
   semiconductor diodes, 245–251
   systems, 264–265
   thyristors, 259–265
   transistors, 251–258
Motor controllers disconnecting means for, 215–216
Motor disconnect switches, 215–216
Motor drives
   AC, 136–139, 275–283
   DC, 119–120, 297–303
Motor ladder diagrams, 17–23
Motor nameplates, 37–42
   ambient temperature, 38
   code letter, 39
   connection diagrams, 41
   current rating, 38
   design letter, 39
   duty cycle, 39
   efficiency, 40
   enclosure, 40
   FLA rating, 214
   frame size, 40
   horsepower rating, 39

insulation class, 38–39
line frequency, 38
manufacturer, 37
NEC requirements for, 37–39
optional information, 39–41
phase rating, 38
power factor, 40
service factor, 39–40
speed, 38
temperature rise, 38
thermal protection, 40–41
VFD, 292
voltage rating, 37
Motor speed, 38, 110
   multispeed motors, 242–243
   wound rotor, 243
Motor starting, 218–231
   DC, 229–230
   full-voltage of AC induction, 218–223
   reduced-voltage of induction motors, 223–229
Motor stator resistance, 292
Motor terminal connections, 28–36
   in AC motors, 30–36
   classification of motors, 28
   in DC motors, 28–30
Motor troubleshooting, 151–156
   charts, 154–156
   guides, 152–154
   ladder/tree, 155–156
Motor-driven timers, 195–196
Motors. *See* Electric motors
Mounting
   motors, 146
   VFD, 285
Multifunction timer, 201–202
Multispeed motors, 242–243
   connections of, 34–36
Multispeed starter, 41
Mushroom-head push buttons, 75
Mutual inductance, and secondary winding, 57

N

N-channel JFET, 254–255
N-type semiconductor material, 245
Nameplate amperes, 140, 214
NAND gate, 271
NAND logic function, 208–209
National Electrical Code (NEC), 6, 13
   Article 240.6, 214
   Article 430 of (*See* Article 430, NEC)
   Article 430.102 of, 287
   on conductor ampacity, 51
   motor nameplate information, 37–39
   motors installation requirements, 211–217
National Electrical Manufacturers Association (NEMA), 14
   code letters, 140
   enclosure types, 74, 75, 171–172
   magnetic contactors ratings, 169–170
   markings/symbols, 182–184
   motor voltages standard, 37
   on physical dimensions of motors, 40
   squirrel-cage induction motors types, 125
National Fire Protection Association (NFPA), 14
   70E, 4, 5
Nationally Recognized Testing Laboratory (NRTL), 14
NC. *See* Normally closed (NC)
NEC. *See* National Electrical Code (NEC)

Negative temperature coefficient (NTC), 94–95
NEMA (National Electrical Manufacturers Association), 14
NEMA C face mount, 146
Neutral point, 66–67
NO. *See* Normally open (NO)
No-load losses, 52
No-load speed, 112
Nonconfigurable alarms, 295
Nonhazardous-location enclosures, 171
Nonlinear loads, 170
Nonregenerative DC drives, 301–302
Nontechnical losses, 52
NOR gate, 271
NOR logic function, 209
Normally closed (NC), 73, 80, 187
Normally open (NO), 73, 80, 187
NOT circuit, 271
NOT logic function, 208
NPN transistor, 252
NRTL (Nationally Recognized Testing Laboratory), 14
NTC. *See* Negative temperature coefficient (NTC)
Numerical cross-referencing, 21–22

O

Occupational Safety and Health Administration (OSHA), 7, 12–13
   on emergency stop switch, 76
ODP motors. *See* Open drip-proof (ODP) motors
Off-delay timer (TOF), 197–200, 309
OL. *See* Overload relay (OL)
On-delay timing/timer (TON), 197, 198, 309
One-shot timer, 200–201
Open drip-proof (ODP) motors, 144
Open enclosures, 143
Open transition, 227–228
Open-loop control, 280
Open-loop mode, servo motors, 102
Operating point, 87–88
Operational amplifier (op-amp) ICs, 266–267
Operator interface, VFD, 285
Operator, push buttons, 74–75
Optocoupler/optoisolator, 250
OR gate, 271
OR logic function, 207–208, 209
Oscilloscope, 152
OSHA. *See* Occupational Safety and Health Administration (OSHA)
OTE (output energize) instruction, 307
Output module, 305
Overcurrent protection, 176–177, 212–214
Overhauling loads, 302
Overheats, motors, 153
Overload protection, 177
Overload relay (OL), 17, 41
   dual-element fuses, 182
   electronic, 181–182
   in motor starters, 175, 178–182
   thermal, 178–181

P

P-type semiconductor material, 245
PAC (Perimeter Access Control), 7
Panelboards, 52–54
   classifications of, 54
   configurations, 54
   grounding, 53
Parallel circuit, 207–208

Parameter programming
    DC motor drives, 302–303
    VFD, 294–295
Parking brakes, 241
Part-winding starters, 226–227
Peak-switching relay, 194
Perimeter Access Control (PAC), 7
Permanent-capacitor motor, 134
Permanent-magnet (PM) DC motors, 110–112
Permanent-split capacitor, 31
Permit-required confined space, 7
Personal protective equipment, 5–7
    face shields, 6
    guidelines for, 5
    hot sticks, 6
    protection apparel, 6
    rubber protective equipment, 5–6
    shorting probes, 6
P.F. *See* Power factor (PF)
Phase angle control, 260
Phases, 124
    diplacement, 131
    rating, 38
Photodiodes, 250
Photoelectric sensor, 89–91
Phototransistors, 253–254
Physical dimensions, motors, 40
Pickup voltage, 164
PID controller. *See* Proportional-integral-derivative (PID) controller
Pilot control devices
    contactors with, 159
    function of, 72–73
Pilot lights, 77–78
PLC. *See* Programmable logic controller (PLC)
Plug-in busways, 51
Plugging, 41, 238–240
PN-junction diode, 245
Pneumatic timers, 196
PNP transistor, 252
Point of Operation Control (POC), 7
Point-to-point wiring, 24
Polarity, 204
    transformers, 62–63
Polarized retroreflective scan sensor, 90
Poles, 190
Position sensors, 95–96
Positive temperature coefficient (PTC), 94–95
Potential transformer, 67
Power circuits, magnetic contactors, 159
Power distribution systems, 47–69
    motor control center, 54–56
    switchboards/panelboards, 52–54
    transformation stages of, 48
    transmission and distribution loss, 51–52
    transmission systems, 47–48
    unit substations, 48–50
Power factor (PF), 40, 126
    poor, 52
Power grid transformers, 48
Power MOSFETs, 256–257
Power rating
    mechanical, 140
    of transformers, 60–61
    VFD, 292
Power riser diagram, 27
Power section, of DC motor drive, 120

Power substations, 48–50
Power supply, of PLC, 304
Preset slow speed, 229
Pressure sensors, 84
Pressure switches, 83–84
Primary control devices, 72–73
Primary full-load current, 60
Primary winding, 57–58, 68
Primary-resistance starters, 224
Program, 306
Programmable logic controller (PLC), 202, 272, 304–315
    fixed, 305
    instructions, 307
    ladder diagrams, 17
    ladder logic programming, 306–309
    modular, 306
    power supply of, 304
    programming timers, 309–310
    sections/configurations, 304–306
Programming device, 305
Programming language, 306
Proper grounding, 286
Proportional-integral-derivative (PID) controller, 294
Protection
    electrical shock, 5–7
    motor, 287–288
    motor branch circuit, 212–214
    overcurrent, 176–177
    overload, 177
    personal equipment for, 5–7
Protective equipment, personal, 5–7
Proximity sensors, 86–89
    capacitive, 88–89
    inductive-type, 87–88
PTC. *See* Positive temperature coefficient (PTC)
Pull-out torque, 143
Pull-up torque (PUT), 143
Pulse width modulation (PWM), 138, 268
    inverter, 278
Push buttons
    defined, 41
    job circuit, 236
Push roller type, limit switches, 81
Push-to-test pilot lights, 78
Pushbutton interlocking, 233–234
Pushbutton station, 74
Pushbutton switches, 73–77
PUT. *See* Pull-up torque (PUT)
PWM. *See* Pulse width modulation (PWM)

R

Rack/slot-based, PLC addressing format, 307
Rails, 17
Ramping, 289
Rated voltage, 164
RC (resistor/capacitor) charge, 196
Read only parameters, 294
Reciprocating machine process, 234–236
Rectifier block, 27
Rectifier diodes, 246–249
Recycle timers, 201
Reduced-voltage starters, 41, 223–229
    autotransformer, 224–225
    open/closed transition in, 227–228
    part-winding, 226–227
    primary-resistance, 224
    soft, 228–229

solid-state, 260–261
wye-delta, 225–226
Regenerative braking, 29, 288, 302
Regenerative DC drives, 301–302
Relay control logic, 207–209
    inputs/outputs, 207
    AND logic function, 207
    NAND logic function, 208–209
    NOR logic function, 209
    NOT logic function, 208
    OR logic function, 207–208, 209
Relay ladder logic (RLL), 306
Relays, 186–209
    actuator, 98–99
    alternating/impulse, 204–206
    control logic for (*See* Relay control logic)
    defined, 41
    double-break, 98
    electromechanical (*See* Electromechanical relay (EMR))
    latching, 203–206
    overload (*See* Overload relay (OL))
    solid-state (*See* Solid-state relay (SSR))
    timing, 195–202
Release point, 87–88
Remote control, 41
Resilient base mounts, 146
Resistance, 2
Resistance check, 69
Resistance temperature detector (RTD), 94, 151
Resistive loads, 170, 190
Resistor/capacitor (RC) charge, 196
Retentive timer (RTO), 309
Retroreflective scanning, 90
Reverse-bias voltage, 246
Reverse-phase relay, 35
Reversing, of motors, 231–234
    AC induction motors, 231–234
    DC motors, 236
Revolutions per minute (rpm), 38
Right-hand motor rule, 108
Rigid base mounts, 146
Riser diagrams, 27
RLL. *See* Relay ladder logic (RLL)
Roller bearings, 147
Rotary solenoids, 99–100
Rotating cam limit switch, 82
Rotating magnetic field, 122–123
Rotation
    direction of, 115–116
    in electric motors, 107–109
RTD. *See* Resistance temperature detector (RTD)
RTO. *See* Retentive timer (RTO)
Rubber protective equipment, 5–6
Rungs, 17
    numbering in ladder diagrams, 21
Running winding, 131

S

Safety Interlock switches, 7–8
Safety laser scanners, 8–9
Safety light curtains, 7
Safety signs, 5
Safety, PLC programs, 308–309
Safety, workplace, 1–15
    electrical codes/standards, 12–15
    electrical shock protection, 1–7
    grounding/bonding, 9–11

lockout/tagout, 11–12
machine, 7–9
Secondary winding, 57–58, 68
Selection criteria, motors, 140–145
    code letters, 140
    current, 140
    design letter, 140
    duty cycle, 143
    efficiency, 140–141
    enclosures, 143–144
    energy-efficiency, 141
    frame sizes, 141
    frequency, 141
    full-load speed, 141
    load requirements, 141–142
    mechanical power rating, 140
    metric motors, 144–145
    temperature rating, 142–143
Selector switches, 41, 78–79, 237
Semiconductor diodes. *See* Diodes, semiconductor
Sensorless open-loop, 282
Sensors, 86–97
    defined, 86
    flow measurement, 96–97
    hall effect, 91–92
    photoelectric, 89–91
    proximity, 86–89
    temperature (*See* Temperature sensors)
    ultrasonic, 92–93
    velocity/position, 95–96
Separate overload, 175
Separate winding motors, 127
Separately derived system, 11
Series circuit, 207
Series-type DC motor, 112
Service entrance, 50, 52
Service factor (SF), 39–40
Service factor amperes, 140
Servo motors, 102–103
Shaded-pole motor, 135
Shading coils/rings, 165
Shielded power cable, VFD, 285, 286
Shock (impact) load, 284
Shock, electrical, 1–3
Short circuits, 4, 231
    protection, 177
Short-circuited capacitor, 133
Shorting probes, 6
Shunt-type DC motors, 113–114
Signs, safety, 5
Silicon-controlled rectifier (SCR), 259–262, 293
    construction styles, types of, 172–173
    shorted, 260–261
    snubber circuit, 174
    in solid-state contactors, 172–174
    SSR and, 193
    testing circuit, 173
Silver contacts, 166
Single-line diagrams, 26
Single-phase AC motors, 131–136, 154
    capacitor-start motors, 133–134
    connections, 30–31
    reversing, 234–236
    shaded-pole motor, 135
    split-phase induction motor, 131–133
    universal motor, 135–136
Single-phase input DC drives, 299–300

Index   I-9

Single-phase transformer, 63–65
Sleeve bearings, 147, 148
Slip, 42, 126
Slip estimator block, 282
Slip-ring motor, 128–129
Snubber circuit, SCR, 174
Soft starters, 228–229
Soft stop, 229
Solenoid, 99–100
    valves, 100–101
Solid-state relay (SSR), 191–195
    defined, 191
    operation of, 191–192
    single-pole configuration, 193
    specifications, 193
    switching methods, 194–195
Solid-state timing relays, 196–197
Speed
    DC motors, 118–119
    full-load, 141
    maximum, 303
    minimum, 303
    of motor (*See* Motor speed)
    no-load, 112
    regulation, 117–118
Speed feedback information, 303
Speed setpoint, 302–303
Speed-regulated drive, 303
Spikes, voltage, 164
Split-phase motor, 31, 131–133
    capacitor-start motors, 133–134
    dual-voltage, 133
Squirrel-cage configuration, AC motors, 30, 124–128
    characteristics of, 125–126
    NEMA-type, 125
SSR. *See* Solid-state relay (SSR)
Stability compensation, 281
Starter auxiliary contact, 309
Starters, 175–184. *See also* specific starters
    autotransformer, 224–225
    combination, 223
    defined, 42
    full-voltage, 218–223
    magnetic motor, 43–45
    manual, 42–43, 218–220
    overcurrent protection in, 176–177
    overload relay in, 175, 178–182
    primary-resistance, 224
    size, 180
    three-phase reversing motor, 231–234
    wye-delta, 225–226
Starting torque, 143
Starting winding, 131
Starting, of motor. *See* Motor starting
Static charge, 270
Stator and rotor resistance losses, 140
Steady-state slip compensation, 281
Step-down transformer, 59, 160
Stepper motors, 101–102
Stopping, of motors, 238–241
    DC injection braking, 240–241
    dynamic braking, 240
    electromechanical friction brakes, 241
    plugging/antiplugging, 238–240
Stray losses, 140–141
Substations, power, 48–50

Subtractive polarity, 62–63, 68
Supply voltage, 292
Switchboards, 52–54
Switches
    centrifugal, 132–133
    E-stop, 222–223
    limit, 80–82, 234
    manually operated, 72–79
    mechanically operated, 80–86
Switchgear, 49
Switching methods, solid-state relay, 194–195
Symbols, 16–23
    electrical control, 182–184
    motor, 16–17
Synchronous condenser, 130
Synchronous motors, 129–130
Synchronous speed, AC motors, 122–123
System grounding, 10, 53, 56

T

Tachometer, 152
Tachometer generators, 95
Tag-based, PLC addressing format, 307
Tagout, 6, 11–12
Target flowmeters, 96–97
TC. *See* Thermocouple (TC)
Technical losses, 51–52
TEFC motors. *See* Totally enclosed, fan-cooled (TEFC) motors
Temperature control devices, 82–83
Temperature ratings, motor, 142–143
Temperature rise, 38, 143
    transformer, 61
Temperature sensors
    integrated circuit, 95
    resistance temperature detector, 94
    thermistors, 95
    thermocouple, 93–94
TENV motors. *See* Totally enclosed, nonventilated (TENV) motors
Thermal contact burns, 3
Thermal dissipation, 194–195
Thermal imagers, 38
Thermal overload relay, bimetallic type, 179
Thermal overload relays, 178–181
    melting alloy-type, 179
Thermal protection, 40–41
Thermal protectors, 151
Thermistors, 95
Thermocouple (TC), 93–94
Thermostats. *See* Temperature control devices
Thermowells, 94
Three-phase AC motors, 122–130, 154
    connections, 31–32
    induction motor, 124–128
    rotating magnetic field, 122–123
    synchronous, 129–130
Three-phase fully controlled rectifier, 300
Three-phase input DC drives, 300
Three-phase reversing motor starter, 231–234
Three-phase transformers, 65–67
Three-pole magnetic contactors, 159
Three-pole manual starter, 219–220
Three-wire control, 44–45, 160
Through-beam scanning, 89–90
Throws, relay contact, 190
Thrust bearings, 147

Thyristors, 259–265
 silicon controlled rectifiers, 259–262
 triac, 262–264
Time-delay fuses, 182
Timer
 defined, 42
 functions of, 197–201
 instruction, 309
 programming, PLC, 309–310
Timing relays, 195–202
 dashpot timers, 196
 motor-driven timers, 195–196
 multifunction, 201–202
 off-delay, 197–200
 on-delay, 197, 198
 one-shot, 200–201
 PLC, 202
 recycle, 201
 solid-state, 196–197
 types of, 196
TOF. See Off-delay timer (TOF)
Toggle switches, 73
TON. See On-delay timing/timer (TON)
Torque
 adequate, 257
 defined, 42
 motors, 108, 1101
 as selection criteria, 143
Torque current estimator, 282
Torque limit, 303
Totally enclosed, 143
Totally enclosed, fan-cooled (TEFC) motors, 144
Totally enclosed, nonventilated (TENV) motors, 144
Tower light indicators, 78
Transformer pilot light, 77–78
Transformer sections, 49
Transformers, 48, 57–61
 autotransformers, 58, 67
 current, 67–68
 instrument, 67–68
 isolation, 58
 operation of, 57–58
 performance of, 61
 polarity of, 62–63
 potential, 67
 power losses in, 52
 power rating of, 60–61
 single-phase, 57, 63–65
 three-phase, 65–67
 voltage/current/turns ratio in, 58–60
Transistors, 251–258
 bipolar junction, 252–254
 Darlington, 253
 field-effect, 254–255
 insulated-gate bipolar, 257–258
 metal oxide semiconductor field-effect, 255–257
Transmission systems, 47–48
Transmission/distribution losses, 51–52
Triacs, 262–264
Triggered SCR, 173
Trip indicator, 178
Troubleshooting
 motors (See Motor troubleshooting)
 VFD, 295–296
Tunable-on-the-fly parameters, 294
Turbine flowmeters, 96

Turns ratio, 58–60
Two-value capacitor, 31
Two-wire control, 160
 circuits, 43

U

UL (Underwriters Laboratories), 145
Ultrasonic sensors, 92–93
Unbalanced motor voltages, 149
Underwriters Laboratories (UL), 145
 logo, 14
Unit substations, 48–50
Universal motor, 135–136

V

V angle, 282
V belts, 147
V/Hz (volts per hertz) drives, 280–281
Vacuum contactor, 168
Variable torque, 35
Variable-frequency AC drive, block diagram of, 26–27
Variable-frequency drive (VFD), 136–139, 229, 276–280
 braking, 288
 bypass contactor, 286–287
 control inputs/outputs, 289–292
 derating, 292
 diagnostics/troubleshooting of, 295–296
 disconnecting means, 287
 electromagnetic interference, 285–286
 enclosures, 284
 grounding requirements for, 286
 installation, 284–296
 line/load reactors, 284
 location, 284
 motor nameplates, 292
 motor protection, 287–288
 mounting technique, 285
 operator interface, 285
 parameter programming, 294–295
 PID control, 294
 ramping, 289
 selection of, 283–284
 shielded power cable, 285, 286
 types of, 293
Variable-speed AC motor drive, 276
Variable-torque loads, 141–142, 284
Variable-voltage acceleration, 230
Variable-voltage inverter (VVI), 293
Vector control, 282
Vector duty, 35
Vector duty motor, 139
Velocity sensors, 95–96
VFD. See Variable-frequency drive (VFD)
Volt-amperes (VA), 60
Voltage
 defined, 2
 dropout, 164
 forward-bias, 246
 hold-in, 164
 levels/balance of motor, 149–150
 pickup, 164
 rated, 164
 rating, 37
 ratio, 58–60
 regulation, 61
 reverse-bias, 246

Voltage (*continued*)
   spikes, 164, 174
   unbalanced motor, 149
Voltage comparator, 267
Voltage-regulated drive, 303
Voltage-source inverter (VSI), 293
Volts per hertz (V/Hz) drives, 280–281
VSI. *See* Voltage-source inverter (VSI)
VVI. *See* Variable-voltage inverter (VVI)

W

Winders, 297
Windings
   primary, 57–58, 68
   running, 131
   secondary, 57–58, 68
   starting, 131
Wire numbering, 22
Wiring diagrams, 24–26

Within sight, defined, 216
Wobble stick, limit switches, 81
Workplace safety. *See* Safety, workplace
Wound-field DC motors, 112
Wound-rotor motor, 30, 128–129
   speed of, 243
Wye-delta starters, 225–226

X

XIC (examine if closed) instruction, 307
XIO (examine if open) instruction, 307

Z

Zener diode, 249
Zener diode regulator circuit, 249
Zero-fired control, 174
Zero-speed switch, 238
Zero-switching relay, 194